Sedimentologic Consequences of Convulsive Geologic Events

Edited by

H. Edward Clifton

229

Published by The Geological Society of America, Inc.
3300 Penrose Place, P.O. Box 9140, Boulder, Colorado 80301

GSA Books Science Editor Campbell Craddock

Library of Congress Cataloging-in-Publication Data

Sedimentologic consequences of convulsive geologic events / edited by
H. Edward Clifton.
p. cm. — (Special paper / Geological Society of America;
229)
Chiefly papers presented at the Inaugural Symposium of the
Sedimentary Geology Division of the Geological Society of America,
held on Oct. 28, 1985, in Orlando, Fla.
Includes bibliographies.
ISBN 0-8137-2229-2
1. Sedimentation and deposition—Congresses. 2. Natural
disasters—Congresses. I. Clifton, H. Edward (Hugh Edward), 1934–
. II. Geological Society of America. Sedimentary Geology Division.
Inaugural Symposium (1985 : Orlando, Fla.) III. Title: Convulsive
geologic events. IV. Series: Special papers (Geological Society of
America) ; 229.
QE57.S419 1988
551.3—dc19 88-25017
CIP

10 9 8 7 6 5 4 3 2

Contents

Preface v

Sedimentologic relevance of convulsive geologic events 1
H. Edward Clifton

Nearshore responses to great storms 7
Robert A. Morton

Origin, behavior, and sedimentology of prehistoric catastrophic lahars at Mount St. Helens, Washington 23
Kevin M. Scott

The Mount Mazama climactic eruption (~6900 yr B.P.) and resulting convulsive sedimentation on the Crater Lake caldera floor, continent, and ocean basin 37
C. Hans Nelson, Paul R. Carlson, and Charles R. Bacon

Sonar images of the path of recent failure events on the continental margin off Nice, France 59
Alberto Malinverno, William B. F. Ryan, Gerard Auffret, and Guy Pautot

The 1929 "Grand Banks" earthquake, slump, and turbidity current 77
David J. W. Piper, Alexander N. Shor, and John E. Hughes Clark

Basin plains; Giant sedimentation events 93
Orrin H. Pilkey

Large-scale bedforms in boulder gravel produced by giant waves in Hawaii 101
George W. Moore and James G. Moore

Sedimentological consequences of two floods of extreme magnitude in the late Wisconsinan Wabash Valley 111
Gordon S. Fraser and Ned K. Bleuer

Deposits of a middle Tertiary convulsive geologic event, San Emigdio Range, southern California 127
Peter G. DeCelles

Cretaceous/Tertiary boundary sediment 143
Kenneth J. Hsü

Index 155

Preface

This volume grew from the Inaugural Symposium of the Sedimentary Geology Division of the Geological Society of America, held on October 28, 1985, in Orlando, Florida. The symposium, entitled "The Sedimentologic Consequences of Convulsive Geologic Events," was stimulated by the geologic community's growing awareness of, and interest in, the effects of infrequent but highly energetic phenomena. Much of this interest arises from the possibility, posed first by Alvarez and others (1980), that the stratigraphic record was profoundly influenced by the impact of a large extraterrestrial body at the end of Cretaceous time. The explosive eruption of Mount St. Helens, although tragic in human terms, provided scientists with an opportunity to study firsthand the effects of such an eruption. Perusal of the recent geological literature shows an increasing number of papers dealing with phenomena that are described in this volume as convulsive.

Seven of the papers in this volume represent contributions to the symposium in Orlando. The additional papers provide considerations that were not possible to cover within the limited time frame of the symposium. I wish to thank all of the authors and the many diligent reviewers of these papers for their cooperation and contributions.

H. Edward Clifton

Geological Society of America
Special Paper 229
1988

Sedimentologic relevance of convulsive geologic events

H. Edward Clifton, *U.S. Geological Survey, 345 Middlefield Road, Menlo Park, California 94025*

ABSTRACT

Convulsive geologic events (extraordinarily energetic events of regional influence) undeniably have occurred repeatedly throughout the history of the Earth. The influence of such events (i.e., explosive volcanic eruptions, large bolide impacts, giant mass failures, catastrophic floods, great earthquakes, large violent storms, giant tsunamis) on the sedimentary record, however, generally is poorly understood. The sedimentary products of many kinds of convulsive events are so seldom well preserved that they are of limited consequence in the depositional record. Moreover, sedimentary processes associated with convulsive events typically are poorly understood, particularly for those types of events that have not occurred within historic time. The difficulty of establishing synchroneity among the products of a large event further impedes their correct interpretation. It is likely that these factors, in combination, explain the enigma posed by the absence of recognized sedimentologic effects of the probably numerous Phanerozoic impacts of major bolides on the Earth. Parsimony demands that we attribute phenomena in the sedimentary record to the most probable explanation, and convulsive geologic events are, by nature, improbable. The recognition of the sedimentologic consequences of convulsive events poses a special challenge to sedimentary geologists. Meeting this challenge will almost certainly demonstrate that convulsive geologic events have greater relevance to the sedimentary record than has been previously recognized.

INTRODUCTION

Sedimentary geologists have become increasingly aware of the importance of extraordinary events in the stratigraphic record (Dott, 1987). The recognition in the 1950s of turbidity currents as important depositional processes demonstrated the potential for abrupt, infrequent events of rapid deposition in an otherwise tranquil environment. Floods have become accepted as a major contributor to fluvial deposits, and the influence of storms on shelf deposits has received much attention in recent years (Aigner, 1985). Most sedimentary geologists today would probably agree with Dott's (1983) contention that the sedimentary record is mostly a record of episodic events.

Much of the preoccupation with these events is directed toward large events that recur at intervals of 100 yr or less. Wolman and Miller (1960) suggest that events recurring at relatively short intervals (1 to 5 yr) have a greater effect on landform development than do more infrequent events of greater intensity, and the same may be true, to a degree, of the development of the depositional record. Relatively little attention has been directed, however, to the sedimentologic consequences of events of the scale that herein is referred to as "convulsive." The questions regarding the possible biological evolutionary significance of phenomena such as large bolide (asteroid or comet) impacts can be appropriately extended to their sedimentologic significance. It is that consideration that forms the basis for this volume.

WHAT IS A CONVULSIVE EVENT?

The term "convulsive event" as used here refers to an extraordinarily energetic event of regional influence. The term "catastrophic," which has traditionally been applied to such events, is undesirable because of its connotation of disaster (Gretener, 1984) and its doctrinal implications (Dott, 1983, 1987). Geologic catastrophes, in terms of human toll, are all too familiar. Some, such as a mudslide that engulfs a village, may be considered catastrophic without being convulsive in a regional sense. The term "cataclysmic" has been suggested as appropriate for convulsive geologic events (Dott, 1983), but that term is also applied in other ways as well: to deluges of the biblical kind (Hsü, 1983), in a hierarchical sense to the events of the greatest magnitude (Benson, 1984), and in planetary science to the last heavy bombardment of planets in the solar system about 3.9 to 4.0 Ga (Short, 1975). The term "convulsive" is neutral in the sense of disaster, carries no doctrinal implications, and has solid precedent in the older geological literature (see, for example, Giekie, 1875, p. 79).

The convulsive events considered in this volume span a range of processes: great storms, giant mass failures on continental slopes, jokulhaups, explosive volcanic eruptions, great earthquakes, giant tsunamis, and large bolide impacts. The events considered here range—using Gretener's (1984) classification based on the time interval required for a 95-percent probability of at least one event—from "common" (10^2 yr) or "regular" (10^3 yr) events such as great storms, explosive eruptions, and great earthquakes, to something between "recurrent" (10^6 yr) and "occasional" (10^8 yr) events such as major bolide impacts. None are "rare events," defined by Gretener (1984) as requiring 10^9 yr for a 95 percent probability of at least one occurrence.

SEDIMENTOLOGIC INFLUENCES OF CONVULSIVE EVENTS

Much discussion in recent years has focused on the influence of convulsive geologic events on the record of biological change (as for example in the collection of papers in Silver and Schultz, 1982; and Berggren and Van Couvering, 1984). Far less consideration has been given to other consequences of such events in the stratigraphic record. Wolman and Miller (1960) argue that "forces of moderate magnitude and frequency have greater net effect on landform development than do intense, short-lived forces associated with short-lived events." In contrast, Ager (1981, p. 56) postulates a "phenomenon of quantum sedimentation" whereby the episodic catastrophic event may have more effect than vast periods of gradual evolution. Very likely each viewpoint is correct for specific depositional settings. Floods of the magnitude that produced the Channeled Scabland of the Columbia Plateau (Bretz, 1923, 1969) are so infrequent over geologic time as to contribute only a small fraction to the overall accumulation of fluvial deposits in any one area. On the other hand, Pilkey (this volume) argues that abyssal plains, major physiographic features of the ocean floor (Heezen and Schneider, 1966), result to a large extent from giant mass failures of the continental margin.

A primary key to the importance of convulsive events is the degree to which their sedimentologic products are preserved. Dott (1983) notes that the effects of many rare events may be temporarily spectacular but are so unlikely to be preserved that they are of limited consequence to the sedimentary record. Dott further argues convincingly that the most important single factor in preservation is base level. Deposits formed at or above a hypothetical base level, such as those produced by glacial outburst floods, are less likely to be preserved than those that form below base level, such as those produced by giant mass failure of the continental margin. Nonmarine and shallow-marine products of convulsive events are likely to be preserved only where subsidence carries the product below erosional base level. Deep-sea products of such events may be tectonically eroded and reconstituted through subduction.

Another problem lies in the ability of sedimentary geologists to recognize the effects of convulsive events. Because of the magnitude and infrequency of such events, our understanding of the processes involved are likely to be inadequate. Convulsive events may, in a general way, be placed in three categories. The first category includes those types of events that have occurred somewhere on the planet in historic time at more or less their maximum intensity. We have a reasonable probability of understanding the sedimentary processes attendent to such events (which might include phenomena such as great earthquakes and giant storms) (see Gretener, 1967). A second category includes those events that have occurred during historic time but at much smaller scales compared to those that may have occurred in the past. Meteorites strike the earth yearly but produce no effects comparable to those generated by the impact of a 10-km bolide. It may be difficult, if not impossible, to extrapolate processes accurately from the present to the past in such cases. The third category includes possible events for which there is no occurrence in historical time. Events of this type might include the effects of supernovae in the vicinity of the solar system (Salop, 1977; Schindewolf, 1977), or variations in the universal gravitational constant (Aspden, 1977). Most are highly conjectural and are given little credence in mainstream geology, although one such phenomenon, the episodic reversal of the earth's magnetic field, is well documented and generally accepted. Sedimentologic processes associated with events of the third type are largely unknown. Although such events must at least be regarded as philosophically possible, they are not considered in this volume.

Because convulsive events as defined herein are of regional consequence, chronologic correlation becomes critical. This is particularly true where different consequences of the same event are postulated, as in the paper by DeCelles in this volume, which attributes a huge nonmarine breccia deposit, a zone of large-scale deformation in shoreline deposits, and a thick submarine mass-flow unit to the same Eocene earthquake or series of earthquakes. The synchroneity of the Cretaceous–Tertiary boundary clay, described herein by Hsü, is fundamental to its attribution to a convulsive event.

Convulsive geologic events are essentially instantaneous in geologic time, but their products may linger for some time within the realm of active sedimentary processes. The character of the products may change as they are weathered, eroded, transported, and redeposited. A major explosive volcanic eruption can choke a drainage system with ash and rubble (see Nelson, this volume). Sediment yields in these fluvial systems become immense (Janda and Meyer, 1985), and any resulting pulse of sediment working its way through the system is likely to undergo textural and compositional evolution. The final deposition of this material may reflect its convulsive origins only in subtle ways.

LESSONS FROM A PARADOX

An example of the difficulty of identifying the consequences of convulsive events lies in the consideration of the effects of great bolide impacts in the geologic past. The suggestion that the biological extinctions at the Cretaceous–Tertiary boundary resulted from a bolide impact (Alvarez and others, 1980) has generated much interest, discussion, and controversy within the geological community. The primary evidence for the impact is the widespread occurrence of a thin layer of clay that is enriched in iridium and other siderophile elements and contains grains of shocked quartz. This clay is also the only documented physical evidence, which creates something of a paradox.

The scenario following the impact of an extraterrestrial body similar to that presumed to have struck the Earth at the end of the Cretaceous (diameter, 10 km; density, 3.0; velocity, 25 km/s) is summarized by Silver (1982). On land, the explosive impact would result in a crater 66 km (Schmidt and Holsapple, 1982) to 150 km in diameter (O'Keefe and Ahrens, 1982). For an oceanic impact, models indicate that a transient crater as much as 150 km across would exist in the water, and a crater 50 to 70 km across and about 500 m deep would be excavated on the sea floor (Melosh, 1982). A 500-km^3 steam bubble of vaporized rock and ocean water would explode from the crater with an energy release equivalent to 10^7 megatons (Melosh, 1982). Waves generated by the explosion and the subsequent collapse of the crater are estimated to be of similar scale to the water depth near the impact (4 to 5 km high in the deep ocean) (Gault and Sonett, 1982). Such waves would still have a height of 10 m after traveling 5,000 km from the impact site. For comparison, most earthquake-generated tsunamis are less than a meter high as they propagate over the deep ocean (Weigle, 1970). Impact into average ocean depths would vaporize a mass of water equivalent to about one-tenth of that presently in the earth's atmosphere. Most of this water would return to the earth's surface as rain or snow in a matter of weeks to months (Croft, 1982).

Some of the consequences of such an impact are almost entirely model-dependent. The greatest explosion recorded in historic time was that of Krakatoa (Jones and Kodis, 1982). The atmospheric pressure waves generated by this explosion traversed the globe at least three times, and tide gauges around the world recorded fluctuations of sea level that locally reached nearly 2 m (Simken and Fiske, 1983). The sea waves were coincident with the arrival of the atmospheric pulse (Ewing and Press, 1955) and can be explained by resonant coupling of the barometric disturbance and the sea surface (Press and Harkrider, 1966). The energy released by the Krakatoa explosion is estimated by Press and Harkrider (1966) to have been in the range of 100 to 150 megatons and by Yokoyama (1981) to be about 33 megatons. The explosive yield of a 10-km bolide impact is estimated to have been 10^6 times greater (Toon and others, 1982). The magnitude and consequences of the resulting atmospheric shock wave and consequent sea-level fluctuations due to resonant coupling with the atmospheric waves of such an explosion deserve consideration.

The foregoing discussion indicates that the sedimentologic consequences of the impact of a large extraterrestrial body are potentially very dramatic, yet none are documented for the postulated impact at the end of the Cretaceous or for any other time in Earth history. Current evidence, however, suggests that large impacts should be geologically significant phenomena. The flux of asteroids is such that an impact of a 10-km body could presently be expected at intervals that average about 40 m.y. (Wetherill and Shoemaker, 1982). The possibility of large cometary impacts (Weissman, 1982) further reduces the postulated cratering interval.

The record of cratering on the Earth supports the notion that it has been struck repeatedly by large bodies throughout geologic time. Data presented by Grieve (1987) show that five craters 50 km or more in diameter can be documented in Phanerozoic time. This almost certainly is a fraction of the total of equivalently large craters that have actually occurred during this interval. The distribution of known craters on land is very uneven; 65 percent of the known structures occur on less than 11 percent of the land surface (Grieve, 1982). They are found predominantly in geologically stable areas that have been the subject of intensive crater research. Moreover, the distribution of craters is uneven with respect to time. About 35 percent of the known structures are less than 100 m.y. old (Silver, 1982). Of the five large Phanerozoic craters, only one is older than 210 Ma. Therefore, one could conservatively estimate that, because of the biases imposed by preservation and recognition, the five known large craters represent only one-half of the large land impacts of the Phanerozoic and but one-sixth of those that have occurred in the oceans over the same time. The average rate for major oceanic impacts during the Phanerozoic, therefore, could be considered to be at least once every 20 m.y., a frequency that is reasonably consistent with astronomers' estimates based on the combined flux of asteroids and comets (Wetherill and Shoemaker, 1982).

Why is there no physical evidence for such impacts other than the Cretaceous–Tertiary boundary clay? As Silver (1982) noted, "The lack of documentation of the anticipated range of consequences of even one major impact event must somehow be reconciled with the probabilities for such events that are currently offered." Part of the problem undoubtedly lies with preservation. As noted earlier, nonmarine and shallow-marine effects are likely to be preserved only in areas of rapid subsidence. But the effects of a major bolide impact are likely to be of such broadly regional, even global, extent that one would intuitively expect at least fragmentary preservation of its consequences.

More likely the paradox arises from our lack of understanding of the processes resulting from a major impact, and the difficulty of demonstrating that possible consequences resulted from a large impact rather than from more mundane local causes. Establishing synchroneity of phenomena among widely separated depositional basins, possibly at the intercontinental level, and linking these phenomena with a known impact presently are difficult, if not impossible, tasks.

In every case, responsible scientific procedure dictates that we accept the most probable, generally simplest explanation for any phenomenon in the geologic record. An anastomosing web of large sandstone dikes that cut hundreds of meters of basin-slope sediment might be due to the seismic shock generated by an impact, but it also could result from a great earthquake, resulting from tectonic stresses that mobilized overpressured turbidite sands buried beneath the slope deposits. The probability of occurrence of such an earthquake in a tectonically active region is 4 to 5 orders of magnitude greater than a 10-km bolide impact. Parsimony demands the more probable explanation.

It may be that in virtually every case, phenomena that could be attributed to a bolide impact are more appropriately attributed to convulsive events of lesser magnitude, operating at a more local scale. It seems likely that the consequences of large bolide impacts do reside in the sedimentary record, and that many have been observed and noted by sedimentary geologists. Scientific protocol, however, almost requires that they be attributed to less exotic origins. If we cannot improve our ability to establish synchroneity of phenomena and linkage with impacts, we likely will never be capable of identifying their effects in the sedimentary record. The paradox does not lie in the nature of things, but in our own procedural methodology. If so, the true relevance of bolide impacts to the sedimentary record is greater than is now, and possibly ever will be, realized.

The lesson from this paradox extends to other, less exotic, convulsive geologic events. The burden of proof correctly lies with the scientist who wishes to invoke such phenomena. In the absence of convincing documentation, their consequences must be attributed to more common events of lesser magnitude. I have observed a great many ancient progradational shoreline sequences in different parts of the world. Virtually everything I have noted seemingly can be explained by a combination of everyday processes and typical seasonal storms—there is no need to invoke truly giant waves or tsunamis. Logic tells me that these convulsive phenomena are probably recorded here and there in the sequences I have seen. My inability to establish their origin obscures their actual importance.

Convulsive events thus pose a challenge to sedimentary geologists. To realize more fully the sedimentologic relevance of these events, we must retain open minds, considering all possibilities in our quest for explanations of features in the depositional record. Careful observation is required, and a willingness to ponder those aspects that are not fully explained by conventional processes. In situations where the products of convulsive events might reasonably be expected, such as the effects of tsunamis on shallow-marine sediment in tectonically active settings, we can ask ourselves how those products might differ from those of more conventional processes and seek these differences. We can explore the possibility of multiple effects from a single event and develop the capability of perceiving the origin of sediment that evolved from a convulsive event. Our understanding of the processes associated with such events is subject to vast improvement.

Meeting this challenge carries significance beyond a better understanding of the stratigraphic record. Most of the convulsive events described in this volume bear a potential for catastrophic toll in a human sense. The ability to recognize the sedimentologic effects of ancient convulsive events will enhance our capability to assess the potential for future occurrences of these events in developed areas and to predict more accurately their human consequences. Herein may lie the greatest relevance of analyzing the sedimentologic consequences of convulsive geologic events.

ACKNOWLEDGMENTS

This paper profited from thoughtful comments and suggestions by Robert H. Dott, Jr., Lee J. Suttner, and Abhijt Basu.

REFERENCES CITED

Aigner, T., 1985, Storm depositional systems; Lecture notes in earth sciences no. 3: Berlin, Springer-Verlag, 174 p.

Ager, D. V., 1981, The nature of the stratigraphical record, 2nd ed.: New York, John Wiley and Sons, 122 p.

Alvarez, L. W., Alvarez, W., Asaro, F., and Michael, H. V., 1980, Extraterrestrial cause for the Cretaceous-Tertiary extinction: Science, v. 208, p. 1095–1108.

Aspden, H., 1977, Galactic domains, G fluctuation and geomagnetic reversals: Catastrophist Geology, v. 2, no. 2, p. 42–47.

Benson, R. H., 1984, Perfection, continuity, and common sense in historical geology, *in* Berggren, W. A., and Van Couvering, J. A., eds., Catastrophes and earth history: Princeton, New Jersey, Princeton University Press, p. 35–75.

Berggren, W. A., and Van Couvering, J. A., 1984, Catastrophes and earth history: Princeton, New Jersey, Princeton University Press, 464 p.

Bretz, J. H., 1923, The Channelled Scablands of the Columbia Plateau: Journal of Geology, v. 31, p. 617–649.

——, 1969, The Lake Missoula floods and the Channelled Scabland: Journal of Geology, v. 77, p. 505–543.

Croft, S. K., 1982, A first-order estimate of shock heating and vaporization in oceanic impacts, *in* Silver, L. T., and Schultz, P. H., eds., Geological implications of impacts of large asteroids and comets on the Earth: Geological Society of America Special Paper 190, p. 143–152.

Dott, R. H., Jr., 1983, 1982 Society of Economic Paleontologists and Mineralogists Presidential Address: Episodic sedimentation; How normal is average? How rare is rare? Does it matter?: Journal of Sedimentary Petrology, v. 53, p. 5–23.

——, 1987, An episodic view of shallow marine clastic sedimentation, *in* deBoer, P. L., van Gelder, A., and Nio, S. D., eds., Tide-influenced sedimentary environments and facies: Amsterdam, Riedel Publishing Company, 530 p.

Ewing, M., and Press, F., 1955, Tide-gauge disturbances from the great eruption of Krakatoa: EOS Transactions of the American Geophysical Union, v. 36, p. 53–60.

Gault, D. E., and Sonett, C. P., 1982, Laboratory simulation of pelagic asteroidal impact; Atmospheric injection, benthic topography, and the surface wave radiation field, *in* Silver, L. T., and Schultz, P. H., eds., Geological implications of impacts of large asteroids and comets on the Earth: Geological Society of America Special Paper 190, p. 69–92.

Giekie, A., 1875, Life of Sir Roderick I. Murchison, BART [v. II]: London, J. Murray, v. II.

Gretener, P. E., 1967, Significance of the "rare event" and related concepts in geology: American Association of Petroleum Geologists Bulletin, v. 51, p. 2197–2206.

——, 1984, Reflections on the "rare event" and related concepts in geology, *in*

Berggren, W. A., and Van Couvering, J. A., eds., Catastrophes and earth history: Princeton, New Jersey, Princeton University Press, p. 77–89.

Grieve, R.A.F., 1982, The record of impact on Earth; Implications for a major Cretaceous/Tertiary impact event, *in* Silver, L. T., and Schultz, P. H., eds., Geological implications of impacts of large asteroids and comets on the Earth: Geological Society of America Special Paper 190, p. 25–37.

—— , 1987, Terrestrial impact structures: Annual Review Earth and Planetary Sciences, v. 15, p. 245–270.

Heezen, B. C., and Schneider, E. D., 1966, Abyssal plains, *in* Fairbridge, R. W., ed., The encyclopedia of oceanography: New York, Reinhold Publishing Corporation, p. 4–6.

Hsü, K. J., 1983, Actualistic catastrophism; Address of the retiring President of the International Association of Sedimentologists: Sedimentology, v. 30, p. 3–9.

Janda, R. J., and Meyer, D. F., 1985, Fluvial sedimentation following Quaternary eruptions of Mt. Saint Helens, Washington: Geological Society of America Abstracs with Programs, v. 17, p. 618.

Jones, E. M., and Kodis, J. W., 1982, Atmospheric effects of large body impacts, *in* Silver, L. T., and Schultz, P. H., eds., Geological implications of impacts of large asteroids and comets on the Earth: Geological Society of America Special Paper 190, p. 175–186.

Melosh, H. J., 1982, The mechanics of large meteroid impacts in the Earth's oceans, *in* Silver, L. T., and Schultz, P. H., eds., Geological implications of impacts of large asteroids and comets on the Earth: Geological Society of America Special Paper 190, p. 121–127.

O'Keefe, J. D., and Ahrens, T. J., 1982, Impact mechanics of the Cretaceous-Tertiary extinction bolide: Nature, v. 298, p. 123–127.

Press, F., and Harkrider, D., 1966, Air-sea waves from the explosion of Krakatoa: Science, v. 154, p. 1325–1327.

Salop, L. J., 1977, Glaciations, biological crises and supernovae: Catastrophist Geology, v. 2, no. 2, p. 22–41.

Schindewolf, O. H., 1977, Neocatastrophism?: Catastrophist Geology, v. 2, no. 2, p. 9–21.

Schmidt, R. M., and Holsapple, K. A., 1982, Estimates of crater size for large-body impact; Gravity-scaling results, *in* Silver, L. T., and Schultz, P. H., eds., Geological implications of impacts of large asteroids and comets on the Earth: Geological Society of America Special Paper 190, p. 93–102.

Short, N. M., 1975, Planetary Geology: Englewood Cliffs, New Jersey, Prentice-Hall, Inc., 361 p.

Silver, L. T., 1982, Introduction, *in* Silver, L. T., and Schultz, P. H., eds., Geological implications of impacts of large asteroids and comets on the Earth: Geological Society of America Special Paper 190, p. xiii–xix.

Silver, L. T., and Schultz, P. H., 1982, Geological implications of impacts of large asteroids and comets on the Earth: Geological Society of America Special Paper 190, 528 p.

Simken, T., and Fiske, R. S., 1983, Krakatoa, 1883; The volcanic eruption and its effects: Washington, D.C., Smithsonian Institution Press, 464 p.

Toon, O. B., Pollack, J. B., Ackerman, T. P., Turco, R. P., McKay, C. P., and Liu, M. S., 1982, Evolution of an impact-generated dust cloud and its effects on the atmosphere, *in* Silver, L. T., and Schultz, P. H., eds., Geological implications of impacts of large asteroids and comets on the Earth: Geological Society of America Special Paper 190, p. 187–200.

Weigle, R. L., 1970, Tsunamis, *in* Weigle, R. L., ed., Earthquake engineering: Englewood Cliffs, New Jersey, Prentice-Hall, Inc., p. 253–306.

Weissman, P. R., 1982, Terrestrial impact rates for long- and short-period comets, *in* Silver, L. T., and Schultz, P. H., eds., Geological implications of impacts of large asteroids and comets on the Earth: Geological Society of America Special Paper 190, p. 15–24.

Wetherill, G. W., and Shoemaker, E. M., 1982, Collision of astronomically observable bodies with the Earth, *in* Silver, L. T., and Shultz, P. H., eds., Geological implications of impacts of large asteroids and comets on the Earth: Geological Society of America Special Paper 190, p. 1–13.

Wolman, M. G., and Miller, J. P., 1960, Magnitude and frequency of forces in geomorphic processes: Journal of Geology, v. 68, p. 54–74.

Yokoyama, I., 1981, A geophysical interpretation of the 1883 Krakatoa eruption: Journal of Volcanology and Geothermal Research, v. 9, p. 359–378.

MANUSCRIPT ACCEPTED BY THE SOCIETY APRIL 4, 1988

Printed in U.S.A.

Geological Society of America
Special Paper 229
1988

Nearshore responses to great storms

Robert A. Morton, *Bureau of Economic Geology, The University of Texas at Austin, P.O. Box X, University Station, Austin, Texas 78713*

ABSTRACT

Great storms originating in both tropical and extratropical regions represent upper limits in the continuum of physical forces that affect sedimentation in shallow marine basins. The energy expended and sediment transported during a few storm hours may equal many years of non-storm work, or they may surpass thresholds unequaled by less energetic processes; consequently, the volume of storm sediment preserved in the basin fill is disproportionate to the frequency of these convulsive events.

Storms can produce large geomorphic features and exaggerated bedforms, especially on low-energy coasts and adjacent shelves where they commonly exert their greatest influence. During peak intensity, combined wind-driven geostrophic currents and wave-induced oscillatory currents flow alongshore and slightly offshore at up to 2 m/s in relatively shallow water (<50 m). The coarsest nearshore sediments (gravel and sand) are transported and deposited by these strong currents, which have a principal unidirectional component. As the unidirectional current component diminishes, wave orbital velocities assume greater importance, and the subequal oscillatory and unidirectional motion molds fine sand that settles from suspension; in the final phase of deposition, mud accumulates below fair-weather wave base in slack water. The resulting graded beds constitute a single but common class of shelf storm deposit.

Thicknesses of graded storm beds typically range from a few decimeters to a few meters, depending directly on wave and current energy and flow duration, and inversely on distance from the shoreline (water depth). Volumetric estimates suggest that each graded bed is not necessarily a product of sand eroded from the beach and shoreface during a single storm. Instead, they may also result from sand being transported offshore by a series of moderately intense events and later selectively sorted during a single extreme storm lasting several days.

Most amalgamated storm beds are preserved in regressive sequences deposited on broad, moderately stable shelves of passive continental margins; however, some accumulated in thick transgressive sequences deposited on narrow, rapidly subsiding and tectonically active shelf margins associated with converging plates. Episodic storm deposition also accounts for coastal and shelf aggradation under static sea-level conditions. Similarities between ancient storm deposits and their modern analogs suggest that extant marine processes adequately explain the observed sedimentologic properties, textural patterns, and vertical sequences of sedimentary structures. Super storms, which are predicted by analysis of extreme values (e.g., wave heights or wind speeds) may be statistical artifacts or, if they exist, their influence may be unrecognized or unimportant in the geological record.

INTRODUCTION

Satellite images from outer space confirm what voyagers and explorers have known since the Middle Ages: our planet is constantly besieged by storms. Geologists are fascinated with storms because they can accomplish in hours the same amount of sedimentological work that normally takes years. Many authors, including Hayes (1967) and Swift and others (1971), have put storms in proper perspective by recognizing their geological significance and the fact that storms approach steady-state conditions when averaged over long periods of time. Most coarse-grained nearshore sediments are deposited during episodic events that dissipate their energy by means of torrential rains, large waves, widespread overwash, and powerful wind-driven currents. The same storm may either simultaneously or sequentially cause stream flooding, shoreline erosion, coastal inundation, and sediment movement on the shelf that would produce coeval storm deposits or erosional surfaces in diverse depositional environments that are widely separated. Vast differences in environmental response to the same storm make the recognition and stratigraphic correlation of these signatures difficult.

The nearshore responses to great storms are best known from studies of low-energy microtidal coasts where surficial features are shaped by events causing the greatest disequilibrium with fair-weather processes (Brown, 1939; Vermeer, 1963; Ball and others, 1967; Hayes, 1967; Morton, 1979; Morton and Paine, 1985). Increased water depths and accelerated flow during storms may produce such large depositional features that they remain inactive until the next storm of equal or greater magnitude. Despite the number of well-documented investigations of modern coastal storms, the size and morphology of some obvious storm deposits cannot always be explained. An example is the washover fan studied by Andrews (1970), which is much larger and exhibits features that are different from any known active fan. The morphologic discrepancies raise questions concerning whether the washover was created by events with magnitudes beyond our experience and perhaps beyond our comprehension, or whether it was formed by a combination of processes operating when the geological setting was somewhat different, such as a slightly higher sea level.

The sedimentologic consequences of great storms are still debated because we almost always lack direct evidence linking the observed depositional features with extremely rare meteorologic events that have poorly defined characteristics other than the implied condition of equaling or exceeding storms of historical record. Additional uncertainties are introduced by speculations regarding the intensity of storms during earlier periods when global temperatures were both warmer and colder and when continents and adjacent shelves occupied different latitudes. The following discussion is limited to the geologic period when the speed of Earth's rotation, proximity to the moon (tidal effect), and composition of the atmosphere were similar to present-day conditions.

Although the important role of storms is now well established, there have been few attempts to compare the characteristics of great storms from different latitudes or to summarize the sequence of storm processes affecting the nearshore marine environment where most data are available. In addition, the sedimentologic attributes of nearshore storm deposits are rarely compared with those of some ancient shelf sequences to determine whether or not they are products of storms having energy levels greatly different from those presently observed.

WIND SYSTEMS AFFECTING THE MARINE ENVIRONMENT

Several different types of atmospheric disturbances can produce extreme winds. The largest and most violent storms are tropical and extratropical cyclones that revolve around centers of low barometric pressure. Although both categories of cyclones share some common attributes, such as direction of rotation, they represent opposite ends of the storm spectrum. Extratropical cyclones are large, mid- to high-latitude, winter storms that form in a baroclinic environment around a cold core and travel from west to east. Their strongest winds occur along frontal boundaries between different air masses having steep density (temperature and moisture) gradients. In contrast, tropical cyclones, which include tropical storms, hurricanes, and typhoons, are much smaller, more intense storms that originate between 30°N and 20°S latitudes and travel from east to west (Gray, 1968). They form during the summer when sea-surface temperature, rain center precursors, and upper-level circulation favor cyclogenesis (Riehl, 1979). The well-organized hurricane vortex forms around the warm-core, low-pressure eye and is driven by convection in a barotropic environment where only minor differences in air temperature (2 to 3°C) are needed to maintain cyclonic circulation (Simpson and Riehl, 1981). The thermodynamics of the two principal cyclonic systems are quite different, and yet their influence on shorelines and continental shelves is similar; they both generate high waves and strong currents that are capable of eroding, transporting, and depositing large volumes of sediment in relatively short periods of time.

High wind speeds and extreme turbulence are also characteristics of thunderstorms and tornadoes. Neither of these atmospheric disturbances is geographically or seasonally constrained and each shares a common attribute with extratropical and tropical cyclones; thunderstorms are driven by convection and tornadoes have vertical vortices. Although thunderstorms and tornadoes can cause extreme wind shears and some of the highest measured wind velocities, both types of storm systems are normally too small and too brief to have any significant influence on nearshore sedimentation.

COUPLING OF STRONG WIND AND MARINE WATER

Variations in solar heating and attendant differences in the reflection and absorption of solar radiation cause strong winds

(Greeley and Iverson, 1985). The transfer of kinetic energy from the atmosphere to the sea is accomplished by tangential shear stresses that form waves and generate wind-drift currents in the upper layers of the ocean. The shear stresses depend directly on air density, wind velocity, and the drag coefficient; the latter term is a measure of the boundary roughness at the wind-water interface and an indicator of the expected transfer of momentum between the two fluids (Roll, 1965). Apparently the drag coefficient increases at wind speeds above hurricane force (33 m/s), possibly because of heavy rainfall (Caldwell and Elliott, 1971) as well as decapitation of wave crests and increasing sea spray and foam (Kraus, 1972).

Our knowledge of shelf currents and sediment transport during storms is improving with the acquisition of data in the bottom boundary layer. Most nearshore data sets show the effects of efficient coupling between wind and water creating a single-layer system where both fluids move in the same direction, or a two-layer system may develop where the bottom currents flow at a high angle or in a direction opposite to that of the wind. Even when the wind blows alongshore, there is a rotation in the water column due to the Coriolis effect, known as Ekman veering. Water depths of about 90 m are necessary for complete rotation, so in shallow water, only a partial spiral develops and bottom currents flow at a slight angle to the wind-driven surface currents.

Contrary to the opinion offered by Duke (1985), coupling of hurricane wind and water need not be less efficient than coupling by winter storms, either because of brief hurricane duration or possible density stratification of the water column. Density stratification, due to differences in temperature, salinity, and suspended sediment in the water, can reduce the near-bottom effect of wind-driven currents by causing multilayer flow and retarding the downward transfer of energy in deep water. However, coastal water masses are normally well mixed by storm waves and currents in water depths of a few tens of meters where most storm beds are deposited.

LIMITS OF STORM SIZE AND INTENSITY

The magnitude and strength of tropical and extratropical cyclones are controlled by different atmospheric and oceanic interactions. As a result, storm diameters and wind velocities cover a wide range of possibilities. For the purposes of this paper, however, attention is focused on the conditions causing the largest and strongest storms for which data are available.

Atmospheric conditions

Most cyclonic wind systems take several days or longer to mature into extreme storms because they rely on positive feedback mechanisms to become larger and stronger. Extratropical storms receive their energy from the sharp horizontal contrast in temperature and barometric pressure between adjacent air masses. As the different air masses collide, ascending warm air and descending cold air transform the available potential energy into kinetic energy. Tropical cyclones require low-level horizontal convergence of wind that is balanced by upper-level divergence, causing a reduction in barometric pressure at the storm center. Storm intensification involves latent heat released by organized convection; the inward-directed pressure gradient causes additional inflow of warm, moist air that ascends and condenses, contributing to further deepening of the storm. The requisite transfer of kinetic energy to the surrounding environment is aided by cold air descending from an upper-level air mass drawn into the storm system from higher latitudes; these processes may cause further reduction of barometric pressure and strengthening of the storm (Simpson and Riehl, 1981). Extremely low barometric pressure is maintained for only 12 to 24 hr, although there may be subsequent periods of intensification (Riehl, 1979).

Oceanic conditions

Extratropical cyclones derive their energy mainly from the atmosphere although instabilities caused by differences in air and ocean temperatures can contribute to their turbulence. In contrast, the ocean is the principal energy source for tropical cyclones. The intensity of tropical cyclones depends on sea-surface temperature and optimum storm speed, assuming favorable conditions in the upper atmosphere. Slowly moving or stalled tropical cyclones may lose strength because upwelling brings cooler water near the surface, whereas fast-moving storms may encounter a land mass or cooler water before reaching unusual strength. Another factor controlling storm size is geographic position. Tropical cyclones grow larger and may intensify as they travel away from the equator; they normally reach their maximum strength just before recurving poleward (Riehl, 1979).

The locations and orientations of continental margins relative to the principal storm trajectories are important because they determine the severity of storm impact and the potential sedimentologic effects. Monthly records kept by the U.S. Department of Commerce indicate that extratropical cyclones are consistently more severe in the northern Atlantic than in the northern Pacific. Furthermore, the northward recurving trajectories of tropical cyclones in the northern hemisphere means that they normally reach their peak when crossing the Gulf of Mexico, but are in their terminal stages before reaching Japan or the northeast, U.S. Atlantic coast. The orientations of shorelines and shelves with respect to the paths of major storms controlled storm dominance in the geologic past, just as they do today.

Paleogeographic and paleoclimactic implications

Speculations regarding the occurrence of super storms in the geologic past (Gretener, 1967; Dott, 1983) are based on probabilities of extreme events rather than on dynamics of the atmosphere. Even meteorologists disagree as to the likelihood of exceptionally intense storms either in the present or in the past when global temperatures were warmer. They point out that counterbalancing forces tend to maintain equilibrium by preventing development of extreme atmospheric conditions. For example, the carbon dioxide and ammonia-rich composition of the Precambrian atmosphere would have increased global temperatures, but these conditions would also have caused offsetting reactions such as increased cloudiness (Schopf, 1980).

Significantly increased global temperatures, like those during the Late Cretaceous (Savin, 1977), would have shifted the zone of tropical cyclogenesis into higher latitudes and would have extended the season of ice-free ocean, thus providing greater opportunity for storm influence. But higher global temperatures also would have decreased the thermal gradient between the poles and equator (Fischer, 1982) and possibly reduced the sharp temperature contrast along frontal boundaries that is so critical to intensification of extratropical storms. The pole-to-equator temperature contrast during the Cretaceous was in the range of 17° to 26°C, as compared to a difference of 41°C at present (Barron, 1983).

Higher atmospheric temperatures would also cause higher vapor pressure and increased evaporation, thereby possibly setting an upper limit on sea-surface temperatures. The exponential increase in evaporation becomes so large at temperatures greater than 25°C that the 28°C isotherm is the highest that can be mapped over tropical oceans (Simpson and Riehl, 1981). The theoretical upper limit of ocean temperature is about 33°C because the cooling effect caused by evaporation increases faster than the warming effect caused by the vertical heat flux (Schopf, 1980).

Regardless of the effects higher temperatures may have had on storm intensity, reduction in the average sea-surface temperature by a few degrees during the summer could have drastically limited the number of hurricanes. Tropical cyclones do not form where ocean temperatures are less than 26°C (Simpson and Riehl, 1981); thus, minor lowering of average sea temperature may have prevented tropical cyclones from forming while allowing extratropical storms to persist.

The speed of Earth's rotation is another variable that may have influenced storms in the geologic past. Variations in rotational speed could have changed storm size, storm trajectory, and possibly storm intensity without changes in atmospheric conditions. Changes in the speed of rotation would alter the Coriolis force, which in turn, would alter preferential direction of movement and the radius of maximum winds.

The number of annual growth rings of Devonian corals suggests that the rotation speed of the Earth was approximately 10 percent faster then than today (Scrutton and Hipkin, 1973). An increased Coriolis force accompanying slightly faster rotation may have resulted in larger storms that tended to recurve more easily toward the poles. However, the increase in Coriolis effects away from the equator (paleolatitude) probably has been a more significant storm modifier because the changes in rotational speed since the Paleozoic have not been exceptionally large. During the Precambrian, before tidal friction slowed the rotational speed and caused the moon to move away from the Earth, tidal ranges and storm frequencies probably were much greater than they are today.

QUANTIFICATION OF GREAT STORMS

Quantifying great storms is difficult because observation stations are rarely located near the storm centers. Available data are biased toward areas that are densely populated or where research has been conducted in the past few decades. Despite the limited monitoring of great storms, measurements from several epicontinental seas and ocean basins (Table 1), including some where storm beds have been reported, show that values of most storm parameters fall within the range observed for a Gulf of Mexico hurricane and a western Atlantic extratropical storm. A noticeable difference between storm parameters is that extratropical cyclones affect much larger areas and normally last several times longer than tropical cyclones, but tropical cyclones typically have larger waves, greater wind velocities, and much higher storm surges.

Storm dimensions and durations

Extratropical storms are normally slow-moving, large-scale systems with wind circulation patterns covering more than 1,000 km. Tropical cyclones, on the other hand, begin as relatively small storms because they form close to the equator and their winds revolve around a narrow center of low pressure. The diameter of cyclonic circulation for most western Atlantic and Gulf Coast hurricanes is between 100 and 300 km, whereas organized circulation for northern Pacific typhoons may be up to 2,000 km in diameter. Typhoons are often larger and stronger than other hurricanes because their paths can cross greater unobstructed expanses of warm water. Diameter of circulation and radius of maximum winds are both measures of storm size, but length of coastline affected (Table 1) is perhaps a better expression of the geologic extent of a storm. This parameter, which is a function of storm path and duration, suggests that the most violent extratropical storms can affect an entire depositional basin and possibly create a chronostratigraphic marker.

Storm systems at any latitude can last for several weeks; however, most of that time is spent over deep water where the influence on bottom currents and sediment transport is negligible. Extratropical storms may move slowly or become nearly stationary, and therefore, they may affect shelf processes for more than two days. Tropical cyclones, on the other hand, typically cross the shelf in less than a day unless their path is parallel to the coast or looped; even then their influence on a given area is normally of limited duration because of the fast forward motion.

Wind velocities and wave heights

Maximum wind velocity recorded during storms or estimated from structural damage is about 100 m/s. The convergence of data toward this value suggests that there may be a physical upper limit to wind speed owing to obtainable pressure gradients and the rapidity of energy dissipation in the atmosphere. Typhoon Tip (Table 1) set a record for lowest measured baro-

TABLE 1. CHARACTERISTICS OF SOME EXTREME NORTHERN HEMISPHERE STORMS

Ocean basin	Gulf of Mexico	Atlantic Ocean	North Sea	Bering Sea	Pacific Ocean	Gulf of Alaska
Type of storm	H†	E	E	E	H‡	E
Storm date	Aug. 1969	Mar. 1962	Nov. 1965	Nov. 1974	Oct. 1979	Jan. 1952
Coastline influenced (km)	150	2,400	250+	700	1,650	1,300
Shelf duration (hr)§	12	48	24	48	36	30
Maximum wave height (m)	22	12	—	—	15*	15
Significant wave height (m)	13	10	6	4	—	—
Peak wind velocity (m/s)	94	41	42	36	85	31
Open coast surge (m)	8	4	2	7	—	—
Bottom current velocity (cm/s)	200	—	150	—	—	—
Depth of erosion (cm)	250	—	—	—	—	—
References	1,2,3,4	5,6	7	8,9	10	11

H = hurricane; E = extratropical.
*Swell during extratropical stage.
†Hurricane Camille.
‡Typhoon Tip.
§Shelf duration is the time the storm spent crossing the continental shelf.
References: 1. Simpson and others (1970); 2. Bea (1974); 3. Bea and Bernard (1973); 4. Murray (1970); 5. Cooperman and Rosendal, (1962); 6. Bretschneider (1964); 7. Gienapp (1973); 8. Fathauer (1975); 9. Brower and others (1977); 10. Dunnavan and Dericks (1980); 11. Danielsen and others (1957).

metric pressure (870 mb) and largest circulation pattern (2,220 km), yet its fastest surface winds were only 85 m/s (Table 1). Although storm-induced currents on the continental shelf largely depend on wind stress, there is no simple relationship between wind speed and bottom-current velocity.

Wave height depends on wind speed, direction, duration, and fetch. Generally the significant waves having long periods travel faster than the forward motion of the storm. These forerunners inflict severe damage on nearby beaches and initiate beach erosion. Extreme wave heights are short-lived and located near the centers of both tropical and extratropical storms. Those generated in the right rear quadrant of tropical cyclones travel in the same direction as the storm, propagate under the influence of the strongest winds, and have the longest effective fetch (Riehl, 1979).

Bea (1974) and Ward and others (1977, 1978) used hindcast methods to estimate extreme and significant wave heights for hurricanes and winter storms in the Gulf of Mexico and the western Atlantic. Comparison of their data indicates that in both areas hurricanes generate the highest waves, and significant wave heights are comparable (10 to 12 m) for both winter and summer storms, but maximum wave conditions last 2 to 3 times longer for winter storms in the western Atlantic than for hurricanes. Simpson and Riehl (1981) estimated that the upper limit of significant deep-water wave heights in extreme hurricanes is about 15 to 18 m. Some extratropical storms may generate even larger waves because extreme winds can persist over nearly unlimited fetch. But because of their overlapping paths and significant wave heights, hurricanes may transport and rework coarse sediment farther out on the shelf and in deeper water than extratropical cyclones.

Quayle and Fulbright (1975) estimated return periods of extreme wind speeds and wave heights for all coasts of the United States using the computational procedures of Thom (1973a, 1973b). Their calculations generally show that atmospheric and oceanic extremes are caused by extratropical cyclones and that wind velocities up to 66 m/s, significant wave heights up to 25 m, and extreme wave heights up to 46 m might be expected every 100 yr (.01 probability) along the stormiest coastal regions such as western Alaska and the southeast coast of the United States. The fact that estimated wind velocities are within the range of historical observations, but that significant or extreme wave heights are not, simply underscores the statistical basis for analysis of extremes which may provide unrealistic estimates. Simpson and Riehl (1981) argued that interference of wave trains generated from different sources would cause premature breaking of extreme waves; therefore, they gave lower estimates than those provided by Quayle and Fulbright (1975).

Surge heights and durations

Abnormal sea levels during storms are governed by atmospheric pressure, wind stress, waves, Coriolis effect, and rainfall, as well as the local influences of coastal configuration and offshore profile. The combination of low barometric pressure, strong onshore winds, large breaking waves, abundant runoff, converging shorelines, and broad, shallow shelves tends to increase storm surge elevations near the coast.

Another component of storm surge is the normal astronomical tide, which increases or decreases the potential storm surge depending on the tidal cycle. The March 1962 storm that affected the U.S. Atlantic coast had a surge of 4 m (Table 1); it was particularly destructive because it lasted more than four to five spring high tides (Cooperman and Rosendal, 1962), prolonging the processes of beach erosion and coastal overwash. Peak surge for hurricanes normally lasts only a few hours (less than one tidal cycle) except for some unusual circumstances such as slow forward movement or a track parallel to the coast. Nevertheless, the highest storm surges are almost always associated with hurricanes such as Camille, which flooded the low-lying coast of Mississippi with an 8-m surge (Table 1).

Bottom currents

Nearshore bottom currents include both unidirectional and orbital motions that incorporate principal components of tides, wind, and waves. In shallow water, the storm wind and wave components dominate and are augmented by the tidal component, which can be important or negligible depending on the local range of astronomical tides. In deeper water, the wave component diminishes substantially. Because of the relationship between wave height and water depth, wave energy dissipation is concentrated on the shoreface.

According to calculations by Herbich (1977), maximum wave orbital velocities at the sea floor would occur in water 10 to 20 m deep, which is the depth range of most known storm deposits. Significant storm waves such as those of Camille (Table 1) would create bottom velocities under the wave crests of about 500 cm/s in 20 m of water and about 300 cm/s in 45 m of water. Even if these calculations slightly overestimate actual conditions, they clearly show that wave orbital velocities can equal or exceed the wind-driven currents at comparable depths.

Near-bottom velocities of 50 to 100 cm/s in up to 80 m of water are reasonably common (Smith and Hopkins, 1972; Caston, 1976; Channon and Hamilton, 1976; Sternberg and Larsen, 1976; Lewis, 1979; Niedoroda and Swift, 1981; Tryggestad and others, 1983; Cacchione and Drake, 1982; Schroeder and McGrail, 1982). The similarity of these measurements is not surprising considering that in the nearshore marine environment, bottom currents reflect the balance between applied wind stresses and bottom friction.

Depths of erosion and deposition

Estimated depths of nearshore erosion and deposition during extreme storms are best obtained from outcrops or cores. The few direct measurements of storm-related erosion are from shelves where exposure of buried pipelines indicate local sea-floor scour in the range of 1 to 2 m (Blumberg, 1964; Herbich, 1977). These estimated depths of erosion are anomalously high and probably are not representative of lowered sea-floor elevations across broad expanses of the shelf. Instead, it seems likely that erosive shelf currents would rapidly entrain sediment and reach their transporting capacity after a few decimeters of shelf sediment were eroded. The temporal and spatial variability of shelf currents during a storm suggests that the depth of erosion also is highly variable from one location to another.

Bathymetric surveys conducted immediately before and after hurricane Camille (Bea and Bernard, 1973) record elevation changes ranging from more than 2 m decrease to 10 m increase. Additional sea-floor surveys of the modern Mississippi Delta (Coleman and others, 1983) indicate that cyclic wave loading and other storm processes may be responsible for redeposition of even larger volumes of sediment. These rapid bottom changes are related primarily to slumping and mass movement on the steep, unstable delta-front slope rather than erosion and deposition that would occur on a more stable shoreface and shelf of an abandoned delta lobe or along an interdeltaic embayment.

Number of storms and frequency of recurrence

Great storms are common compared to major earthquakes, tsunamis, catastrophic volcanic eruptions, and other convulsive events. Of the 50 to 100 tropical cyclones and several thousand extratropical cyclones that form each year (Riehl, 1979), some die before making landfall and others weaken before crossing a continental shelf, but many storms enter relatively shallow water and are strong enough to be considered significant geologic events, especially in terms of shoreline and shelf sediment transport.

Simpson and Lawrence (1971) calculated recurrence intervals of severe hurricanes (wind speeds >56 m/s) based on storms striking the U.S. mainland since the late 1800s. They estimated return periods ranging from 14 to 85 yr, depending mainly on shoreline orientation with respect to the tracks of major storms. Improved monitoring techniques using aircraft, satellite, and radar observations indicate some slight biases in estimated frequency because paths and landfall positions are now accurately plotted as compared to determinations more than a century ago. The spatial distribution of storm probabilities clearly demonstrates the substantial differences in storm influence among coastal segments and the frequent recurrence of extreme storms where their preferred paths intersect the coast.

STORM INFLUENCE ON MUDDY SHELVES

Synoptic observations from the Gulf of Mexico

Despite recent studies documenting the physical and biological processes operating in the shelf environment, our understanding of sediment transport and fluid behavior in the bottom boundary layer during storms is imprecise and largely inferred from the characteristics of modern and ancient storm beds deposited on muddy shelves. The presence of coarse clastics in an otherwise heterogeneous or muddy substrate requires explanation that storm processes can easily provide. For this reason, the alternating sand and mud layers are perhaps the best known of the myriad of storm-influenced sedimentary sequences. Graded beds do not represent all shelf sequences, but this one type of storm bed occurs worldwide and is recognized as an indicator of storm deposition.

Hayes (1967) first equated thin graded shelf deposits in the Gulf of Mexico with storm processes that he interpreted as storm-surge ebb, or the seaward flow of elevated water after storm landfall. After Hayes introduced the concept of storm-surge ebb, the term was retained and eventually broadened to encompass a variety of nearshore processes quite different from those originally proposed. Some authors still use storm-surge ebb to explain graded shelf deposits despite the explicitness of the term and the strong evidence favoring waves and wind-driven alongshore currents as the principal mechanisms dispersing sediment on most non-tidal shelves. Tidal currents and reflux of flood waters may locally reinforce or temporarily exceed bottom currents, but they probably are subordinate to the high-energy wind-driven currents that persist over most of the shelf region.

The classical study by Hayes was from the northwestern Gulf of Mexico where modern sediments have traditionally served as facies models for interpreting depositional sequences. Fluvial, deltaic, barrier island, and strandplain systems of this area have been described in much more detail than the subtidal systems (shoreface, shelf, and slope), which have only recently received much attention. These latter depositional environments are better known from other regions, especially those where sand sheets or sand ridges form the sea floor.

The muddy continental shelf west of the Mississippi Delta is a storm-dominated system undergoing both transgression and minor aggradation. The diverse Holocene histories, textural patterns, and depositional sequences of this broad region allow it to serve simultaneously as a model of both nondeposition and sediment accumulation on a microtidal muddy shelf during a relative sea-level rise or stillstand. Because high-energy events are the dominant mechanism of sediment transport on this shelf, it is instructive to present some of the observed effects of recent storms and to interpret their stratigraphic significance.

Hurricane Camille. The most intense Gulf Coast hurricane in modern times (class 5 of Saffir/Simpson scale) had winds up to 94 m/s, waves of 13 m (Table 1), and a diameter of about 650 km as it crossed the continental shelf just east of the Mississippi Delta. A current meter located 100 km east of the eye and in 10 m of water recorded near-bottom current velocities up to 160 cm/s while the storm center was over deep water (Murray, 1970). These high current velocities were generated by sustained winds of only 20 m/s. As the storm moved onshore, the direction of bottom currents changed from alongshore to offshore in response to a change in wind direction.

The shear stresses of powerful bottom currents and large waves near the eye of Camille were not measured and can only be inferred from lowered sea-floor elevations of up to 3 m. Such bathymetric changes signify slope failure and mass movement in water more than 100 m deep (Bea and Bernard, 1973). High water content of these oversteepened prodelta muds undoubtedly facilitated their basinward transport because bottom velocities of only 20 cm/s are sufficient to erode clay with 90 percent water (Schopf, 1980). Mudflows, slumps, and similar types of gravity-driven processes are mechanisms that transfer shallow-water sediments into deep water seaward of actively prograding deltas (Coleman and others, 1983). Storms can initiate these slope instabilities through rapid discharge and sediment loading near distributary mouths as well as through large waves that create pore-pressure fluctuations within the sediment column.

Tropical storm Delia. Large waves and powerful bottom currents are not always associated with extreme storms. Even minor disturbances with sustained wind speeds less than 25 m/s, such as tropical storm Delia (Fig. 1), can produce 7-m waves and bottom currents up to 200 cm/s in 18 m of water (Forristall and others, 1977, 1978). These strong alongshore currents, which lasted for more than a day, reworked and preferentially sorted the bottom sediments, producing a textural pattern parallel to flow direction (Fig. 2) that resembles a longitudinal sand ridge system without topographic relief. Insufficient sediment supply and brief flow duration prevented construction of bedforms having the same spacing as the large-scale alongshelf current patterns. Nevertheless, the selective sorting probably mimics the scale of features that would have formed under equilibrium conditions if ample sand had been available. Dunes and sand waves composed of fine sand are the bedforms normally predicted for the measured depth and sustained current velocities (Rubin and McCulloch, 1980); however, such bedforms rarely appear in fine sand in the marine environment because these size classes commonly travel in suspension.

The spatial distribution of grain sizes indicates both erosion and deposition by contour parallel currents having a large-scale flow structure that may have been locally influenced by sea-floor topography. Entrainment of modern muds left the cohesive relict (Pleistocene) mud exposed at the sea floor and flanked by silt and sandy silt that graded laterally into silty sand. Strong winter currents and biological mixing probably modified the post-storm stratification and sediment textures; however, it appears that the imprint of Delia's currents survived more than a year after the storm.

The spacing of the coarsest deposits (~5 km) is similar to dimensions reported for both modern and ancient sand ridges (Stubblefield and others, 1984; Swift and Rice, 1984). The linear shapes of the sandy deposits are also similar to yardangs, which offer minimum resistance to streamline flow (Greeley and Iverson, 1985). Macroscale flow patterns forming sand ridges in the marine environment have been attributed to slightly oblique flow that experiences accelerations and decelerations due to interactions with bottom topography (Swift and Rice, 1984). The observed textural pattern (Fig. 2) also favors a mechanism that responded to and interacted with sea-floor irregularities (Fig. 2).

Hurricane Alicia. In 1983, a hurricane of moderate strength (class 3 of the Saffir/Simpson scale) crossed the Texas shelf just west of the path taken by Delia (Fig. 1). Plots of composite wind vectors show that Alicia had a diameter of about 200 km and maximum sustained winds of about 60 m/s at peak intensity, which occurred shortly before landfall. Significant wave heights near Alicia's center were not recorded, but they must have been higher than 2 m considering that waves of 1 to 2 m were measured far from the direct influence of the storm (Garcia and Flor, 1984).

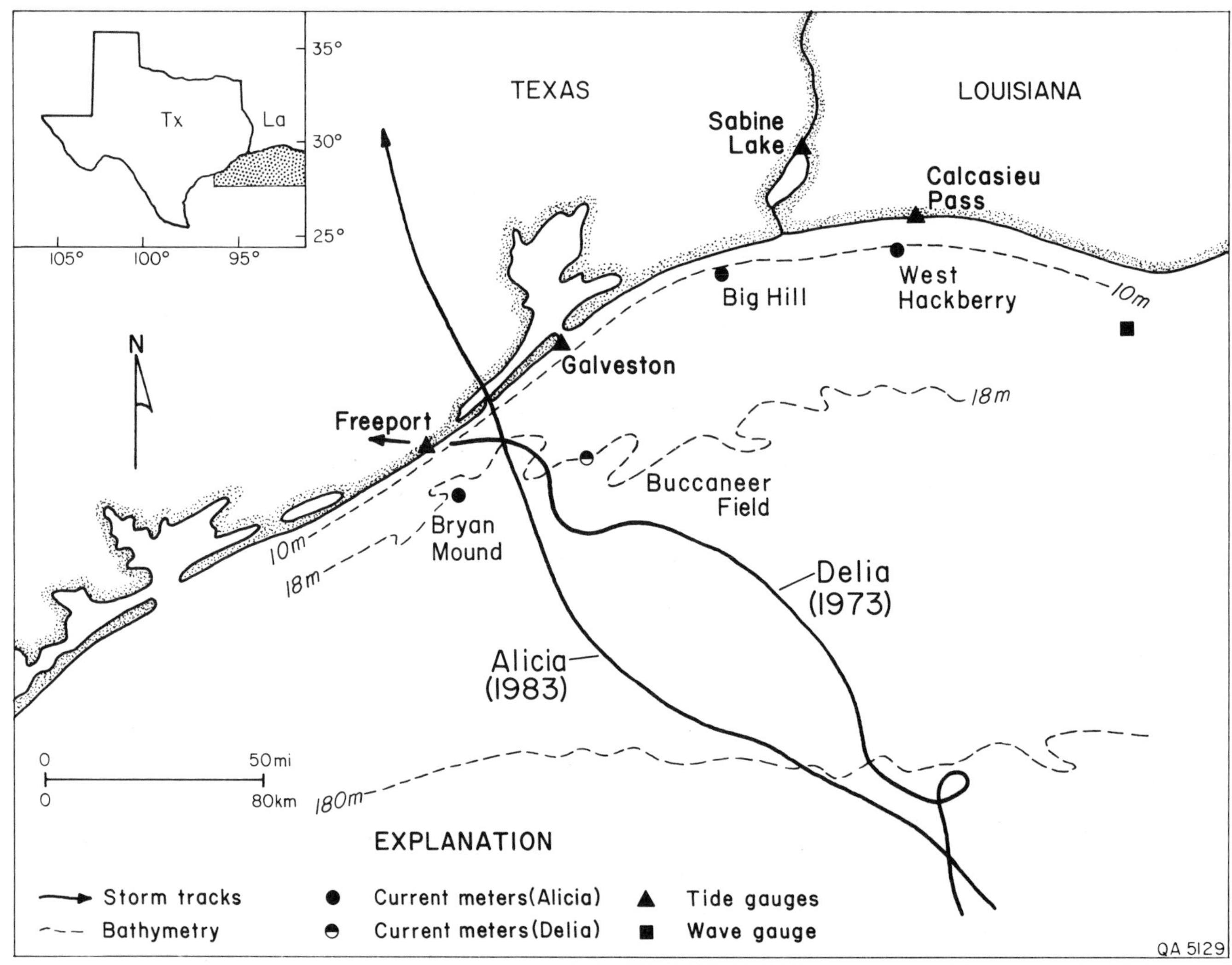

Figure 1. Locations of current meters and tide gauges operating in the northwestern Gulf of Mexico during tropical storm Delia (1973) and Hurricane Alicia (1983). Wind and current velocities for Delia were published by Forristall and others (1977) and those for Alicia were reported by Hann and others (1984). Wave heights and water levels for Alicia were summarized by Garcia and Flor (1984).

Alicia caused surge elevations of about 1 m along most of the coast except at Galveston (Fig. 3), which was near the eye in the right front quadrant (Fig. 1). There the surge rapidly rose to 3 m and dropped abruptly a few hours after landfall (U.S. Army Corps of Engineers, 1983).

Current meters operated by Texas A&M University (Hann and others, 1984) were located on either side of the storm center with the array near Bryan Mound being closest to the eye (Fig. 1). As Alicia crossed the shelf, the unidirectional component of bottom currents flowed alongshore and slightly offshore (Fig. 4) at up to 100 cm/s. Bottom currents of more than 50 cm/s lasted for 2 days, suggesting that the period of greatest sediment transport was of comparable duration.

Comparing Figures 2 and 4 illustrates that bottom currents began accelerating about a day before most tide gauges recorded peak water levels, and nearly two days before the surge reached its maximum elevation. Furthermore, while storm water receded from the barriers and adjacent bays, shelf bottom currents decelerated and flowed alongshore (southwest). These observations are incompatible with the storm-surge ebb model which would predict accelerated or maintained flow in an offshore (south or southeast) direction.

Current meters at three levels at the Bryan Mound site (Fig. 5) recorded well-organized unidirectional flow with slight Ekman veering near the bottom. The flow was highly frictional, causing the entire water column to move at nearly the same speed except when the greatest wind stress was applied; then the bottom currents reached their maximum velocities, which were about 33 percent slower than the surface currents. Peak bottom velocities occurred several hours before landfall and before maximum storm surge was recorded at the coast. The occurrence of maximum bottom currents following the highest wind stress, but

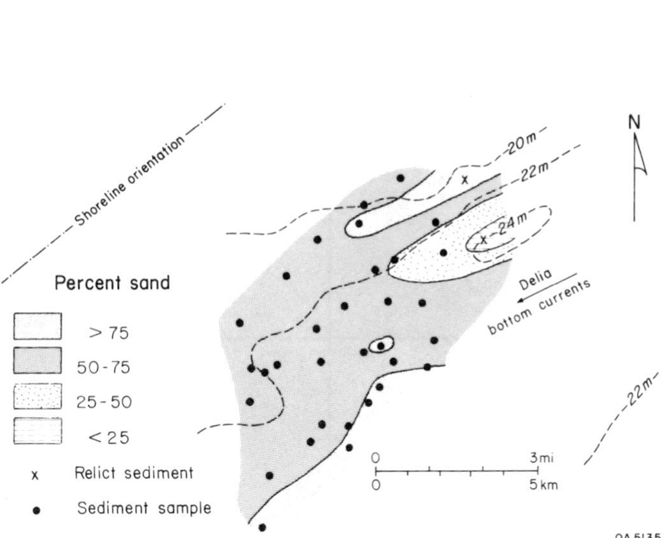

Figure 2. Percent sand in surface sediments collected at Buccaneer Field about one year after tropical storm Delia. Site location is shown in Figure 1. Sediments are composed of dark gray muddy sand and sandy mud with variable amounts of shell and mud clasts eroded from the underlying relict deltaic sediments. Sediment textures and descriptions were reported by Harper and others (1976).

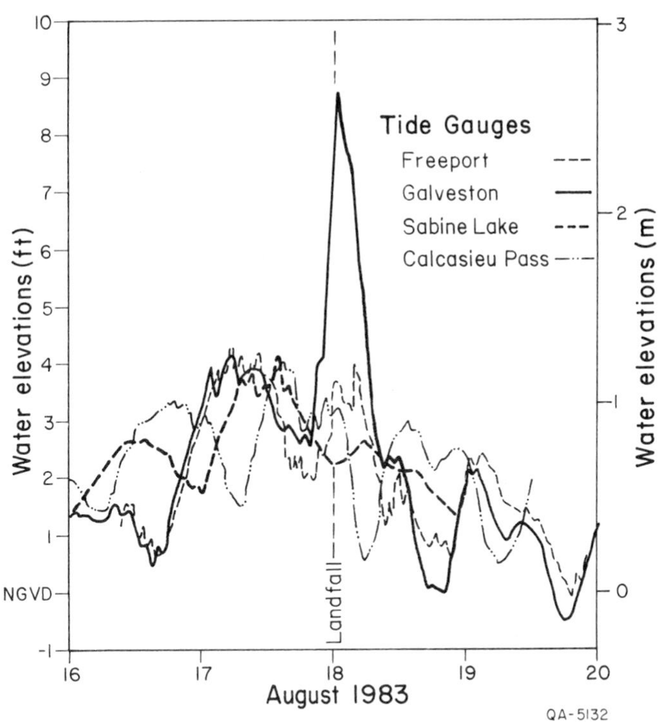

Figure 3. Water levels along the Texas-Louisiana coast during Hurricane Alicia. Data from Garcia and Flor (1984). Tide gauge locations are shown in Figure 1.

preceding the maximum coastal surge, is consistent with the observations made for tropical storm Delia (Morton, 1981).

Before landfall, the periods of bottom current acceleration, stability, and deceleration closely corresponded with changes in wind velocity and direction (Fig. 5). Bottom current velocities periodically increased and decreased in response to wind gusts and lulls lasting several hours. These short-term velocity fluctuations may result in small-scale interlaminations of sand and mud observed in some shelf deposits. After landfall, surface currents continued to flow southwestward even though the wind was blowing northeast and onshore. At the same time, water at middepths was nearly stationary while bottom water flowed to the northeast (Fig. 5) in response to the alongshore pressure gradient that preceded landfill. Momentum of the water column and brief duration of reversed currents prevented a reversal in bottom flow and transport of sediment toward the beach.

Sediment samples also were collected at the Bryan Mound site before and after Alicia (D. Harper, personal communication, 1985). The fact that observed changes can be explained by several different processes complicates interpreting the textural data. For example, a decrease in mean grain size or percent sand could result from either deposition of mud or selective removal of sand. Comparing sediment descriptions with the textural changes helped improve the sedimentological interpretations.

Before the storm, the surface sediments were nearly homogeneous with little or no evidence of layering. The coarsest sediment was sand, exhibiting cross-shelf trends in which the sand interfingered with and graded into mud (Fig. 6a). After the storm, the area was blanketed by mud (Fig. 6b), and the coarsest sediment was mostly sandy mud representing sand transported alongshelf (parallel to the bottom currents) and deposited where mud previously existed (Fig. 6c). Nearly all of the post-storm deposits were graded and typically consisted of 0.5 to 5 cm of silt over 1 to 2 cm of sand over muddy sand or mud. The uppermost sand and mud couplet, which was not present before Alicia, represents the storm-graded layer, whereas the underlying fine-grained sediments are the older shelf deposits. Mostly ripples would have formed at the Bryan Mound site if moveable fine sand had been present; instead of ripples, it is likely that irregular mounds and depressions formed because the bottom sediments are dominantly mud not sand.

Alicia eroded nearly 1.5 million m^3 of sand from a 30-km stretch of west Galveston Island while beach erosion elsewhere was negligible. Two years after the storm, about 60 percent of the total eroded volume was accounted for in beach recovery and washover deposits. The remaining sand volume, representing about 20 m^3/m of beach, was mainly transported southwestward along the shoreface and is presumably nourishing adjacent barrier beaches (Morton and Paine, 1985).

The Bryan Mound site is about 50 km downdrift from west Galveston Island and about 35 km offshore from a segment of the coast where Alicia surge channels drained the barrier flats and adjacent bay. It was ideally situated to receive the fine sand eroded from the shoreface, and there was ample time to travel the required distance if the sand was primarily transported in suspension or by a turbidity current, as is commonly proposed (Dott,

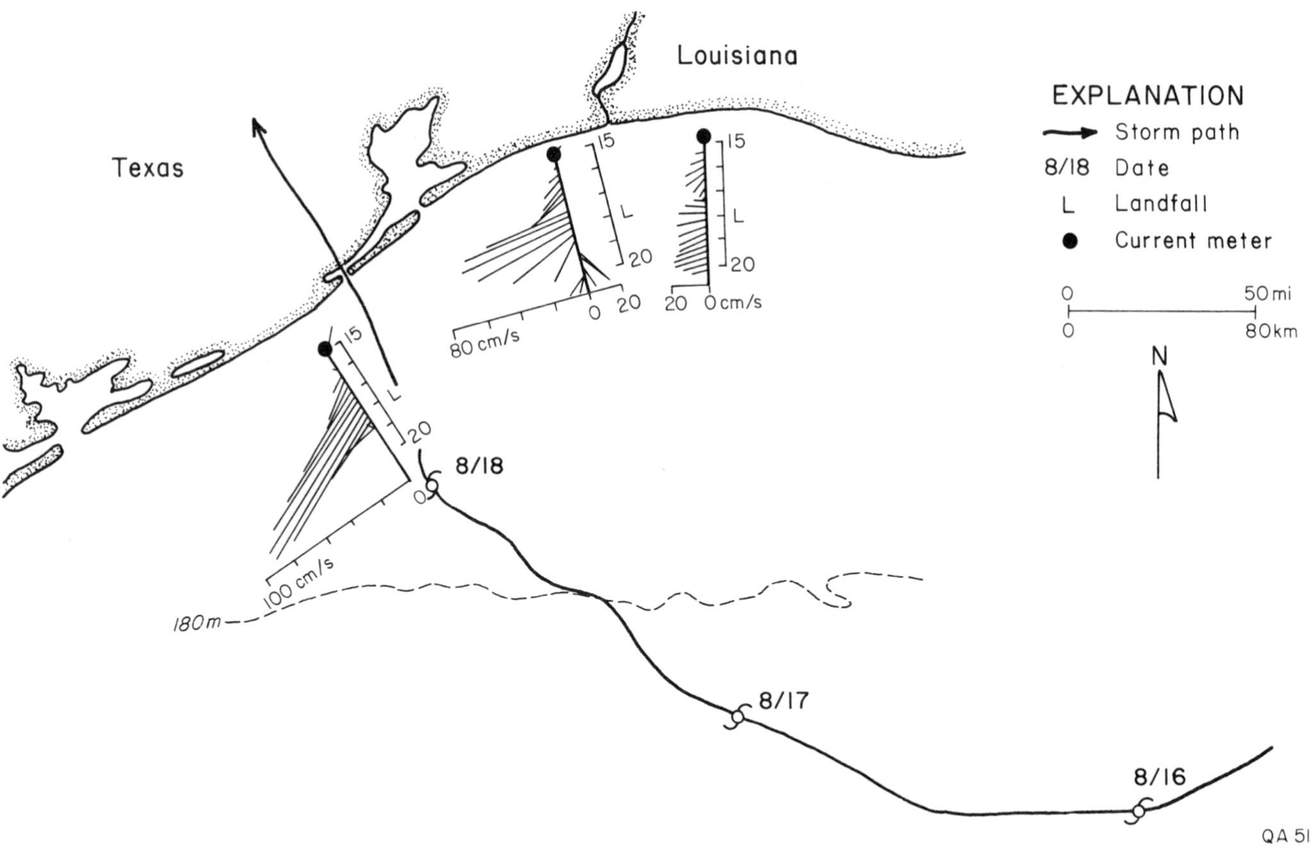

Figure 4. Nearshore bottom current velocities and directions along the Texas-Louisiana coast during Hurricane Alicia, August 15–20, 1983. Stick vectors presented in Hann and others (1984) were provided by Frank Kelly (personal communication, 1985).

1983; Walker and others, 1983). Yet the evidence suggests simple in situ reworking and size fractionation of preexisting shelf sediments rather than the introduction of allochthonous material from nearby beaches. The strongest evidence supporting this conclusion is based on calculations involving the storm sediment budget. These calculations indicate that the volume of sand contained in a layer 1 cm thick covering the area influenced by storm currents greatly exceeds the volume of sand available from beach and shoreface erosion. Furthermore, temporal variations in surface sediment textures at the Bryan Mound site show that sand and mud couplets can form without importing sand from adjacent beaches.

Descriptions of sediment samples collected monthly for a year after Alicia show that the post-storm graded bed was greatly modified within 3 months. The most common change was an increase in percent sand and mean grain size caused by erosion and concomitant reduced thickness of the uppermost layer of silt. Subsequent changes during the winter months included physical mixing and obliteration of the graded bed. By the end of the first post-storm year (August 1984), bottom sediment reworking had formed another graded bed having thickness and composition similar to the post-Alicia deposit. If sampling had been conducted a year after the storm, the second graded bed undoubtedly would have been attributed to Alicia rather than to the subsequent reworking and redeposition by winter storms.

The persistence of sand or mud at a particular location is an indicator of sediment mobility and the resulting interaction of bottom flow and available shelf sediments. Time averaging of sediment textures six months before and after Alicia (Fig. 7) show markedly different patterns that reflect the dominant shelf processes. These data suggest that before the storm the principal direction of sediment reworking was cross-shelf or oblique to the bathymetric contours (Fig. 7a), whereas after the storm, preferential sorting was primarily alongshelf (Fig. 7b). The coarsest post-storm sediments were about 4 km apart, which is comparable to the spacing observed after Delia (Fig. 2). The differences in these textural patterns (Figs. 7a and 7b) emphasize the importance of sequential events and the time-dependent construction of depositional units.

Sediments of the Texas shelf

Although the textural patterns after Delia and Alicia provide a basis for analyzing short-term shelf processes, they are not necessarily predictors of long-term shelf sedimentation. In fact, these observations were made where rates of sedimentation are

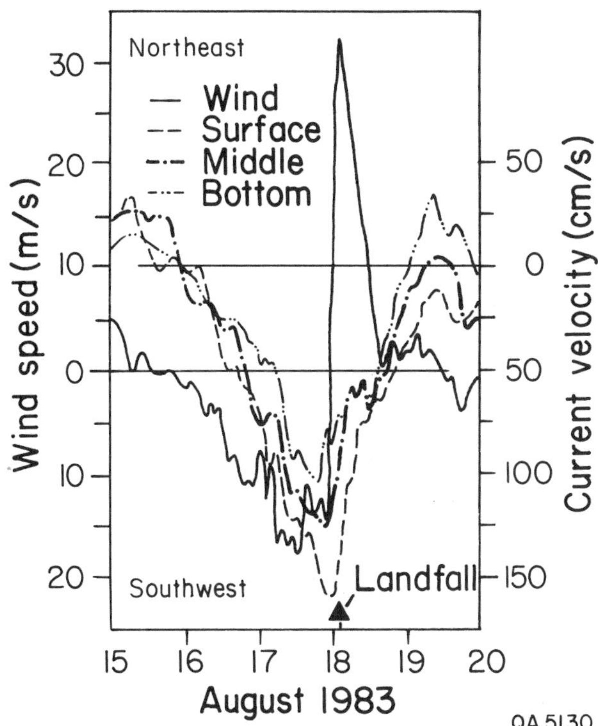

Figure 5. Alongshore current velocities at three levels and wind speeds recorded at the Bryan Mound site during Hurricane Alicia. Upper meter was 3.6 m below the surface, middle meter was 11.0 m below the surface, lower meter was 20.0 m below the surface or 1.8 m above the sea floor. Location of the Bryan Mound site is shown in Figure 1.

extremely low. The shelf of the upper Texas coast is composed of less than 2 m of soft gray clay (60 to 80 percent water content) with variable amounts of sand and shell overlying stiff tan to gray clay (<50 percent water content); the stiff relict sediments were deposited by Pleistocene fluvial-deltaic systems (Curray, 1960; Morton and Winker, 1979) during a period of lower sea level. Together the low rates of sediment influx, suspended nature of the available sediment, and relatively deep storm-wave base have kept the relict sediments at or near the sea floor. This part of the shelf has received little of the sediment eroded from the beach and shoreface as the Holocene Brazos–Colorado Delta retreated landward more than 16 km. The thin veneer of modern shelf sediments is continuously reworked; consequently, the preservation potential of storm deposits on this broad, shallow segment of the shelf is extremely low.

On the adjacent shelf sector between the Holocene Brazos–Colorado and Rio Grande Deltas, post-Wisconsinan sediments are up to 43 m thick (Berryhill, 1976), and at least the upper 5 m is composed of storm deposits interlayered with shelf mud. The modern shelf deposits aggraded as the flanking deltas retreated landward and adjacent barriers prograded seaward. These interbedded sand and mud sequences exhibit characteristics recognized in both modern and ancient sediments that are interpreted as having been deposited on the shelf below storm wave base.

The distal storm deposits have erosional bases that commonly scoured into mud. The basal shelly sand layer grades upward into sand, which in turn, grades into mud. Other graded beds lack the basal shell and are composed of sand with parallel laminations that sometimes pass into ripple cross laminations overlain by mud (Morton, 1981). The proximal and distal relationships of storm-bed thickness, number, and composition in the Gulf of Mexico are essentially identical to the trends described by Aigner and Reineck (1982) for the North Sea.

SEDIMENTOLOGIC CONSEQUENCES OF STORMS

Interaction of waves and currents

There is a growing awareness that slowly varying currents combined with waves initiate and maintain sediment motion more efficiently than either force can independently (Curray, 1960; Grant and Madsen, 1979; Niedoroda and others, 1984). When current and orbital velocities are combined, the shear stress in the direction of the current is three to ten times that of the current alone (Silvester, 1974).

Calculations using actual or estimated wave parameters indicate that orbital velocities of waves rarely exceed the threshold of sediment motion at mid-shelf or near the outer shelf (Curray, 1960; Hadley, 1964; Draper, 1967; Silvester and Mogridge, 1970). Nevertheless, bottom current velocities on the outer shelf of 30 to 100 cm/s lasting for several days occur several times a year and are not necessarily related to storms (McGrail, 1979; Ebbesmeyer and others, 1982). Rather they can be caused by intruding semipermanent currents as well as by combined tidal, wind-driven, or density-driven currents. It appears that such strong currents may be capable themselves of reworking shelf sediments and producing thin graded beds on the outer shelf, even in the absence of large storm waves.

Storm bed characteristics

A few sedimentologic features such as preferentially oriented shells, hummocky cross stratification, and oscillation ripples appear to be diagnostic of some marine storm deposits. These features, together with the upward fining textures, vertical sequence of sedimentary structures, and spatial relationships of inferred storm deposits, have led to an idealized stratigraphic sequence and integrated hydrodynamic model of nearshore storm processes (Walker, 1979; Morton, 1981; Walker and others, 1983). The preferred alignment of shells at some exposures proves that peak velocities are associated with strong combined flow having a dominant unidirectional component that occurs as the storm passes over the shelf. The overlying parallel laminated sands are also deposited under similar conditions, but hummocky cross stratification is thought to be the product of currents having subequal unidirectional and reversing components. As the local currents decrease, they interact with storm waves having orbital velocities of about the same magnitude as the currents. With continued current decay, the combined flow eventually is domi-

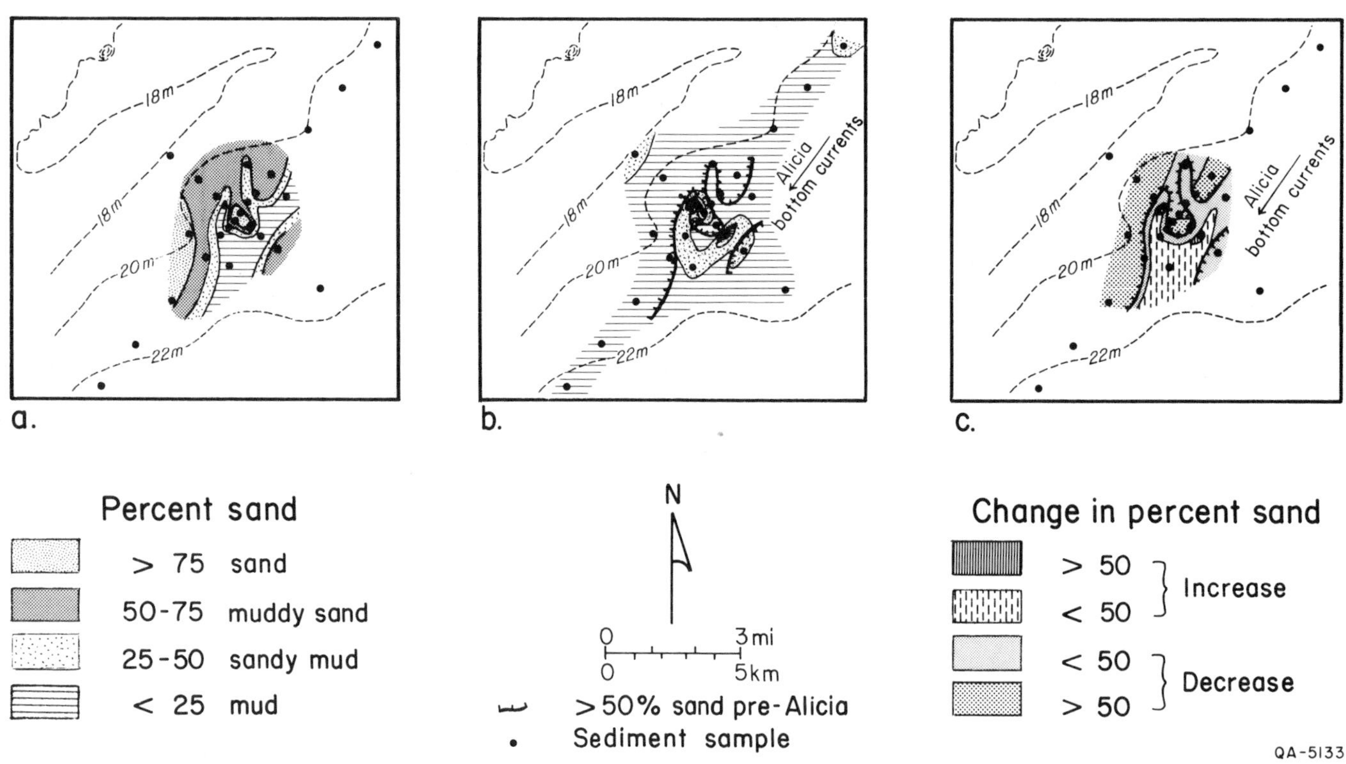

Figure 6. Percent sand in surface sediments collected at the Bryan Mound site. (a) July 1983 (pre-Alicia). (b) August 1983 (post-Alicia). (c) The change in percent sand before and after the storm. Mapping and interpretation were based on sediment descriptions and textural analyses provided by Don Harper (personal communication, 1985).

nated by the oscillatory motion that produces the nearly symmetrical ripples. Later, mud settles out of suspension during the quiescent period following the storm. This depositional model explains the geographic (modern) and stratigraphic (ancient) position of thin graded beds as well as thicker units containing hummocky cross stratification, but it also implies that sand eroded from the beach and surf zone is deposited on the inner shelf during the same high-energy event.

Sand eroded from the beach during a storm is temporarily stored in an offshore bar system on the upper shoreface; what happens on the lower shoreface is less precisely known (Niedoroda and others, 1984). Normally a large volume of the eroded sand returns to the beach shortly (weeks and months) after the storm. This recovery process can occur even during periods of high wave energy, such as the winter, because the raised elevation of the upper shoreface and lowered beach profile promote onshore transport (Morton and Paine, 1985). Subsequent beach recovery diminishes as seasonal erosion-accretion cycles temporarily disrupt the restoration process and the surplus of sand on the shoreface is exhausted, balancing the sediment budget.

Niedoroda and others (1984) stated that converging currents during storms caused bedload deposition on the upper shoreface, lower shoreface, and inner shelf. If this statement is generally correct and if equilibrium shoreface profiles are eventually restored, then the net loss of sand after beach recovery (several years) should be a measure of the volume transported onto the shelf below fair-weather wave base and incorporated into the shelf sediment.

Sand volumes of modern storm beds attributed to single depositional events were calculated from descriptions and illustrations presented by Hayes (1967), Morton (1981), Nelson (1982), and Aigner and Reineck (1982). Although storm deposits are commonly less than 10 cm thick, they cover broad areas and their minimal sand volumes (respectively estimated from the data of Aigner and Reineck, 1982; and Nelson, 1982) range from 6 million to nearly 400 million m^3 just in the areas where they have been described. When compared to the length of adjacent coastline, these volumes would require erosion and seaward transport of about 400 m^3/m to more than 2,000 m^3/m of coast. This range of normalized sand volumes is one to two orders of magnitude greater than the volumes of sand typically eroded from the beach and surf zone during a single storm.

Discrepancies in the estimated volume of sand eroded from beaches versus the volume stored in storm beds suggest that either (1) sand also is removed from the lower shoreface, causing a temporary disequilibrium profile; or (2) sand is transferred to the shelf from the beach and shoreface by a series of moderately intense events and is subsequently sorted from the muddy matrix by large waves and strong bottom currents. This step-wise repetitive progression and reworking would explain the presence of

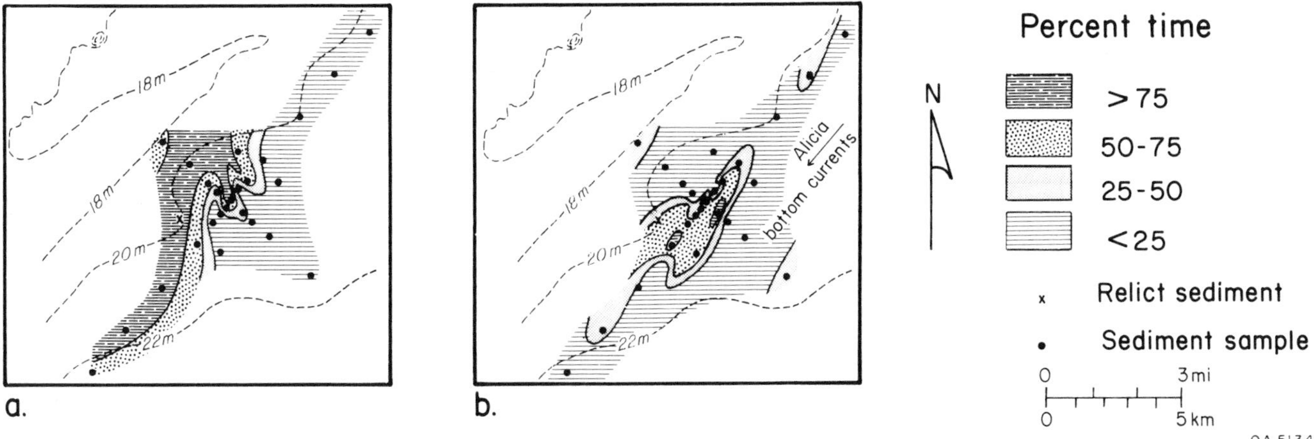

Figure 7. Results of repetitive sediment sampling at the Bryan Mound site during two periods. (a) Six months before, and (b) six months after Hurricane Alicia. The calculated value (percent) at each sample station is a ratio of the number of samples composed predominantly of sand compared to the total number of samples collected at that station during the six-month period. Mapped values illustrate the time-averaged effects of shelf processes influencing depositional patterns. Highest frequencies of sand occurrence signify the areas of most intense bottom currents.

graded beds that appear to have been deposited as a single unit but are far removed from the shoreface. However, this alternative explanation still would not account for the large number of vertically stacked storm beds, each containing more sand than is probably eroded from the beach and surf zone during a single storm. Both proposed mechanisms (net loss of sand from the nearshore zone and intermittent reworking) may operate simultaneously but at different places during the same storm, as suggested by the current histories, textural analyses, beach profiles, and sediment budget calculations for hurricane Alicia (Hann and others, 1984; Morton and Paine, 1985). These data indicate that flow expansion in slightly deeper water (10 m) across the lower shoreface and inner shelf caused downdrift shoreface and shelf accretion, which was supplied by beach and shoreface erosion along an updrift coastal sector. At the same time in deeper water (20 m) on the inner shelf, resuspension and deposition of preexisting heterogeneous sediments formed a graded layer that was probably unrelated to the sediment transported along the lower shoreface and shelf to the northeast.

The sand eroded from the nearshore zone and deposited on the inner shelf as a graded bed represents a net loss to the littoral drift system if the sand is not replenished either from onshore transport by waves or from external sources such as updrift coastal erosion or fluvial supply. Some littoral-drift cells lack a supply of "new" sediment, consequently, cumulative losses averaged over years or decades result in long-term shoreline and shoreface retreat. Considering the nearshore zone from a regional perspective and for various periods of time (one storm, centuries of shoreline movement, or coastal transgression), it is clear that some shorefaces experience net erosion at all time scales. Other littoral-drift cells, however, may gain from the repetitive action of high waves and wind-driven currents because of their orientation and location with respect to storm trajectories. For these cells, a surplus of sand results in progradation of the shoreline and shoreface and aggradation of the inner shelf.

Conditions favoring preservation

Outcrop observations and field measurements (Blumberg, 1964; Herbich, 1977) indicate that the depth of shoreface and shelf erosion commonly exceeds the thickness of non-storm deposits (mud caps) or of entire storm beds; therefore many storm deposits are amalgamations resulting from several events. Moreover, preservation depends largely on the sequence of events rather than on the magnitude of a single event. The common practice of estimating storm frequency by calculating the ratio of time to the number of storm beds within a stratigraphic interval, actually measures preservation potential and greatly underestimates the number of high-energy events in a given period of time.

Storm beds can form on either passive or tectonically active continental margins; they are most commonly preserved in the middle of upward-coarsening sequences that were deposited on broad, slowly subsiding shelves. On these relatively stable platforms, storm wave base periodically exceeded the shallow water depths. This generalization appears to be true regardless of whether the regressive sequence is measured in tens or hundreds of meters. Shelf storm deposits can also be preserved in transgressive sequences of some rapidly subsiding basins along converging plate boundaries where rates of sedimentation are high (Hunter and Clifton, 1982). In transgressive sequences, they commonly overlie an erosional unconformity, depositional hiatus, or compressed stratigraphic horizon.

DISCUSSION AND CONCLUSIONS

The four interactive models proposed by Figueiredo and

others (1982) to explain the origin of graded beds (stratigraphy, liquefaction, bedform migration, and waning energy) can be tested using data from the Texas coast and shelf. Their stratigraphic model, resulting from landward displacement of the shoreline, is inconsistent with the specific setting on the central Texas shelf where storm beds are preserved. There the adjacent shoreline prograded as storm beds accumulated, but where the shoreline continues to retreat, storm beds are not preserved. Bottom liquefaction by waves is undoubtedly an important mechanism enhancing the turbulent resuspension of sediment, but gravity settling from wave suspension would not explain the sedimentary structures or oriented clasts that are commonly preserved in storm beds. Bedform migration may be a plausible explanation for size-graded layers on sand-rich shelves, but it seems an inappropriate model for shelves composed dominantly of silt and clay that prevent development of migratory transverse bedforms. Only the model of waning storm energy consistently explains the graded beds on the muddy shelf of the western Gulf of Mexico.

A question that remains unresolved concerns the importance of shallow-water density or turbidity currents as suggested by Walker and others (1983). Whether or not the criteria for autosuspension (Pantin, 1979) is satisfied by storm processes on most shelves is still uncertain; however, theoretical considerations (D.J.P. Swift, personal communication, 1986) suggest that the requirements cannot be easily obtained. Steep vertical concentration gradients would be necessary to generate turbidity currents on relatively flat muddy shelves, but the fine-grained sediment composing these shelves is resuspended high in the water column, thus diluting the sediment concentration near the sea floor and preventing formation of a density underflow. Considering the hydrodynamics and physical setting of modern shelves, it seems unlikely that turbidity currents are responsible for the storm-graded beds. The predominant alongshore transport directions and extremely low cross-shelf gradients currently observed—and also inferred for many ancient shelves—are compelling evidence that shelf sediment transport mechanisms do not involve turbidity currents. On the other hand, the suspension of sandy mud by advective processes and selective sorting by an initially strong, decelerating current are capable of producing size-graded beds.

The magnitude and frequency of episodic geologic events determine their effectiveness in transporting sediment and shaping the Earth's surface (Wolman and Miller, 1960). These characteristics are independent, even though large, extremely violent events rarely occur. Great storms are good examples of events that meet the criteria of being extreme in terms of size and intensity and rare on the basis of elapsed time, but they are frequent events in terms of global experience and when averaged over geologic time. Truly catastrophic events having regional or global impact originate deep within the Earth's crust (earthquakes, volcanoes) or outside our atmosphere (asteroids) where potential energy is nearly infinite. They manifest themselves as near-surface phenomena (faults, eruptions, impact craters) that instantaneously alter the morphology, and commonly the biota and stratigraphy, of large areas around the epicenter. Even great storms are incapable of modifying the terrane like other high-energy episodic events.

In conclusion, great storms are rare historical events for any given nearshore region, but it appears that less intense storms occurring more frequently are also capable of producing many of the sedimentologic features commonly observed in the rock record. The fact that graded beds are found on the Brazilian shelf (Figueiredo and others, 1982) where hurricanes do not occur is ample evidence that relatively weak storms can form sand and mud couplets. Furthermore, the lack of compelling stratigraphic evidence of super storms suggests that they may be a consequence of probability rather than reality, or if they exist, their sedimentologic influence may be unimportant or still unrecognized.

ACKNOWLEDGMENTS

I am particularly grateful to Frank Kelly and Don Harper of Texas A&M University for, respectively, providing velocity plots and textural data obtained before and after Hurricane Alicia. Appreciation is also extended to R. H. Simpson for discussing the origins of storms. The reviews by Don Swift and Ed Clifton greatly improved the manuscript. Publication was authorized by the Director, Bureau of Economic Geology, The University of Texas at Austin.

REFERENCES CITED

Aigner, T., and Reineck, H-E., 1982, Proximality trends in modern storm sands from the Helgoland Bight (North Sea) and their implications for basin analysis: Senckenbergiana Maritima, v. 14, p. 183–215.

Andrews, P. B., 1970, Facies and genesis of a hurricane-washover fan, St. Joseph Island, central Texas coast: Austin, University of Texas Bureau of Economic Geology Report of Investigations no. 67, 147 p.

Ball, M. M., Shinn, E. A., and Stockman, K. W., 1967, The geological effects of Hurricane Donna in south Florida: Journal of Geology, v. 75, p. 583–597.

Barron, E. J., 1983, A warm, equable Cretaceous; The nature of the problem: Earth-Science Reviews, v. 19, p. 305–338.

Bea, R. G., 1974, Gulf of Mexico hurricane wave heights: Offshore Technology Conference, v. 2, p. 791–810.

Bea, R. G., and Bernard, H. A., 1973, Movements of bottom soils in the Mississippi Delta offshore, *in* Offshore oil and gas fields: Lafayette Geological Society, p. 13–28.

Berryhill, H. L., Jr., ed., 1976, Environmental studies, south Texas outer continental shelf; geology: U.S. Geological Survey report prepared for the Bureau of Land Management, 270 p.

Blumberg, R., 1964, Hurricane winds, waves, and currents test marine pipeline design: Pipeline Industry, v. 20, p. 43–72.

Bretschneider, C. L., 1964, The Ash Wednesday East Coast storm, March 5–8, 1962; A hindcast of events, causes, and effects: Proceedings 9th Coastal Engineering Conference, p. 617–659.

Brower, W. A., Jr., Diaz, H. F., and Prechtel, A. S., 1977, Marine and coastal climatic atlas of the outer continental shelf waters and coastal regions of Alaska: University of Alaska, Arctic Environmental Information and Data Center, v. 2, Bering Sea, 443 p.

Brown, C. W., 1939, Hurricanes and shoreline changes in Rhode Island: Geographical Review, v. 29, p. 416–430.

Cacchione, D. A., and Drake, D. E., 1982, Measurements of storm-generated bottom stresses on the continental shelf: Journal of Geophysical Research, v. 87, p. 1952–1960.

Caldwell, D. R., and Elliott, W. P., 1971, Surface stresses produced by rainfall: Journal of Physical Oceanography, v. 1, p. 145–148.

Caston, V.N.D., 1976, A wind-driven near bottom current in the southern North Sea: Estuarine and Coastal Marine Science, v. 4, p. 23–32.

Channon, R. D., and Hamilton, D., 1976, Wave and tidal current sorting of shelf sediments, southwest England: Sedimentology, v. 23, p. 17–42.

Coleman, J. M., Prior, D. B., and Lindsay, J. F., 1983, Deltaic influence on shelfedge instability processes, *in* Stanley, D. J., and Moore, G. T., The shelfbreak; Critical interface on continental margins: Society of Economic Paleontologists and Mineralogists Special Publication 33, p. 121–137.

Cooperman, A. I., and Rosendal, H. E., 1962, Great Atlantic storm, 1962: Mariners Weather Log, v. 6, p. 79–85.

Curray, J. R., 1960, Sediments and history of Holocene transgression, continental shelf, northwest Gulf of Mexico, *in* Recent sediments, northwestern Gulf of Mexico: American Association of Petroleum Geologists, p. 221–266.

Danielsen, E. F., Burt, W. V., and Rattray, M., 1957, Intensity and frequency of severe storms in the Gulf of Alaska: EOS American Geophysical Union Transactions, v. 38, p. 44–49.

Dott, R. H., Jr., 1983, Episodic sedimentation; How normal is average? How rare is rare? Does it matter?: Journal of Sedimentary Petrology, v. 53, p. 5–23.

Draper, L., 1967, Wave activity at the sea bed around northwestern Europe: Marine Geology, v. 5, p. 133–140.

Duke, W. L., 1985, Hummocky cross-stratification, tropical hurricanes, and intense winter storms: Sedimentology, v. 32, p. 167–194.

Dunnavan, G. M., and Diercks, J. W., 1980, Supertyphoon Tip: Monthly Weather Review, v. 108, p. 1915–1923.

Ebbesmeyer, C. C., Williams, G. N., Hamilton, R. C., Abbott, C. E., Collipp, B. G., and McFarlane, C. F., 1982, Strong persistent currents observed at depth off the Mississippi River Delta: Offshore Technology Conference, paper 4322, p. 259–272.

Fathauer, T. R., 1975, The great Bering Sea storms of 9–12 November 1974: Weatherwise Magazine, v. 28, p. 56–84.

Figueiredo, A. G., Jr., Sanders, J. E., and Swift, D.J.P., 1982, Storm-graded layers on inner continental shelves; Examples from southern Brazil and the Atlantic coast of the central United States: Sedimentary Geology, v. 31, p. 171–190.

Fischer, A. G., 1982, Long-term climatic oscillations recorded in stratigraphy, *in* Climate in Earth history: Washington, D.C., National Academy Press, p. 97–104.

Forristall, G. Z., Hamilton, R. C., and Cardone, V. J., 1977, Continental shelf currents in tropical storm Delia; Observations and theory: Journal of Physical Oceanography, v. 7, p. 532–546.

Forristall, G. Z., Ward, E. G., Borgman, L. E., and Cardone, V. J., 1978, Storm wave kinematics: Offshore Technology Conference, paper 3227, p. 1503–1513.

Garcia, A. W., and Flor, T. H., 1984, Hurricane Alicia storm surge and wave data: U.S. Army Corps of Engineers Technical Report CERC 84-6, 89 p.

Gienapp, H., 1973, Stromungen wahrend der Sturmflut vum 2, November 1965 in der Deutschen Bucht und ihre Bedentung fur den Sedimenttransport: Seckenbergiana Maritima 5, p. 135–151.

Grant, W. D., and Madsen, O. S., 1979, Combined wave and current interaction with a rough bottom: Journal of Geophysical Research, v. 84, p. 1797–1808.

Gray, W. M., 1968, Global view of the origin of tropical disturbances and storms: Monthly Weather Review, v. 96, p. 669–700.

Greeley, R., and Iverson, J. D., 1985, Wind as a geological process on Earth, Mars, Venus, and Titan: Cambridge University Press, 333 p.

Gretener, P. E., 1967, Significance of the rare event in geology: American Association of Petroleum Geologists Bulletin, v. 51, p. 2197–2206.

Hadley, M. L., 1964, Wave-induced bottom currents in the Celtic Sea: Marine Geology, v. 2, p. 164–167.

Hann, R. W., Jr., Giammona, C. P., and Randall, R. E., eds., 1984, Offshore oceanographic and environmental monitoring services for the Strategic Petroleum Reserve; Annual report for the Bryan Mound site from September 1982 through August 1983: Texas A&M University report prepared for the Department of Energy, DOE/PO10850-4.

Harper, D. E., Jr., Scrudato, R. J., and Giam, C. S., 1976, A preliminary environmental assessment of the Buccaneer oil and gas field: Galveston, Texas A&M University report prepared for National Marine Fisheries Service, 62 p.

Hayes, M. O., 1967, Hurricanes as geological agents; Case studies of Hurricane Carla, 1961, and Cindy, 1963: The University of Texas at Austin Bureau of Economic Geology Report of Investigations 61, 54 p.

Herbich, J. B., 1977, Wave-induced scour around offshore pipelines: Offshore Technology Conference, v. 4, p. 79–90.

Hunter, R. E., and Clifton, H. E., 1982, Cyclic deposits and hummocky cross-stratification of probable storm origin in upper Cretaceous rocks of the Cape Sebastian area, southwestern Oregon: Journal of Sedimentary Petrology, v. 52, p. 127–143.

Kraus, E. B., 1972, Atmosphere ocean interaction: London, Oxford University Press, 275 p.

Lewis, K. B., 1979, A storm-dominated inner shelf, Western Cook Strait, New Zealand: Marine Geology, v. 31, p. 31–43.

McGrail, D. W., 1979, New observations on factors controlling the vertical extent of and sediment concentration in the nepheloid layer [abs.]: EOS American Geophysical Union Transactions, v. 60, p. 847.

Morton, R. A., 1979, Subaerial storm deposits formed on barrier flats by wind-driven currents: Sedimentary Geology, v. 24, p. 105–122.

——, 1981, Formation of storm deposits by wind-forced currents in the Gulf of Mexico and the North Sea: International Association of Sedimentologists Special Publication 5, p. 385–396.

Morton, R. A., and Paine, J. G., 1985, Beach and vegetation-line changes at Galveston Island, Texas; Erosion, deposition, and recovery from Hurricane Alicia: The University of Texas at Austin Bureau of Economic Geology

Geological Circular 85-5, 39 p.
Morton, R. A., and Winker, C. D., 1979, Distribution and significance of coarse biogenic and clastic deposits on the Texas inner shelf: Gulf Coast Association of Geological Societies Transactions, v. 29, p. 136–146.
Murray, S. P., 1970, Bottom currents near the coast during Hurricane Camille: Journal of Geophysical Research, v. 75, p. 4579–4582.
Nelson, C. H., 1982, Modern shallow-water graded sand layers from storm surges, Bering shelf; A mimic of Bouma sequences and turbidite systems: Journal of Sedimentary Petrology, v. 52, p. 537–545.
Niedoroda, A., and Swift, D.J.P., 1981, The role of wave orbital currents versus wind-driven currents in maintaining the shoreface; Observations from the Long Island Coast: Geophysical Research Letters, v. 8, p. 337–340.
Niedoroda, A. W., Swift, D.J.P., Hopkins, T. S., and Chen-Mean Ma, 1984, Shoreface morphodynamics on wave-dominated coasts, *in* Greenwood, B., and Davis, R. A., Jr., eds., Hydrodynamics and sedimentation in wave-dominated coastal environments: Marine Geology, v. 60, p. 331–354.
Pantin, H. M., 1979, Interaction between velocity and effective density in turbidity flows; Phase-plane analysis with criteria for autosuspension: Marine Geology, v. 31, p. 59–99.
Quayle, R. G., and Fulbright, D. C., 1975, Extreme wind and wave return periods for the U.S. coast: Mariners Weather Log, v. 19, p. 67–70.
Riehl, H., 1979, Climate and weather in the tropics: New York, Academic Press, 611 p.
Roll, H. U., 1965, Physics of the marine atmosphere: New York, Academic Press, 426 p.
Rubin, D. M., and McCulloch, D. S., 1980, Single and superimposed bedforms; A synthesis of San Francisco Bay and flume observations: Sedimentary Geology, v. 26, p. 207–231.
Savin, S. M., 1977, The history of the Earth's surface temperature during the last 100 million years: Annual Review of Earth and Planetary Science, v. 5, p. 319–355.
Schopf, T.J.M., 1980, Paleoceanography: Cambridge, Massachusetts, Harvard University Press, 341 p.
Schroeder, W. W., and McGrail, D., 1982, Shelf currents observed during tropical storms [abs.]: EOS American Geophysical Union Transactions, v. 63, p. 66.
Scrutton, C. T., and Hipkin, R. G., 1973, Long-term changes in the rotation rate of the Earth: Earth-Science Review, v. 9, p. 259–274.
Silvester, R., 1974, Coastal engineering: Amsterdam, Elsevier, 338 p.
Silvester, R., and Mogridge, G. R., 1970, Reach of waves to the bed of the continental shelf: Proceedings of the 12th Coastal Engineering Conference, v. 2, p. 651–668.
Simpson, R. H., and Lawrence, M. B., 1971, Atlantic hurricane frequencies along the U.S. coastline: National Oceanic and Atmospheric Administration Technical Memorandum NWS SR-58, 14 p.
Simpson, R. H., and Riehl, H., 1981, The hurricane and its impact: Baton Rouge, Louisiana State University Press, 398 p.
Simpson, R. H., Sugg, A. L., and staff, 1970, The Atlantic hurricane season of 1969: Monthly Weather Review, v. 98, p. 293–306.
Smith, J. D., and Hopkins, T. S., 1972, Sediment transport on the continental shelf off of Washington and Oregon in light of recent current measurements, *in* Swift, D.J.P., Duane, D. B., and Pilkey, O. H., eds., Shelf sediment transport; process and pattern: Stroudsburg, Pennsylvania, Dowden, Hutchinson, and Ross, p. 143–180.
Sternberg, R. W., and Larsen, L. H., 1976, Frequency of sediment movement on the Washington continental shelf; A note: Marine Geology, v. 13, p. M37–M47.
Stubblefield, W. L., McGrail, D. W., and Kersey, D. G., 1984, Recognition of transgressive and post-transgressive sand ridges on the New Jersey continental shelf, *in* Tillman, R. W., and Siemers, C. T., eds., Siliciclastic shelf sediments: Society of Economic Paleontologists and Mineralogists Special Publication 34, p. 1–23.
Swift, D.J.P., and Rice, D. D., 1984, Sand bodies on muddy shelves; A model for sedimentation in the Western Interior Cretaceous seaway, North America, *in* Tillman, R. W., and Siemers, C. T., eds., Siliciclastic shelf sediments: Society of Economic Paleontologists and Mineralogists Special Publication 34, p. 43–62.
Swift, D.J.P., Stanley, D. J., and Curray, J. R., 1971, Relict sediments on continental shelves; A reconstruction: Journal of Geology, v. 79, p. 322–346.
Thom, H.C.S., 1973a, Distribution of extreme winds over oceans: Journal of Waterways, Harbors, and Coastal Engineering Division, American Society of Civil Engineers, v. 99, p. 1–17.
——, 1973b, Extreme wave height distributions over oceans: Journal of Waterways, Harbors, and Coastal Engineering Division, American Society of Civil Engineers, v. 99, p. 355–374.
Tryggestad, S., Selanger, K. A., Mathisen, J. P., and Johansen, O., 1983, Extreme bottom currents in the North Sea, *in* North Sea dynamics: Springer-Verlag, p. 148–158.
U.S. Army Corps of Engineers, 1983, Report on Hurricane Alicia, August 15–18, 1983: Galveston, Texas, U.S. Army Corps of Engineers District, 44 p.
Vermeer, D. E., 1963, Effects of Hurricane Hattie, 1961, on the cays of British Honduras: Zeitschrift für Geomorphologie, v. 7, p. 332–354.
Walker, R. G., 1979, Shallow marine sands, *in* Walker, R. G., ed., Facies models: Geoscience Canada Reprint Series 1, p. 75–89.
Walker, R. G., Duke, W. L., and Leckie, D. A., 1983, Hummocky stratification; Significance of its variable bedding sequences; Discussion and reply: Geological Society of America Bulletin, v. 94, p. 1245–1251.
Ward, E. G., Evans, D. J., and Pompa, J. A., 1977, Extreme wave heights along the Atlantic coast of the United States: Offshore Technology Conference, p. 315–324.
Ward, E. G., Borgman, L. E., and Cardone, V. J., 1978, Statistics of hurricane waves in the Gulf of Mexico: Offshore Technology Conference, p. 1523–1536.
Wolman, M. G., and Miller, J. P., 1960, Magnitude and frequency of forces in geomorphic processes: Journal of Geology, v. 68, p. 54–74.

Manuscript Accepted by the Society April 4, 1988

Printed in U.S.A.

Geological Society of America
Special Paper 229
1988

Origin, behavior, and sedimentology of prehistoric catastrophic lahars at Mount St. Helens, Washington

Kevin M. Scott, *U.S. Geological Survey, Cascades Volcano Observatory, 5400 MacArthur Boulevard, Vancouver, Washington 98661*

ABSTRACT

A series of four lahars (volcanic debris flows and their deposits) occurred in rapid succession about 2,500 radiocarbon yr ago in the river system that drains the northwest sector of Mount St. Helens. The huge initial lahar had an instantaneous peak discharge near that of the Amazon River at flood stage, and the third in the series was the second largest in the history of the watershed. The deposits of the flows form a widespread terrace underlain by as much as 12 m of deposits, but its distribution reflects the fact that the lahars were the middle segments of flood waves beginning and ending as streamflow surges. These flood waves originated as lake-breakout surges analogous to those that would have been released, without engineering intervention, from the outlet-blocked Spirit Lake and several new lakes formed in 1980.

The lahar deposits are granular and noncohesive, reflecting the origin of the entrained sediment as stream alluvium beyond the base of the volcano. Megaclasts consisting of blocks probably derived from an ancient debris avalanche are abundant in the initial flow and locally form a diamicton with the laharic diamicton as matrix. Although fine-pebble angularity resulting partly from cataclasis during flow is diagnostic of the lahars, a high content of rounded clasts and a channel facies with zones of clast support may explain why similar deposits have not been more widely recognized. The lahar channel facies superficially resembles alluvium.

The depositional record of lahars and their distally evolved, hyperconcentrated lahar-runout flows many tens of kilometers from their source can provide a remarkably detailed history of volcanic activity. It may, in fact, provide the only evidence of some important events, of which the magnitude, frequency, and behavior are vital to the assessment of future volcanic hazards. Lake breakouts, meltwater surges caused by large pyroclastic flows, and forms of catastrophic volcanic ejection are examples of events that can create large downstream lahars but may leave little evidence of their true magnitude preserved on and near a volcano. The largest lahar in the river system, for example, began as a streamflow surge that bulked to a debris flow only after more than 20 km of flow.

INTRODUCTION

In the record of lahars of all ages in the Toutle-Cowlitz River system (Scott, 1985a), which drains the northwestern sector of Mount St. Helens, paleohydraulic analysis shows that a series of four lahars stands out in both magnitude and close association in time. The series formed during the Pine Creek eruptive period, from 3,000 to 2,500 yr B.P. (eruptive history after Mullineaux and Crandell, 1981; and Crandell, written communication, 1983), and is part of the Silver Lake lahar assemblage of Mullineaux and Crandell (1962). The series is interpreted here as an ancient analog of the events that would have occurred naturally within several years following the 1980 eruption. On May 18, 1980, a debris avalanche blocked the outlet of Spirit Lake with up to 170 m of deposits and catastrophically raised the lake level 63 m (Youd and others, 1981; Meyer and Carpenter, 1982). The avalanche also blocked tributary valleys to form Coldwater, Castle, and Jackson Creek Lakes (Fig. 1).

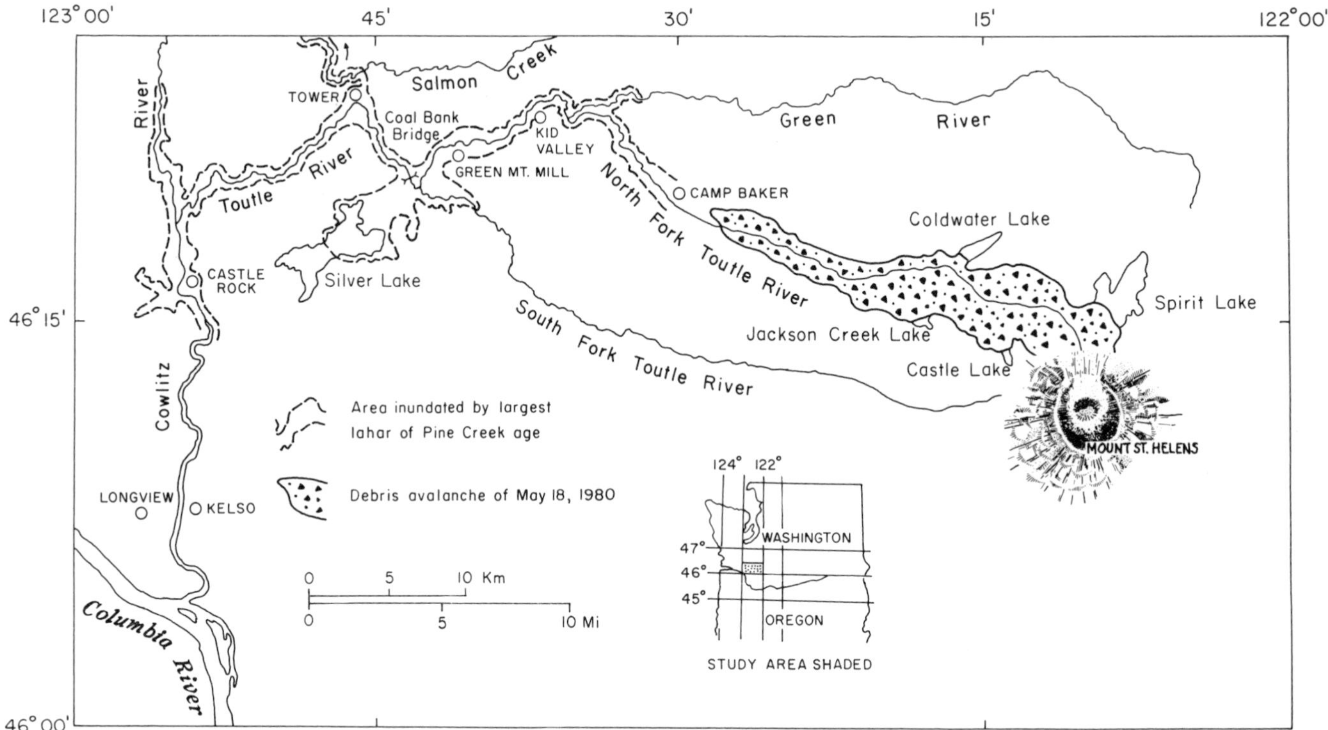

Figure 1. Index map of Toutle-Cowlitz River system, showing area inundated by lahar PC 1.

Complex engineering problems were presented by the stability of the damming deposits (Meyer and others, 1986).

Without engineering intervention, Spirit Lake and each lake formed in 1980 would have enlarged by runoff, and would have overtopped its natural dam or otherwise caused the dam to fail. The behavior of most modern and ancient lahars, as revealed by their proximal deposits, shows that the resulting lake-breakout surges would have transformed to lahars as the flood waves moved downvalley. The largest lahar, from a breakout of Spirit lake, would have inundated flood plains to depths of tens of meters, as did its analog in Pine Creek time. The potential breakout of modern Spirit Lake from overtopping as a result of runoff inflow has been eliminated with a drainage tunnel that lowered and now stabilizes the lake level. Continuing threats from the displacement of lake water by a large volcanic flow, or from a failure of the avalanche dam, are reduced by the lowered lake level.

Lahars may be formed around volcanoes by lake breakouts through natural dams of a variety of materials. These include the deposits of debris avalanches, landslides, large pyroclastic flows (especially lithic types), lahars, lava flows, and ice or snow avalanches. The blocking deposits may be rapidly eroded or covered by later deposits. Debris avalanches create great hazards through the damming of drainages at a volcano because of their size (commonly more than 1 km^3; Siebert, 1984), mobility (median H/L = 0.11; Siebert, 1984), and thick hummocky deposits, features that can result in multiple unstable natural dams. Volcanic debris avalanches are being recognized with increasing frequency at volcanoes in the Cascade Range and elsewhere (for example, Newhall, 1982; Glicken, 1982; Crandell and others, 1984; Siebert, 1984; Vallance, 1985); they have been identified at 59 volcanoes in Japan (Ui and others, 1986). Lahars acting themselves as natural dams do not pose as great a hazard; their role in blocking tributary drainages of the Toutle River is described in a subsequent section. Failures of various types of volcanic and nonvolcanic natural dams are discussed by Costa (1985) and Schuster (1985).

GENERAL CHARACTER AND AGE OF THE LAHAR SEQUENCE

This paper describes the behavior and sedimentology of lahars of catastrophic size that in the Toutle River were formed by lake breakouts. With the exception of flow as a non-Newtonian fluid, the lahars of Pine Creek age are distinctly different from the largest lahars recognized in rivers draining at least two other Cascade Range volcanoes. Lahars originating directly as slope failures are the largest known at Mt. Rainier (Crandell, 1971) and Mt. Baker (Hyde and Crandell, 1978). Such lahars may have a relatively high clay content derived from hydrothermally altered source rock. In such cases the downstream lahar behavior is commonly as a broad, texturally uniform wave of partially cohesive sediment. The flows do not mix readily with streamflow in the channel and do not readily undergo the downstream transformation to streamflow described below.

In contrast to the large lahars at Mt. Rainier and Mt. Baker,

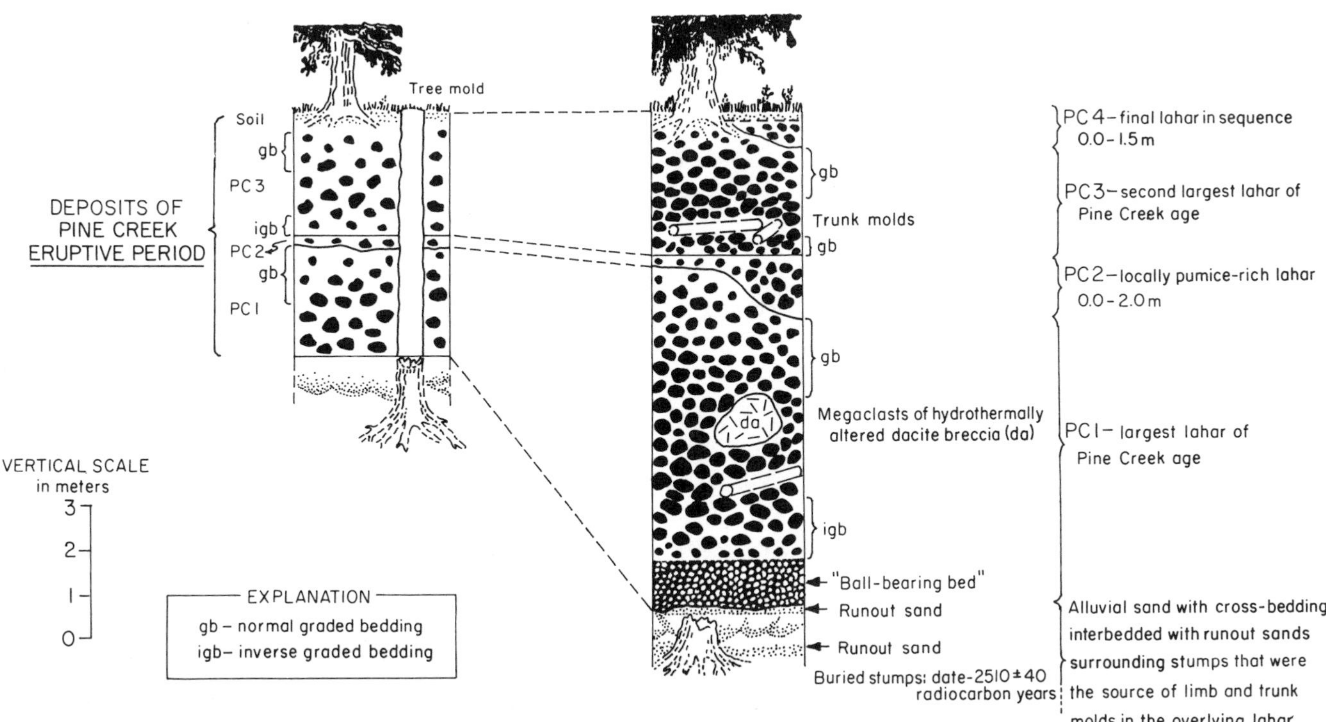

Figure 2. Stratigraphy of deposits of the Pine Creek eruptive period, based on 1980–1983 exposures near the Green Mountain Mill. Note tree mold in the flood plain facies.

the four lahars of Pine Creek age have: (1) particle-supported matrices of silty sand; (2) a low clay content (even less than 1 percent) relative to many other reported lahars and debris flows; and (3) well-defined flood plain and channel facies (Fig. 2), with the latter including bars of clast-supported cobbles and boulders that settled in the flow at sites of rapid energy loss (whaleback bars; Figs. 8 and 9, Scott, 1985a). These features are common to most of the lahars in the river system and in general resulted from the transformation of other flow types (Scott, 1985a and b). Streamflow surges have been the most common proximal lahar-producing flows, but the record includes catastrophically ejected pyroclastic surges that deflated to form lahars, and catastrophically ejected rubbly avalanches that continued to flow as lahars. Meltwater flood surges that transformed to lahars were produced by pumiceous pyroclastic flows during eruptions of Ape Canyon time (about 50 to 36(?) ka), but mainly by lithic pyroclastic flows in later eruptive episodes.

The granularity of most of the lahars of all ages in the Toutle-Cowlitz River system thus reflects the inclusion of sediment that was subject to previous selective sorting, either hydraulically or explosively induced, which caused a significant loss of fine sediment prior to incorporation in the lahar. Of the more than 25 large lahars inundating flood plains more than 50 km from Mount St. Helens, only 2 have a clay content greater than 3 percent, suggesting that they originated by failure of slopes underlain by hydrothermally altered volcanic rocks.

Most of the granular lahars of all ages in the Toutle-Cowlitz River system underwent the distal transformation to hyperconcentrated streamflow (40 to 80 percent sediment by weight; about 20 to 60 percent by volume) described as a lahar-runout flow (Scott, 1985a). The evidence for, and the origin of, this transformation is discussed by Scott (1985a), Pierson and Scott (1985), and Scott and Dinehart (1985). Most of the lahars, including those of Pine Creek age, thus may be visualized as the middle segment of a flood wave beginning and ending as a streamflow surge. In the case of the lahar of March 19, 1982, a distal transformation to hyperconcentrated flow occurred at a sediment content by volume of 57 percent (Pierson and Scott, 1985), close to the originally proposed upper boundary of hyperconcentrated flow (60 percent; Beverage and Culbertson, 1964).

Deposits of the flow sequence of Pine Creek age dominate downstream valley fills in the North Fork and main Toutle River, forming a widespread terrace underlain by as much as 12 m of deposits (Fig. 2). About 30×10^6 m^3 of primary deposits of the first lahar (PC 1) remain in the North Fork and main Toutle River valleys, a minimum volume because it does not include the unknown but probably significant volume of PC 1 deposits underlying Silver Lake. More than 90 percent of the original PC 1

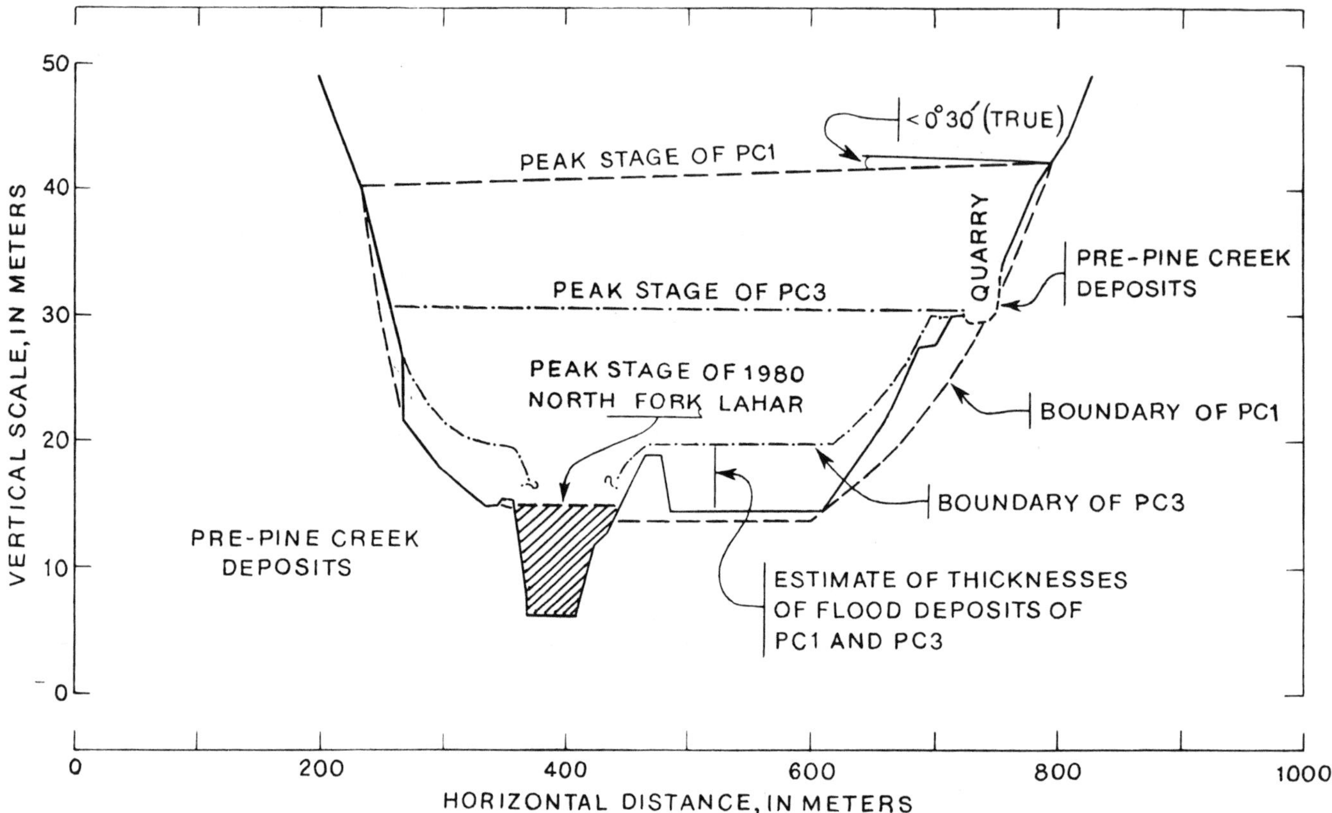

Figure 3. Reconstructed cross sections of the peak discharges (assumed synchroneity with peak stage) of lahars PC 1 and PC 3 at Kid Valley. The flow cross section of the lahar of May 18, 1980, is shown for comparison.

flow volume, estimated at about 1 km^3 in its medial portion, yielded deposits that have since been reworked, or continued in the flood wave to the Columbia River. Mainly before the start of the Castle Creek eruptive period (about 2,200 to 1,700 yr B.P.), a locally widespread alluvial fill as much as 2.5 m thick was formed in response to reworking of the vast influx of laharic sediment.

The series of four lahars originated in the North Fork of the Toutle River about 2,500 radiocarbon yr ago, close to the end of the Pine Creek eruptive period. Stumps of trees sheared off by flow PC 1 yielded uncorrected radiocarbon dates of 2,510 ±40 yr B.P. at an upstream locality in the North Fork Toutle River (Fig. 2), and 2,550 ±145, 2,560 ±155, and 3,290 ±195 yr B.P. at a downstream locality near Tower. Wood beneath flow PC 1 in the Cowlitz River was dated at 2,440 ±40 yr B.P. The 3,290-yr date represents a dead but well-preserved cedar, or the older interior wood of a stump abraded by the intense transport of boulders near the base of the flow.

EVIDENCE OF LAKE-BREAKOUT ORIGIN

Size and dynamics of initial flow

The first (PC 1) and third (PC 3) flows in the sequence are the largest lahars in the history of the Toutle-Cowlitz River system during the approximately 50,000-yr existence of Mount St. Helens. Discharges were determined from measured flow cross sections (like Fig. 3), and mean peak-stage velocities determined from runup measurements (see Scott, 1985a and 1986, for details). The instantaneous peak discharge of flow PC 1 was in the range of 200,000 to 300,000 m^3/s at distances between 30 and 50 km from the modern crater (Fig. 4). In terms of instantaneous flow rate, that discharge is comparable to the mid-course Amazon River at flood stage (Oltman and others, 1964; Oltman, 1968). Mean peak velocities of flow PC 1, 30 to 50 km from the crater, were in the range of 16 to 22 m/s.

The transition interval shown in Figure 4 is the channel segment where the PC 1 debris flow was transforming downstream to hyperconcentrated streamflow. It is described and illustrated in Scott (1985a and b).

The estimate of the volume of the flood wave of lahar PC 1, 1 km^3, is crude; it is smaller than the volume of 2.8 km^3 estimated for a modeled modern lahar and based on a released volume of water from Spirit Lake of about 0.4 km^3 and a peak discharge at release of 15,000 m^3/s (Swift and Kresh, 1983). The shape and volume of a lahar flood wave are equally difficult to estimate for both modern and ancient flows, however. The shape of the initial hydrograph in response to a lake breakout is extremely sensitive to the assumed rate and final width of breach opening, and any estimates of those figures are really guesses. Even if the initial hydrograph could be estimated, the rate and amount of downstream bulking involved in formation of the lahar are also highly subjective.

The smaller volume of lahar PC 1 is based on its probably more highly peaked shape, which is in turn a function of more rapid breakout through a volumetrically smaller ancient natural dam, compared to the broad extent of the 1980 debris avalanche. Comparisons of peak discharges are not helpful in this regard. Highly peaked lahar flood waves may change markedly in volume but only slightly in peak discharge (see tables in Laenen and Hansen, 1988).

A sudden release from an ancestral Spirit Lake is the most likely, if not the only possible, source of the streamflow surge that produced lahar PC 1. A large volume of catastrophically ejected material (see Scott, 1985a) could produce a large lahar, but it would not contain the high percentage of stream alluvium seen in the lahars of Pine Creek age. Comparisons with worst-case estimates of snow and ice melting by pyroclastic flows and the possible emptying of a summit crater lake are discussed by Scott (1986).

It is highly improbable that a debris flow as large as PC 1 could be initiated by a relatively small water surge from the melting of snow or ice. Unlike the moraine-lake breakout in Russia described by Yesenov and Degovets (1979), in which a debris flow formed almost immediately and continued to enlarge downstream, the large modern and ancient lahars at Mount St. Helens have, once fully bulked, attenuated over most of their downstream courses. This behavior probably can be extrapolated to other Cascade Range volcanoes if channel stability and sedimentology are considered. Modern breakout flood waves from moraine-impounded lakes on Cascade Range stratovolcanoes have not been amplified by downstream bulking comparable to the Russian case, including examples where debris flow formed (unpublished data of A. Laenen, J. E. Costa, B. Voight, and K. M. Scott from North Sister, Middle Sister, and Brokentop volcanoes). Eruption-induced lahars and streamflow surges would probably behave in a similar manner at those volcanoes. A debris flow beginning as a slope failure induced by precipitation on Mount Hood bulked rapidly, but thereafter showed little net overall increase in peak discharge over most of its 8-km course (Gallino and Pierson, 1985). The flow occurred in a steeply sloping channel containing much erodible sediment and with unstable valley side slopes.

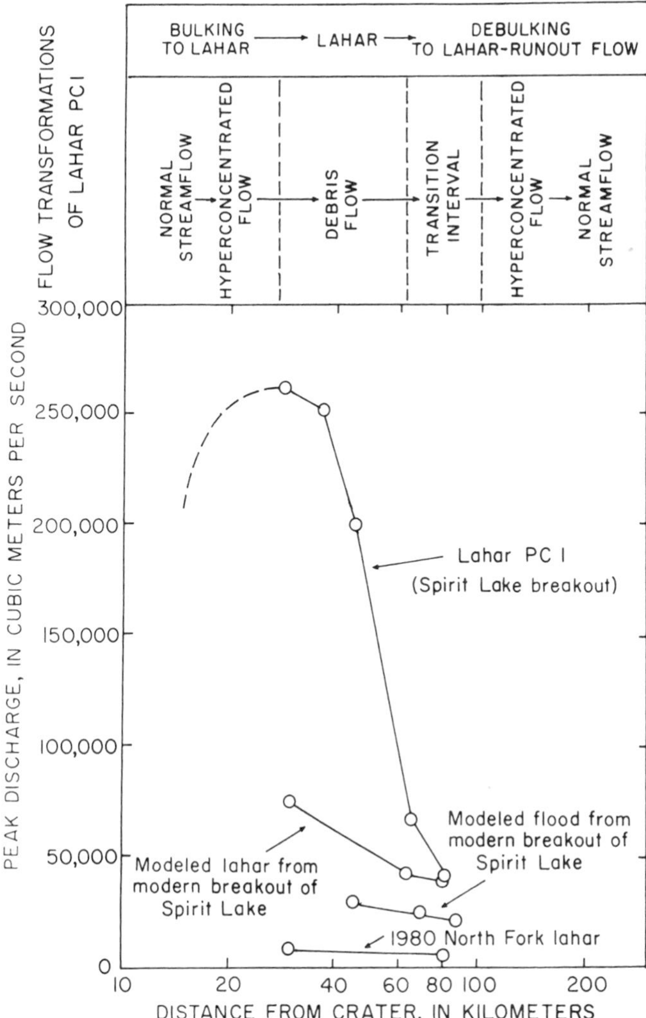

Figure 4. Peak discharges of lahar PC 1, a modeled lahar from a modern breakout of Spirit Lake (Swift and Kresh, 1983), and a modeled flood flow from a modern breakout of Spirit Lake (Bissell and Hutcheon, 1983). Peak discharge (Fairchild and Wigmosta, 1983) of the relatively cohesive (3 to 5 percent clay) lahar of May 18, 1980, is shown for comparison.

Clast roundness and a lahar-bulking factor

The clasts in the lahars of Pine Creek age are notably better rounded (Fig. 5) than those in the typically granular lahar in the river system. The percentage of alluvium bulked into lahars PC 1 and PC 3 was more than 50 percent, shown by a lahar-bulking factor (LBF). LBFs for the smaller flows PC 2 and PC 4, although not calculated, are similar. The LBF is the demonstrable percentage of sediment in a lahar introduced from channels beyond the base of the volcano and is calculated from roundness in size classes common to both channel and flood plain facies (Scott, 1985a). In calculation of the factor, a fractured particle is classified as alluvium if only a small part retains a smooth, stream-rounded surface.

The high LBFs of the large lahars of Pine Creek age illustrate a difference from any potential modern breakout of Spirit Lake. In a modern breakout, the bulking of the flood surge to a lahar would occur mainly, if not entirely, on the surface of the debris avalanche deposit, which extends 22 km down the North Fork Toutle River from the volcano. A breakout of Elk Rock Lake, a small lake on the avalanche surface, occurred on August 27, 1980, and may have formed a lahar; only the distal part of that flow was sampled, and peak measured sediment concentra-

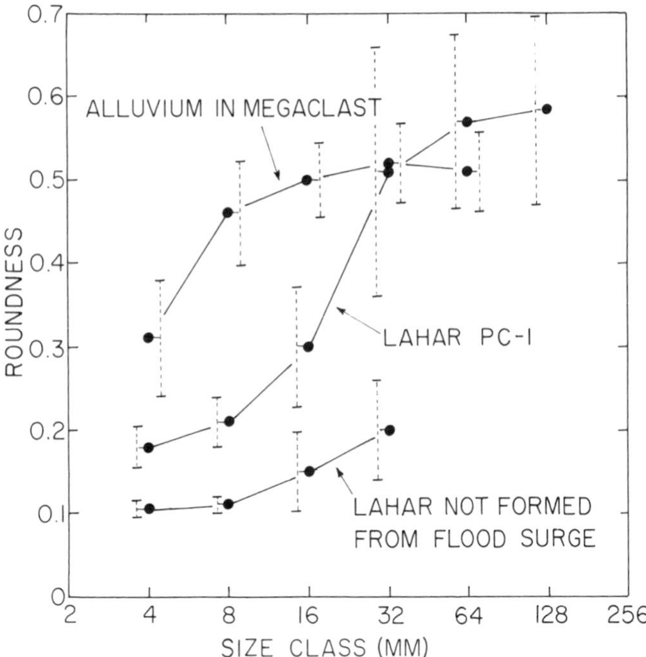

Figure 5. Roundness of clasts in a lahar (PC 1) produced by bulking of a flood surge, a lahar derived from a slope failure (lahar of May 18, 1980, in the North Fork Toutle River), and alluvium in a megaclast in lahar PC 1. The alluvium represents the main source of sediment in lahar PC 1. Confidence limits about the mean values are shown at the 95 percent level. See Krumbein (1941) for techniques.

tion was in the hyperconcentrated range. The upstream behavior of the 1982 (Pierson and Scott, 1985) and 1984 (K. A. Cameron and P. T. Pringle, oral communication, 1985) meltwater surges and lahars indicates that the proximal flows transformed rapidly into lahars on the avalanche surface, or on pyroclastic flow deposits that now overlie the avalanche. The low LBFs of most pre–1980 lahars also indicate the speed with which the bulking process normally takes place. This is a function of the erodibility of the volcaniclastic detritus on the slopes of most Cascade Range stratovolcanoes. The matrix facies and dilated blocks of debris avalanches are especially erodible. Flow PC 1 needed at least 20 km of valley-wide erosion of alluvium to complete its bulking, suggesting that the ancient natural dam was smaller in volume and consequently yielded less sediment directly to the flow than would the 1980 debris avalanche during a breach.

Megaclasts

Masses of hydrothermally altered dacite at least 8 m in intermediate diameter from the core of Mount St. Helens are common in flow PC 1 upstream from the gorge heading at the Coal Bank Bridge (Fig. 6). These megaclasts were blocks in an ancient debris avalanche, whether the lake-impounding deposit had that origin or not. The abundance of the blocks strongly suggests the existence of a debris avalanche large enough to have dammed the valley of the North Fork Toutle River in Pine Creek time, although the pre–1980 distribution of the Y tephra of pre–Pine Creek age is believed by D. R. Crandell (oral communication, 1984) to preclude a large debris avalanche of Pine Creek age in that valley. A smaller, more areally restricted debris avalanche than the 1980 feature could have created an effective natural dam, however, as well as one that could have breached rapidly. At least one debris avalanche of pre–Kalama age (500 to 350 yr B.P.) is exposed on the north side of the mountain.

Some of the dacite megaclasts show a contorted intertonguing of igneous structures and alteration banding like that described by H. Glicken in the 1980 debris avalanche and interpreted as a product of flow within the avalanche (H. Glicken, oral communication, 1982). A similar structure is reported from other debris avalanches by Ui (1983) and Siebert (1984) and was seen by the author in blocks in the 1984 debris avalanche at Mount Ontake, Japan. Most of the dacite megaclasts in lahar PC 1 are finely brecciated, a structure not characteristic of most of the rocks in place in the modern crater, and a probable response to slope failure. Coarser brecciation in blocks in landslides and debris avalanches is described as a jigsaw puzzle effect (Shreve, 1968), or as jigsaw cracks (Ui, 1983). The result is a dilatant response in which the individual pieces have not been sorted or widely separated.

The abundant dacite megaclasts locally form a diamicton within the laharic diamicton, with the latter forming the matrix phase; at other sites a clast-supported deposit of megaclasts occurs within the lahar matrix. The dacite megaclasts are present but smaller below the Coal Bank Bridge, probably as a result of disaggregation during flow through the gorge below the bridge. Similar masses occur in the other flows of the sequence, but they are not as common or as large. Some protrude above the top surface of the deposit (Fig. 7, A), having grounded on the underlying surface like an iceberg, while others may have floated in the lahar (Fig. 7, B).

A single megaclast, about 5 by 12 m in its original exposure, contains a stratigraphic section of flood plain deposits eroded upstream by lahar PC 1 (Fig. 8; Scott, 1985a, Fig. 24). It contains five stratigraphic units, including lahars and well-rounded cobble alluvium (Fig. 5) that was the source of most of the sediment in the flow. The delicate mass, itself rounded during lahar transport, shows no internal deformation. The originally horizontal bedding in the megaclast slopes upstream at an angle of 7°.

Close association in time

The passive flood plain flow of lahar PC 1, especially in backwater areas, surrounded trees and left them standing, rather than shearing them off as did the high-velocity flow along channels (Fig. 2). The trees remained standing as the succeeding flows in the series were emplaced. They are now represented by cylindrical vertical molds known locally as tree "wells" (Fig. 2). The interior surfaces of some molds show that bark was uniformly intact throughout the emplacement of the entire flow sequence, after the trees were killed by the initial flow (lesser depths of lahar

Figure 6. Photograph of a dacite megaclast in lahar PC 1 near the confluence of the North Fork and South Fork of the Toutle River. Note the roundness of the cobbles in the lahar and the apparent roundness of the megaclast. Exposed diameter of the megaclast is 1.5 m.

inundation killed the same species in 1980). This indicates that the sequence was emplaced within 20 yr at most, based on the normal decompositional evolution of a standing old-growth Douglas fir (Franklin and others, 1981, Fig. 16).

The nature of the stratigraphic contacts between the lahars suggests an even shorter time period for emplacement of the sequence. The contacts are either sharp and nearly planar (except where there was primary relief; Fig. 7), or the units are separated by thin intercalations of fluvial sand, deposition of which began either with dewatering of the flow, or with the first rain. There is no evidence of significant regrowth of vegetation. In contrast, the flood plain deposits of the 1980 lahars, including the granular, noncohesive flows, were covered with a nearly impenetrable growth of alders by 1985. This is evidence that the flow sequence of Pine Creek time was emplaced within several years or less.

These results contrast sharply with the mean recurrence interval of lahars (those large enough to inundate flood plains 50 km or more from the volcano) of less than 100 yr during eruptive periods of the last 4,500 yr (Scott, 1986). Other lahars also tended to cluster in time within eruptive periods, but in no known case to the extent of the Pine Creek lahars.

Thus the four lahars of similar type (granular and with high LBFs) occurred in rapid succession, leading to the inference of a similar lake-breakout origin. The largest ancient lahar, PC 1, resulted from a breakout of Spirit Lake and was the first in a sequence of four large lahars. Without engineering intervention and without additional large-scale volcanic activity, a modern breakout of Spirit Lake would have been the last in a sequence and, based on systematic monitoring, would probably have occurred in 1984 (Meyer and others, 1986). This potential modern sequence of events is additional evidence that, although the ancient and modern blockages of the valley were probably of similar composition and produced, respectively, actual and hypothetical (modeled) lahars that were closely similar in upstream inundation area (cf. Scott and Janda, 1982, and Swift and Kresh, 1983), the configuration and areal extent of the blockages were different. It is possible that, unlike the potential modern sequence, the four ancient lahars were derived from breakouts of only Spirit Lake. Repeated large-scale flows or avalanches into an ancestral Spiril Lake may have displaced water that surged across what remained of an initial damming deposit to produce lahars PC 2 through 4.

FLOW BEHAVIOR

Mapping of the lahar deposits shows that the streamflow surge that generated lahar PC 1 required at least 20 km of travel from near Spirit Lake before it bulked enough to become a lahar. Thin lahar deposits accreted to valley side slopes 1 to 3 km below Camp Baker (Fig. 1). Over the next several kilometers the lahar

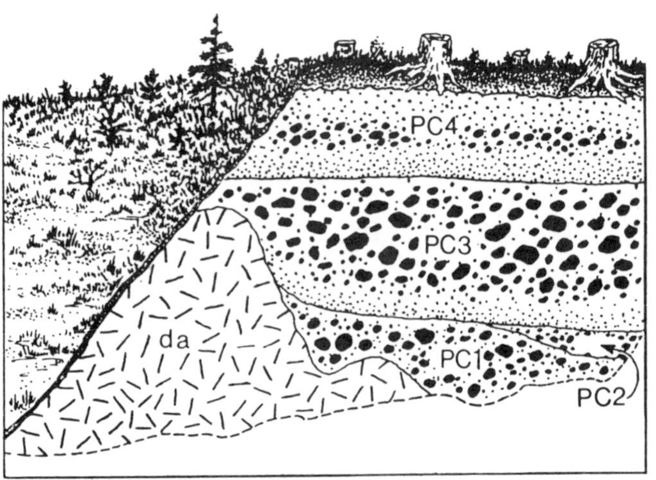

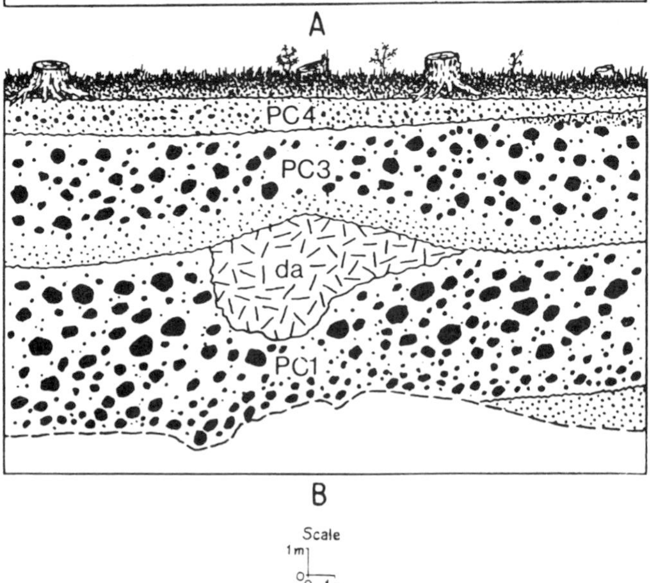

Figure 7. Examples of megaclasts in lahar PC 1 upstream from the confluence of the Green and North Fork Toutle Rivers. Dashed line at base of exposures is limit of talus.

became well developed, forming both the flood plain facies as well as clast-supported boulder bars of the channel facies.

Based on the LBF and the estimated volume of lahar PC 1, erosion of about 32 km^2 of upstream valley flat to a depth of 10 to 20 m yielded much of the sediment that transformed the flood surge to debris flow. Valley walls are mainly of glaciated bedrock and yielded relatively little sediment to the flow. Erosion of 10 to 15 m occurred during enlargement of the smaller debris flow described by Yesenov and Degovets (1979). The megaclast of flood plain deposits (Fig. 8) documents at least 5 m of erosion at a point.

In light of the evidence presented heretofore, the supposition can be made that any attempt to synthesize a modern Spirit Lake breakout should yield results similar to the ancient event. The depths and inundation levels of lahar PC 1 (Scott and Janda, 1982) and a hypothetical modern lahar (Swift and Kresch, 1983), both resulting from Spirit Lake breakouts, are very similar in reaches between 30 and 50 km from the crater. However, the flow dynamics as well as the subsequent downstream attenuation of the lahars are strikingly divergent. The following comparisons with that model and two others show that the behavior of the ancient lahar is the more accurate guide to the dynamics of a modern lahar—reversed uniformitarianism as a guide to modern catastrophic behavior.

The hypothetical lahar of Swift and Kresh (1983) was considerably slower (mean velocities were less) than lahar PC 1, and flow depths were correspondingly higher for the lesser discharges (Fig. 4) yielded by the hypothesis. These differences reflect partly the modeled assumptions of a texturally uniform wave of viscous debris and, in effect, too high a resistance to flow, either or both as channel friction or internal resistance to flow (see discussion of Manning's "n" values in Laenen and Hansen, 1988, and a rheologic consistency factor in Chen, 1983). The resulting behavior was more like that of the 1980 North Fork lahar (Fig. 4), one of two lahars of significant clay content and size (but small relative to PC 1) in the history of the river system. Velocities of the modeled flow were locally similar to or even less than those of the relatively small 1980 lahar (6 to 12 m/s; Fairchild and Wigmosta, 1983). That lahar developed mainly by dewatering and slumping of fine-grained surficial deposits of the 1980 avalanche, contained from 3 to 5 percent clay, and consequently was more cohesive than the Pine Creek–age lahars and most other 1980 and pre–1980 lahars.

Although much of the sediment in a modern lake-breakout lahar would come from the debris avalanche, its net composition probably would fall in the low-clay range, as is seen in the post–1980 lahars developed on its surface. Consequently, behavior of a modern lake-breakout lahar would be like that of the typically more granular flows in the system, of which the transformations and attenuation of lahar PC 1 shown in Figure 4 are an example. The critical transformation-limiting clay content of lahars in the river system is about 3 percent. Preliminary sampling of the avalanche deposits, from surficial parts of which the 1980 lahar was derived, showed an average clay content of 4 percent (Voight and others, 1983). By the time of any probable breakout, part of the residuum would be depleted in fine sediment by hydraulic sorting. More recent sampling of the avalanche deposits, possibly at greater depth in the now-incised unit, indicates a clay content of less than 3 percent (H. Glicken, written communication, 1985).

In part as a result of its granular, noncohesive character, flow PC 1 attenuated more rapidly than the 1980 North Fork lahar, the modeled lahar, and a modeled flood surge. The differences are illustrated in Figure 4. The models did not include deposition losses, which caused substantial attenuation. In addition, however, a significant part of the attenuation in the ancient flow was due to the progressive downstream change to lahar-runout flow and the textural changes that accompanied that transformation. The transformation in three, much smaller, modern examples in the river system required distances from 9 to 19 km.

Figure 8. Oblique aerial photograph (from height of 40 m) of megaclast (egg-shaped mass, approximately 5 by 12 m, in center of photo) in lahar PC 1. The megaclast consists of a delicate section of upstream flood plain deposits.

Transformation of even flow PC 1 to a lahar-runout flow probably was mainly complete at the Cowlitz-Columbia River confluence. The effects at that point would have been those of a devastating flood, albeit one highly charged with sediment.

The model of a modern Spirit Lake breakout flood surge did not generate sediment concentrations above 15 percent by volume (Bissell and Hutcheon, 1983), so the largest of several modeled flood waves routed downstream was relatively small (Fig. 4). Yet another model assumed formation of hyperconcentrated flow from a modern Spirit Lake breakout, based on the notion that the only sediment available to the flood surge would be from nearly saturated debris avalanche deposits (Sikonia, 1985). The high LBFs of the Pine Creek lahars show that these results are unrealistic, even if transformation of the modern flow to a lahar was not completed on a saturated debris avalanche. The LBFs indicate that bulking to a lahar would continue below the 1980 avalanche terminus. That bulking would be accelerated by post-1980 sediment-retention structures now filled with potentially mobile sediment.

TEXTURE

The reasons for the granular, noncohesive character of the lahar deposits were reviewed above. Cumulative size distribution curves and histograms illustrate the significant differences in texture between the source material and the flood plain facies of lahar PC 1 (Fig. 9). The flood plain facies has no clast support, and records generally passive flow with features such as vertical tree molds and overrun forest litter. The channel facies is coarser, more poorly sorted, and more likely to have a clast-supported framework than the flood plain facies. The coarse mode of the channel facies of the Pine Creek lahars is commonly in the cobble size range (64 to 256 mm), whereas that of the flood plain facies is commonly in the pebble range (4 to 64 mm).

Inverse grading is common in the basal part of both facies. This part of each flow unit may be overlain by a poorly graded or ungraded zone, that in the channel facies of lahar PC 1 contains extensive regions of clast support. The normal tripartite subdivision (see Fig. 7, lahars PC 3 and 4) ends with an uppermost normally graded zone. The basal inverse grading is interpreted as a particle-size-dependent response to near-boundary shear stress, and the upper normal grading represents settling of coarser clasts in the zone of plug flow (see Denlinger and others, 1984; Scott, 1985a).

The granular lahars in the river system have much variation in gravel content (more than 2 mm)—in size, grading, and degree of dispersal. The texture of the silty sand matrix is comparatively constant, although change does occur over many tens of kilome-

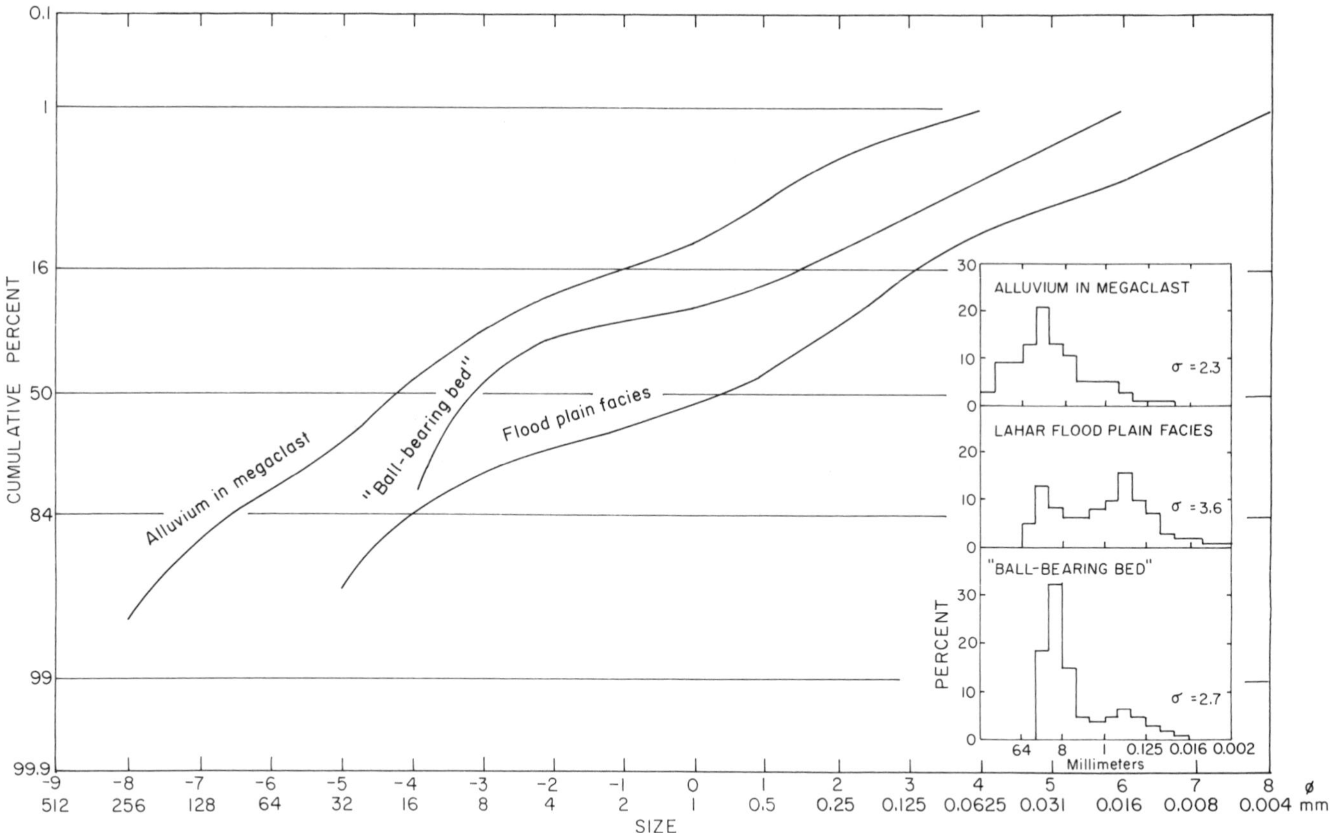

Figure 9. Cumulative curves and histograms illustrating the size distributions of alluvium in a megaclast in lahar PC 1, the flood plain facies of lahar PC 1, and the "ball-bearing bed." Megaclast location is upstream from the confluence of the Green and North Fork Toutle Rivers; lahar localities are near the Green Mountain Mill. Sorting is the inclusive graphic sorting of Folk (1980).

ters. Most of the normal grading and part of the inverse grading is coarse-tail grading, in which size variation is concentrated in the gravel-size clasts. Thus the critical diameter (term introduced for pyroclastic flows by Sparks and others, 1973) that separates these phases is about 2 mm. The same variation with size was recognized by Fisher (1971) and Crandell (1971); Major (1984) has documented a critical diameter of 4 mm in coarse rubbly lahars. Because the larger clasts commonly form an intact framework in the channel facies, however, the two size divisions generally should be considered as a matrix phase and a coarse-grained phase, rather than as the continuous or matrix phase and the dispersed phase of Fisher (1971). The latter terminology is accurate for the flood plain facies, which, because of its obviously dispersed coarse-grain phase and its original wide distribution, is the lahar facies most commonly seen.

The clast-supported deposits of the channel facies of a lake-breakout lahar and those of alluvium are texturally very similar where exposures are poor, and where the massive graded character of the lahar may not be clear. A distinctive feature of the lahars, apparent in even a small sample, is the degree of angularity in the finer pebble size classes (Fig. 5). This angularity is produced in part by cataclasis during flow.

BOUNDARY FEATURES

Boundary features of the large lahars in the river system include inverse grading and a lahar-abraded pavement closely resembling that produced by glacial scour (Scott, 1985a). The latter feature is formed at sites of rapid energy loss where boulders, supported away from the flow boundary by particle-to-particle contact forces during sustained flow, sink to the bottom of the flow and abrade the underlying surface. This commonly occurs over a short channel segment prior to deposition of the coarse clasts in clast-supported whaleback bars. Boundary features also include sole layers, defined as the texturally distinct basal sublayers of lahars (Scott, 1985a), which allow the correlation of lahar boundaries and thus the reconstruction of flow cross sections like Figure 3. Some sole layers may be accretionary during the flood wave rise and thus act to reduce resistance to flow (boundary roughness) during peak flow.

A unique variety of sole layer is the "ball-bearing bed" at the base of the channel facies of lahar PC 1 (Figs. 2 and 10). This unit is a 1-m-thick concentration of marble-size pebbles, is clast supported where best developed, and contains the same silty sand matrix as the remainder of the lahar. Described as a Type III sole

Figure 10. Photograph of the basal part of the channel facies of lahar PC 1 near the Green Mountain Mill (Fig. 2). Note the clast-supported pebble-dominated texture of the "ball-bearing bed," which is about 1.0 m thick at this locality. The contact between the "ball-bearing bed" and the overlying part of the lahar may be transitional or sharp.

layer by Scott (1985a), its size distribution is truncated and positively skewed (no coarse tail; Fig. 9). More than 50 percent of the unit is concentrated in 2 or 3 pebble size classes. The unit itself is poorly graded or ungraded, but can be considered a part of the inversely graded basal part of the channel facies of lahar PC 1.

The distribution of the layer (Fig. 11) shows its thickness and sorting to correlate directly with the amount and duration of shear stress at channel boundaries. It is a diagnostic feature of the channel facies of lahar PC 1 where boundary shear was intense and sustained. It is absent in the passively emplaced flood plain facies of that flow. A thinner, finer-grained equivalent is present at the base of PC 1 deposits high on steep valley side slopes within several meters of peak stage where flow depth exceeded 30 m. At such locations, shear was significant but was not sustained because of the transitory passage of the flood wave at that height.

The truncated size distributions of the layer show that the layer is the result of near-boundary shear-induced sorting in which clasts larger than 32 mm were either moved up into the body of the flow, or were broken. Locally derived boulders were observed to have moved rapidly up several meters into flow PC 1 to associate with clasts of similar size. The velocity gradient that developed within the matrix surrounding the larger particles near the base of the lahar caused them to rotate and, in effect, to climb up the gradient (R. P. Denlinger, written communication, 1985). As particles larger than 32 mm moved upward or were broken, and the remaining pebbles approached an intact framework, kinetic sieving (Middleton, 1970) probably moved the smaller particles preferentially downward. However, the "ball-bearing bed" itself is at best poorly graded and in all cases is the basalmost deposit of the flow. The primary forces in its formation are clearly size-dependent particle interactions that moved the coarser clasts away from the boundary.

Although the sorting response to the extreme duration and intensity of near-boundary shear was the primary factor producing the unusual size distribution of the "ball-bearing bed," cataclasis was also a factor. The bed developed at the base of the channel facies of lahar PC 1, where cataclasis was probably a result of the dilatant response to shear being suppressed by vertical loading in the deep (more than 30 m) and high-velocity flow. This effect in producing cataclasis is described from experimentally produced shear zones (Mandl and others, 1977). Broken clasts constitute 38 percent of two pebble size classes in the "ball-bearing bed," 32 percent of the same classes in the channel facies, but only 16 percent of those classes in the source material. Cataclasis in the "ball-bearing bed" was accompanied by coarse abrasion that actually contributed to clast rounding. The coarsely abraded surfaces of some pebbles in the sole layer are in sharp

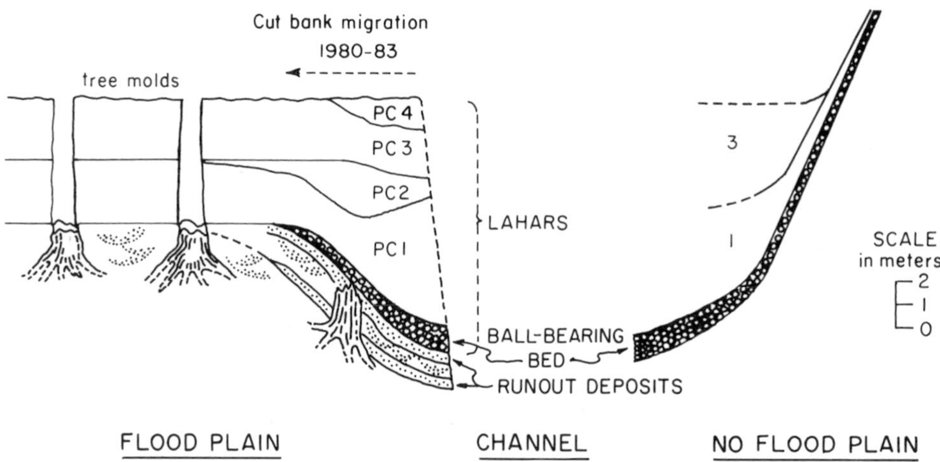

Figure 11. Diagrammatic cross section of deposits of Pine Creek age near the Green Mountain Mill, showing distribution of the "ball-bearing bed."

contrast to the smooth, mainly stream-rounded surfaces of coarser clasts in the main body of the lahar.

The three succeeding lahars do not have a sole layer like the "ball-bearing bed." Each has a locally well-developed sandy sole layer, seen at the bases of lahars PC 3 and PC 4 in Figure 7. This type, described as a Type I sole layer, lacks pebble-size or coarser clasts and is typically a feature of the flood plain facies of some granular lahars. Most features of the sandy sole layer, which has a texture similar to that of the matrix in the main body of the lahar, support its origin as a boundary layer from which the coarse clasts moved upward. It did not thin appreciably when crossing large roughness elements (Fig. 7) and was seen on vertical flow boundaries, so in general the layer cannot be the deposit of an early dilute part of the flood wave. A sole layer with that origin occurs, however, within the transition intervals where distal lahars were transforming downvalley to hyperconcentrated flow (see Scott, 1985b, Fig. 4).

The differences in sole layers of facies of the Pine Creek lahars relate both to surface roughness and flow depth, two other factors that influence shear-induced sorting. A sole layer did not form at the base of the PC 1 flood plain facies when that huge lahar moved passively across a maturely forested flood plain, in contrast to the smooth, sparsely vegetated surfaces that were more rapidly overrun by the succeeding lahars having pronounced sandy sole layers. Thus the presence or absence of sole layers in the flood plain facies can be explained by variations in roughness of the underlying surface and the consequent effects on shear-induced sorting. The effects are probably achieved through the differences in duration and shape of the vertical velocity distribution. Differences at higher levels of near-boundary shear explain whether a channel facies is marked by a sandy sole layer or by a coarser unit like the "ball-bearing bed."

LAHAR-MARGIN LAKES

Silver Lake (Fig. 1) formed in a tributary blocked by the lahars of Pine Creek age (Mullineaux and Crandell, 1962). General characteristics of lahar-margin lakes, based on Silver Lake as the type example and several others of 1980 and Pine Creek age, are a shallow depth and a wide blockage. Because of their fluid character, lahars flow into areas of backwater channel storage, commonly leaving a broad deposit surface with a low lateral slope. Breakouts of the resulting shallow lakes consequently are not a hazard generally comparable to that of the initiating lahar. Breakout may be rapid or may not occur, depending on the size of the blocked tributary as well as the lahar. The behavior of the 1980 lahar in the North Fork Toutle River as it briefly dammed the larger Cowlitz River is described by Scott (1985a). Silver Lake, with a surface area of more than 15 km^2, has a maximum present depth of only 3 m. The original depth probably was similar. The natural outlet, not modified until 1971, was entirely stable. These characteristics contrast markedly with the commonly deeper lakes impounded by the less-stable deposits of debris avalanches.

SUMMARY

With their distal transformation to hyperconcentrated lahar-runout flow, most modern and ancient lahars in the watershed draining the northwest sector of Mount St. Helens were the middle segment of a flood wave beginning and ending as a streamflow surge. Evidence that a series of four lahars was derived from lake-breakout flood surges about 2,500 radiocarbon yr ago includes (1) the magnitude of the initial gigantic lahar relative to a size probable with other origins, (2) a high content of alluvium introduced into the flood surge beyond the base of the volcano, (3) abundant megaclasts derived from a debris avalanche in the initial flow, and (4) occurrence in a time-clustered series. The series is an analog of the catastrophic modern lahars that would have resulted from breakouts of lakes formed or enlarged by the 1980 eruption, had not lake levels been stabilized by emergency engineering works.

The study of the sequence of Pine Creek age verifies that the dynamics of ancient lahars can be a better guide to forecasting the character and behavior of large modern flows than any model to date. The approach of using the past as a guide to the future also has practical potential for predicting lahar magnitude and frequency in the Toutle River system. Throughout most of that watershed the peak discharge for inundation-level design purposes is not a flood from precipitation runoff but a volcanically induced lahar and its proximal or distal streamflow phases.

The sedimentology and stratigraphy of lahars and lahar-runout flows beyond the base of a volcano may be the only, as well as the most complete, evidence of important volcanic events. Lake-breakout lahars may be revealed only in that way. Their granular deposits probably are present downstream from many volcanoes but may not have been recognized because of their resemblance to alluvium, particularly where the content of rounded clasts is high or the deposit consists of the clast-supported channel facies. Significant angularity in the fine pebble size classes is diagnostic of large granular lahars and results from cataclasis during flow, even where the sediment source is rounded stream alluvium. The existence of now-eroded debris avalanches and other natural-dam deposits may be revealed by these downstream deposits and their included megaclasts.

ACKNOWLEDGMENTS

The manuscript benefited from review by J. E. Costa, A. Laenen, D. R. Crandell, K. A. Cameron, H. Glicken, and J. J. Major. Most of the other acknowledgments noted in Scott (1985a) and Scott (1986) apply to this paper as well.

REFERENCES CITED

Beverage, J. P., and Culbertson, J. K., 1964, Hyperconcentrations of suspended sediment: Proceedings of American Society of Civil Engineers, Journal of Hydraulics, v. 190, HY6, p. 117–128.

Bissell, V. C., and Hutcheon, R. J., 1983, Forecasting and data collection preparedness for streams affected by the Mt. St. Helens volcano: Portland, Oregon, National Weather Service, Northwest River Forecast Center, 72 p.

Chen, C., 1983, On frontier between rheology and mudflow mechanics; Proceedings of Conference on Frontiers in Hydraulic Engineering, Cambridge, Massachusetts, August 1983: American Society of Civil Engineers and Massachusetts Institute of Technology, p. 113–118.

Costa, J. E., 1985, Floods from dam failures: U.S. Geological Survey Open-File Report 85-560, 54 p.

Crandell, D. R., 1971, Postglacial lahars from Mount Rainier Volcano, Washington: U.S. Geological Survey Professional Paper 677, 73 p.

Crandell, D. R., Miller, C. D., Glicken, H., Christiansen, R. L., and Newhall, C. G., 1984, Catastrophic debris avalanche from ancestral Mount Shasta, California: Geology, v. 12, p. 143–146.

Denlinger, R. P., Scott, K., and Pierson, T., 1984, Granular resorting and its effect on resistance to flow in grain-supported lahars [abs.]: EOS Transactions of the American Geophysical Union, v. 65, p. 1142.

Fairchild, L. H., and Wigmosta, M., 1983, Dynamic and volumetric characteristics of the 18 May 1980 lahars on the Toutle River, Washington; Proceedings of the Symposium on Erosion Control in Volcanic Areas, Seattle and Vancouver, Washington, July 6–9, 1982: Ministry of Construction, Government of Japan, Technical Memorandum of Public Works Research Institute no. 1903, p. 131–153.

Fisher, R. V., 1971, Features of coarse-grained, high-concentration fluids and their deposits: Journal of Sedimentary Petrology, v. 41, p. 916–927.

Folk, R. L., 1980, Petrology of sedimentary rocks: Austin, Texas, Hemphill Publishing Company, 182 p.

Franklin, J. F., Cromack, K., Jr., Denison, W., McKee, A., Maser, C., Sedell, J., Swanson, F., and Juday, G., 1981, Ecological characteristics of old-growth Douglas-fir forests: U.S. Forest Service, Pacific Northwest Forest and Range Experiment Station General Technical Report PNW-118, 48 p.

Gallino, G. L., and Pierson, T. C., 1985, Polallie Creek debris flow and subsequent dam-break flood of 1980, East Fork Hood River Basin, Oregon: U.S. Geological Survey Water-Supply Paper 2273, 22 p.

Glicken, H., 1982, Criteria for recognition of large volcanic debris avalanches [abs.]: EOS Transactions of the American Geophysical Union, v. 63, p. 1141.

Hyde, J. H., and Crandell, D. R., 1978, Postglacial volcanic deposits at Mount Baker, Washington, and potential hazards from future eruptions: U.S. Geological Survey Professional Paper 1022-C, 17 p.

Krumbein, W. C., 1941, Measurement and geologic significance of shape and roundness of sedimentary particles: Journal of Sedimentary Petrology, v. 11, p. 64–72.

Laenen, A., and Hansen, R. P., 1988, Using a one-dimensional unsteady-state streamflow model to simulate debris flows: American Society of Civil Engineers, Journal of the Hydraulics Division (in press).

Major, J. J., 1984, Geologic and rheologic characteristics of the May 18, 1980, southwest flank lahars at Mount St. Helens, Washington [M.S. thesis]: University Park, Pennsylvania State University, 225 p.

Mandl, G., de Jong, L.N.J., and Maltha, A., 1977, Shear zones in granular material; An experimental study of their structure and mechanical genesis: Rock Mechanics, v. 9, p. 95–144.

Meyer, W., and Carpenter, P. J., 1982, Filling of Spirit Lake, Washington, May 18, 1980, to July 31, 1982: U.S. Geological Survey Open-File Report 82-771, 19 p.

Meyer, W., Sabol, M. A., and Schuster, R. L., 1986, Landslide-dammed lakes at Mount St. Helens, Washington, *in* Schuster, R. L., ed., Landslide Dams; Processes, Risk, and Mitigation; Proceedings of Symposium: American Society of Civil Engineers Geotechnical Special Publication, no. 3, p. 21–41.

Middleton, G. V., 1970, Experimental studies related to problems of flysch sedimentation, *in* Lajoie, J., ed., Flysch sedimentology in North America: Geologic Association of Canada Special Paper 7, p. 253–272.

Mullineaux, D. R., and Crandell, D. R., 1962, Recent lahars from Mount St. Helens, Washington: Geological Society of America Bulletin, v. 73, p. 855–870.

——, 1981, The eruptive history of Mount St. Helens, *in* Lipman, P. W., and Mullineaux, D. R., eds., The 1980 eruptions of Mount St. Helens, Washington: U.S. Geological Survey Professional Paper 1250, p. 3–15.

Newhall, C. G., 1982, A prehistoric debris avalanche from Mount St. Helens [abs.]: EOS Transactions of the American Geophysical Union, v. 63, p. 1141.

Oltman, R. E., 1968, Reconnaissance investigations of the discharge and water quality of the Amazon River: U.S. Geological Survey Circular 552, 16 p.

Oltman, R. E., Sternberg, H. O'R., Ames, F. C., and Davis, L. C., Jr., 1964, Amazon River investigations, reconnaissance measurements of July 1963: U.S. Geological Survey Circular 486, 15 p.

Pierson, T. C., and Scott, K. M., 1985, Downstream dilution of a lahar; Transition from debris flow to hyperconcentrated streamflow: Water Resources Research, v. 21, p. 1511–1524.

Schuster, R. L., 1985, Landslide dams in the western United States; Proceedings of the International Conference and Field Workshop on Landslides, IV, Tokyo, 1985: The Japan Landslide Society, Committee for International Technique, p. 411–418.

Scott, K. M., 1985a, Lahars and lahar-runout flows in the Toutle-Cowlitz River system, Mount St. Helens, Washington; Origins, behavior, and sedimentology: U.S. Geological Survey Open-File Report 85–500, 202 p. (also in press as Professional Paper 1447–A).

——, 1985b, Lahars and flow transformations at Mount St. Helens, Washington, U.S.A.; Proceedings of the International Symposium on Erosion, Debris Flow and Disaster Prevention, 1985, Tsukuba, Japan: The Erosion-Control Engineering Society, Japan, p. 209–214.

——, 1986, Lahars and lahar-runout flows in the Toutle-Cowlitz River system, Mount St. Helens, Washington; Magnitude and frequency: U.S. Geological Survey Open-File Report 86–500, 95 p. (also in press as Professional Paper 1447–B).

Scott, K. M., and Dinehart, R. L., 1985, Sediment transport and deposit characteristics of hyperconcentrated streamflow evolved from lahars at Mount St. Helens, Washington; Proceedings of International Workshop on Flow at Hyperconcentrations of Sediment, September 10–14, 1985, Beijing, China: International Research and Training Center on Erosion and Sedimentation, 33 p.

Scott, K. M., and Janda, R. J., 1982, Preliminary map of lahar inundation during the Pine Creek eruptive period in the Toutle-Cowlitz River system, Mount St. Helens, Washington: U.S. Geological Survey Water Resources Investigation Report 82–4067.

Shreve, R. L., 1968, The Blackhawk landslide: Geological Society of America Special Paper 108, 47 p.

Siebert, L., 1984, Large volcanic debris avalanches; Characteristics of source areas, deposits, and associated eruptions: Journal of Volcanology and Geothermal Research, v. 22, p. 163–197.

Sikonia, W. G., 1985, Impact on the Columbia River of an outburst of Spirit Lake: U.S. Geological Survey Water-Resources Investigations Report 85–4054, 29 p.

Sparks, R.J.S., Self, S., and Walker, G.P.L., 1973, Products of ignimbrite eruptions: Geology, v. 1, p. 115–118.

Swift, C. H., and Kresh, D. L., 1983, Mudflow hazards along the Toutle and Cowlitz Rivers from a hypothetical failure of Spirit Lake blockage: U.S. Geological Survey Water-Resources Investigations Report 82–4124, 10 p.

Ui, T., 1983, Volcanic dry avalanche deposits; Identification and comparison with nonvolcanic debris stream deposits: Journal of Volcanology and Geothermal Research, v. 18, p. 135–150.

Ui, T., Yamamoto, H., and Suzuki-Kamata, K., 1986, Characterization of debris avalanche deposits in Japan: Journal of Volcanology and Geothermal Research, v. 29, no. 1–4, p. 231–243.

Vallance, J. W., 1985, Late Pleistocene lahar assemblage near Hood River, Oregon [abs.]: EOS Transactions of the American Geophysical Union, v. 66, p. 1150.

Voight, B., Janda, R. J., Glicken, H., and Douglass, P. M., 1983, Nature and mechanics of the Mount St. Helens rockslide-avalanche of 18 May 1980: Geotechnique, v. 33, p. 243–273.

Yesenov, U. Y., and Degovets, A. S., 1979, Catastrophic mudflow on the Bol'shaya Almatinka River in 1977: Soviet Hydrology, Selected Papers, v. 18, p. 158–160.

Youd, T. L., Wilson, R. C., and Schuster, R. L., 1981, Stability of blockage in North Fork Toutle River, *in* Lipman, P. W., and Mullineaux, D. R., eds., The 1980 eruptions of Mount St. Helens, Washington: U.S. Geological Survey Professional Paper 1250, p. 821–828.

Manuscript Accepted by the Society April 4, 1988

Printed in U.S.A.

Geological Society of America
Special Paper 229
1988

The Mount Mazama climactic eruption (~6900 yr B.P.) and resulting convulsive sedimentation on the Crater Lake caldera floor, continent, and ocean basin

C. Hans Nelson, Paul R. Carlson, and Charles R. Bacon, *U.S. Geological Survey, 345 Middlefield Road, Menlo Park, California 94025*

ABSTRACT

The climactic eruption of Mount Mazama and the resulting sedimentation may have been the most significant convulsive sedimentary event in North America during Holocene time. A collapse caldera 1,200 m deep and 10 km in diameter was formed in Mount Mazama, and its floor was covered by hundreds of meters of wall-collapse debris. Wind-blown pyroclastic ash extended 2,000 km northeast from Mount Mazama and covered more than 1,000,000 km^2 of the continent. On the Pacific Ocean floor, Mazama ash was transported westward 600 to 700 km along deep-sea channels by turbidity currents.

The initial single-vent phase of the climactic eruption, a Plinian column, emptied over half of the magma erupted. Debris from this phase accumulated as a pumice deposit 10 m thick at the rim to 50 cm thick as much as 100 km from the vent. This deposit created a mid-Holocene stratigraphic marker over the continent and the continental margin of western North America. A ring-vent phase followed as a second part of the climactic eruption and produced highly mobile pyroclastic flows. These flows covered the mountain for at least 14 km from the vent, continued down the valleys nearly 60 km, and deposited as much as 100 m of pumiceous ignimbrite.

After the caldera collapsed as a result of the eruption of more than 50 km^3 of magma, heat of the climactic eruption apparently created phreatic explosion craters along the ring fracture zone of the caldera floor. Initially, explosion debris and sheetwash of pyroclastics off highlands seems mainly to have filled the local craters with bedded volcaniclastics. This basal, generally flat-lying unit, was quickly covered by wedges of chaotically bedded debris flow and avalanche-type deposits that thin inward from the caldera walls. These deposits may have formed in response to seismic activity associated with postcaldera volcanism that apparently began soon after the caldera collapsed. The lower two units of non-lacustrine beds (50 to 60 m) make up the majority of the postcaldera sedimentary deposits and seem to have deposited rapidly after the climactic eruption. Twenty to 25 m of lacustrine sediment has been accumulating more slowly over the subaerial debris during the past 6,900 yr.

Some Mazama ash probably was transported by rivers to the sea immediately after the climactic eruption because significant amounts of this ash appear in mid-Holocene turbidites of Cascadia Basin. The presence of Mazama ash mixed with Columbia River sand in texturally and compositionally graded turbidites shows that Mazama ash periodically was moved by sediment-gravity flows down the canyons and through channels to deposition sites in the Astoria Fan and the Cascadia Channel. The coarsest and thickest tuffaceous turbidites were deposited on channel floors, and the ash-rich suspension

flows that overtopped the levees were deposited as thin-bedded turbidites in interchannel areas.

Study of the Mount Mazama climactic eruption shows that such an event in the Cascade Mountains has the potential to: (a) cause major destruction within 100 km of the vent, (b) severely affect biota as far as 2,000 km downwind, and (c) disrupt commercial river and marine transportation or natural sedimentation as far as several hundred kilometers in the opposite direction from wind-blown debris. Present geologic characteristics on the Crater Lake caldera floor suggest that geologic hazards from a significant volcanic event appear to be minimal for the next few thousand years.

INTRODUCTION

The climactic eruption of Mount Mazama and the formation of a caldera about 6,900 yr B.P. resulted in significant deposition across the floor of Crater Lake, Oregon, over the continental area of the Pacific Northwest (Fryxell, 1965; Mullineaux, 1974; Sarna-Wojcicki and others, 1983), and on the sea floor of the northeast Pacific Ocean basin (Fig. 1; Nelson and others, 1968). The depositional history of this widespread sedimentary deposit can be well documented because of the recency of this event and the large volume of juvenile material erupted (about 50 km^3 of magma, or 100 times that of the Mount St. Helens 1980 eruption) (Fig. 1; Bacon, 1983; Lipman and Mullineaux, 1981). Disruption of the natural sedimentary system over extensive continental and marine areas indicates possible geologic hazards of immense proportions in the Cascade Range of western North America.

The geologic history of Mount Mazama and its climactic eruption have been documented by Williams (1942) and Bacon (1983); the general distribution of Mazama ash over western North America has been mapped by Fryxell (1965), Lidstrom (1971), Kittleman (1973), Mullineaux (1974), and Sarna-Wojcicki and others (1983); and the deposits of the tuffaceous turbidites of the northeast Pacific Ocean basin have been studied previously (Nelson and others, 1968).

The purpose of this chapter is to summarize all phases of this convulsive sedimentary event by presenting new information on seismic stratigraphy of the caldera floor and correlating it with the history of continental and marine sedimentation. First we will interpret the acoustic stratigraphy of the deposits that accumulated rapidly on the caldera floor following the cataclysmic event. Then we will characterize the pyroclastic-flow deposits of the caldera's subaerial slopes and outline the deposits of air-fall pumice and ash on the continent. Finally, we will describe the flushing of some of this continental pyroclastic sediment down the Columbia River, through submarine canyons, and across the abyssal sea-floor channels.

METHODS OF STUDY

Data for the subsurface lake floor deposits were collected in 1979 using a high-resolution seismic-reflection profiling system along tracklines spaced about 200 to 500 m apart (Fig. 2). Position of the boat was controlled by precision (±50 m) range-range transponder navigation. Sound sources used for the survey included a 16.4-cm^3 (1-in^3) airgun, fired once every 4 s, and a 300-joule Uniboom transducer, triggered every 0.25 or 0.5 s. Depths and thicknesses of the acoustic stratigraphic units were later calculated with the use of a computer by digitizing reflection times to these units every 0.25 to 2.0 min along the trackline.

To determine thickness, seismic velocities (two-way travel time) had to be assumed for each unit. The velocity selected for Crater Lake water (1,435 m/s) is based on the measured velocity of similar Lake Geneva water at 9°C (Hodgeman and others, 1961, p. 2538). The velocity assumed for unit II bedded volcaniclastics is based on recent work by Hill and others (1985) in Long Valley Caldera where the velocity of nonwelded to partly welded tuff ranges from 2,500 to 4,000 m/s, and that of unconsolidated volcaniclastic beds varies from 1,200 to 1,800 m/s. Because of the water saturation, the strong reflectors, and the possible induration of mixed deposits of tuff and unconsolidated pyroclastic debris, we selected a velocity of 2,500 m/s for unit II. We have arbitrarily chosen 1,525 m/s for the unit III debris wedges, 1,500 m/s for unit IV undifferentiated lake beds, and 1,450 m/s for unit V upper lake beds, based on typical velocities for nearsurface volcaniclastic lake sediment (Otis and others, 1977).

GEOLOGIC SETTING

Mount Mazama is a composite volcano made up of several overlapping shield and stratovolcanoes that are composed of andesite and dacite lava flows and pyroclastic layers (Bacon, 1983). The cone-building stage of volcanism occurred between ~420,000 and ~50,000 yr B.P. A shallow magma chamber subsequently developed and erupted catastrophically 6,845 ±50 yr B.P. The caldera collapsed during this climactic eruption and now forms a closed 8 × 10-km basin ~1200 m deep (Fig. 2). The lake has filled the caldera to a maximum depth of 622 m. The caldera walls rise 150 to 600 m at an average angle of 45° above the lake surface. Below the lake, the basin walls slope about 30° to a sediment-smoothed floor (Nelson, 1967; Nelson and others, 1986b).

Summarizing events of the Mount Mazama climactic eruption helps clarify the contemporaneous sedimentary history of the lake-floor, continental, and marine depositional environments. Bacon (1983) has divided the climactic eruption into single-vent and ring-vent phases, with the onset of collapse causing the switch

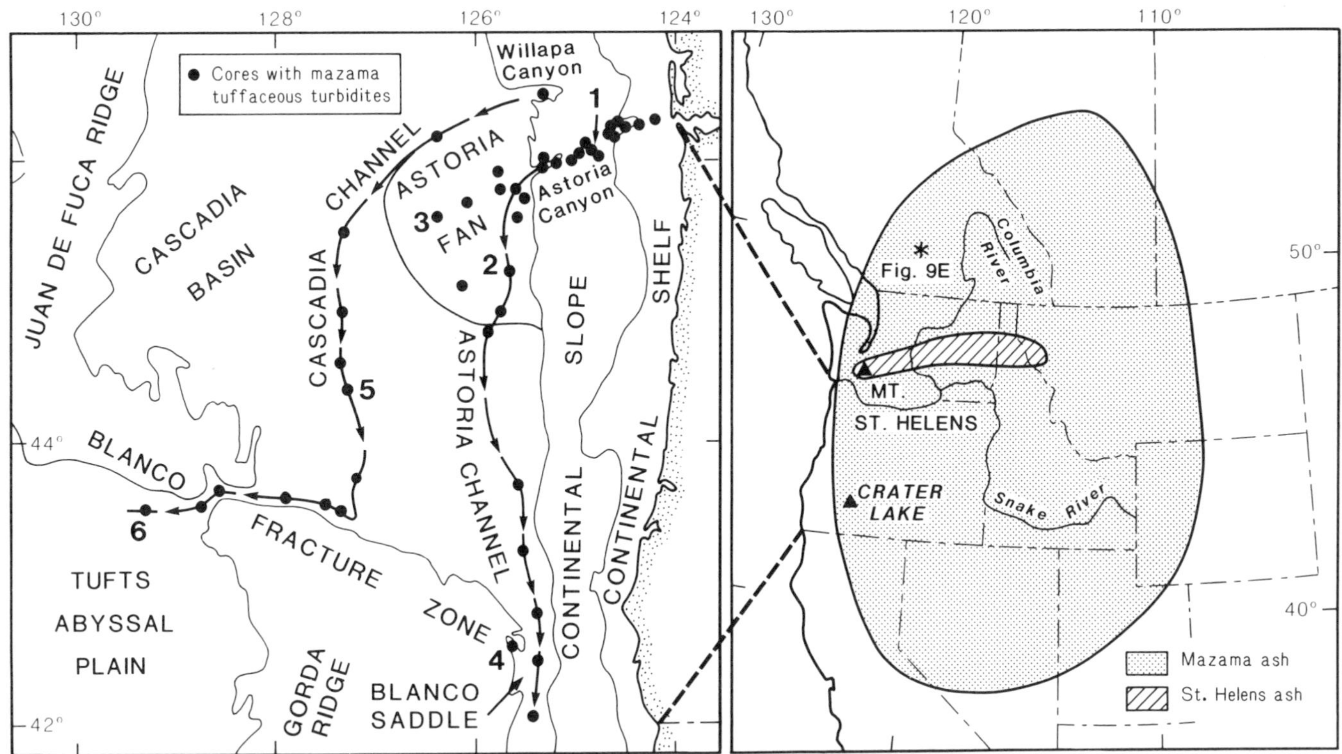

Figure 1. Occurrence of ash from the Mount Mazama eruption in the Pacific Northwest and on the sea floor of the northeast Pacific Ocean (continental limits modified from Sarna-Wojcicki and others, 1983; sea-floor locations modified from Carlson, 1967, and Nelson and others, 1968). Numbered cores are shown on Figure 10. Distribution of ash from the May, 1980, Mount St. Helens eruption to the 0.5-cm isopach (uncompacted)—approximately the same thickness as the limit of Mazama ash occurrence—is shown for comparison.

from the first to the second phase. In the first and main part of the single-vent eruptions, voluminous airfall pumice and ash derived from a high Plinian eruption column spread to the north and northeast over the Pacific Northeast; in the second part, the Wineglass Welded Tuff (Williams, 1942) deposited only locally on the north and east rim and valleys of the caldera flank (Bacon, 1983).

During the single-vent phase, more than one-half of the magma was released in a climactic eruption that created a subterranean void and caused the caldera to collapse. During this collapse, highly mobile pyroclastic flows derived from the ring-vent phase descended all flanks of the mountain and deposited ignimbrite in stream valleys (Bacon, 1983; Druitt and Bacon, 1986). Limited ash deposits on the caldera rim suggest that the ring-vent phase was followed by a period of phreatic(?) eruption of intracaldera debris, much of which probably fell within the caldera.

Shortly after the end of the climactic eruption and caldera collapse, renewed volcanic activity formed the central platform, Wizard Island, and Merriam Cone—the three volcanic edifices that rise from the caldera floor (Fig. 2; Williams, 1961; Nelson and others, 1986b). The cinder cone of Wizard Island forms the top of a pile of andesite flows that extend east as a central platform from the base of the Wizard Island edifice. Merriam Cone is a submerged andesite cone north of the lake center. These three features divide the lake floor into three sub-basin areas: one in the northwest, one in the southwest, and the largest and deepest on the east side of the lake (Fig. 2).

CALDERA FLOOR DEPOSITS

Acoustic stratigraphy

Five acoustic units have been identified on high-resolution seismic reflection profiles of the caldera floor (Fig. 3). The five units are: (I) the acoustic basement; (II) a moderate- to well-layered sedimentary sequence that overlies the acoustic basement and generally exhibits continuous flat-lying to slightly dipping strong reflectors; (III) a poorly to moderately well layered sedimentary sequence that thickens toward the caldera walls and is composed of interbedded chaotic and discontinuous to continuous reflectors; (IV) a well-layered sedimentary sequence of flat-lying strong reflectors; and (V) a well-layered sed-

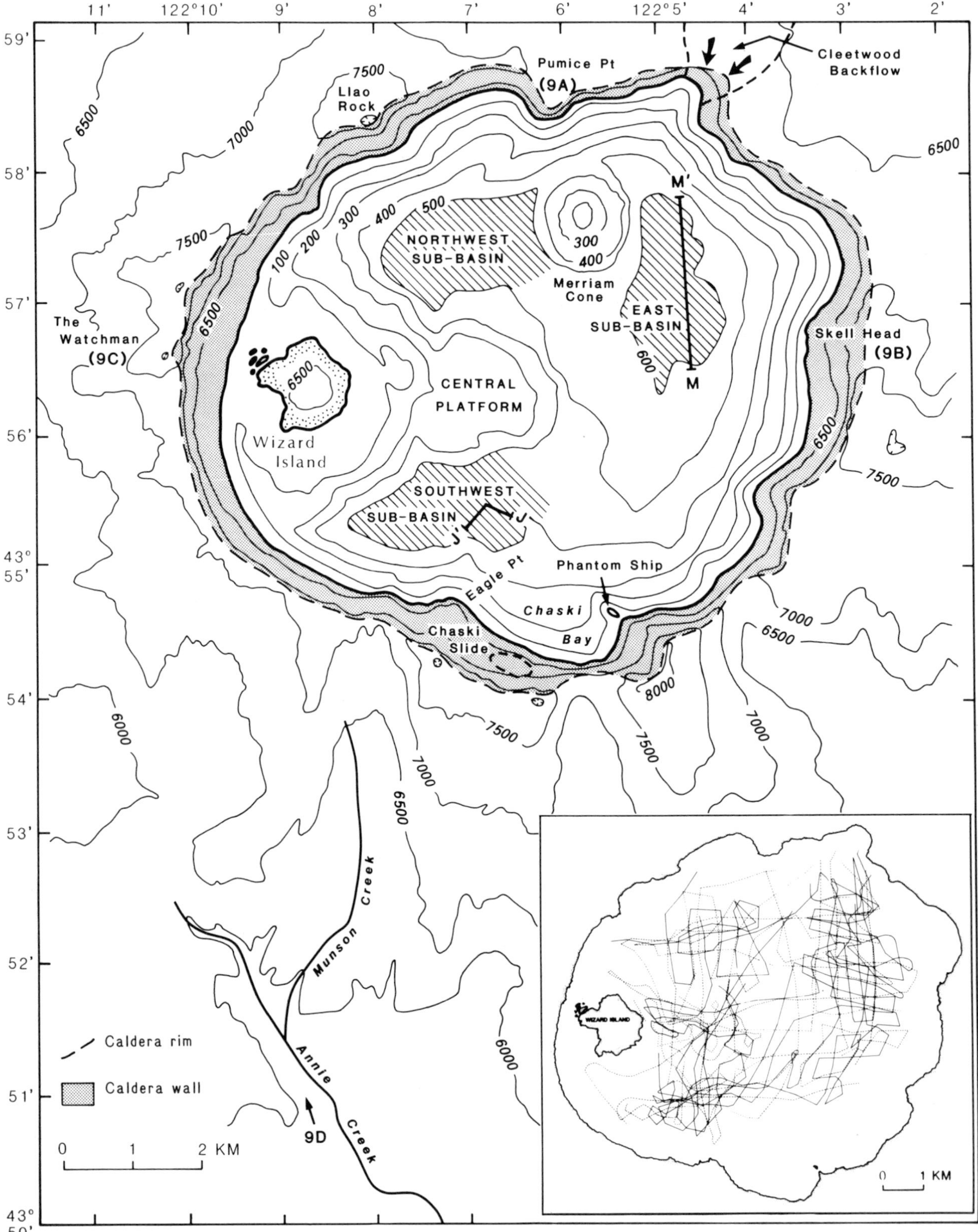

Figure 2. Map showing bathymetry of Crater Lake (from Byrne, 1962) and surrounding topography (from Williams, 1956). Lake shoreline is at 6,176 ft (1,882 m). Bathymetric contour interval below the shoreline is 100 m, topographic contour interval above the shoreline is 500 ft (152.4 m). Profile M-M′ is shown in Figure 3; J-J′ shown in Figure 4. Inset shows geophysical tracklines used in this study.

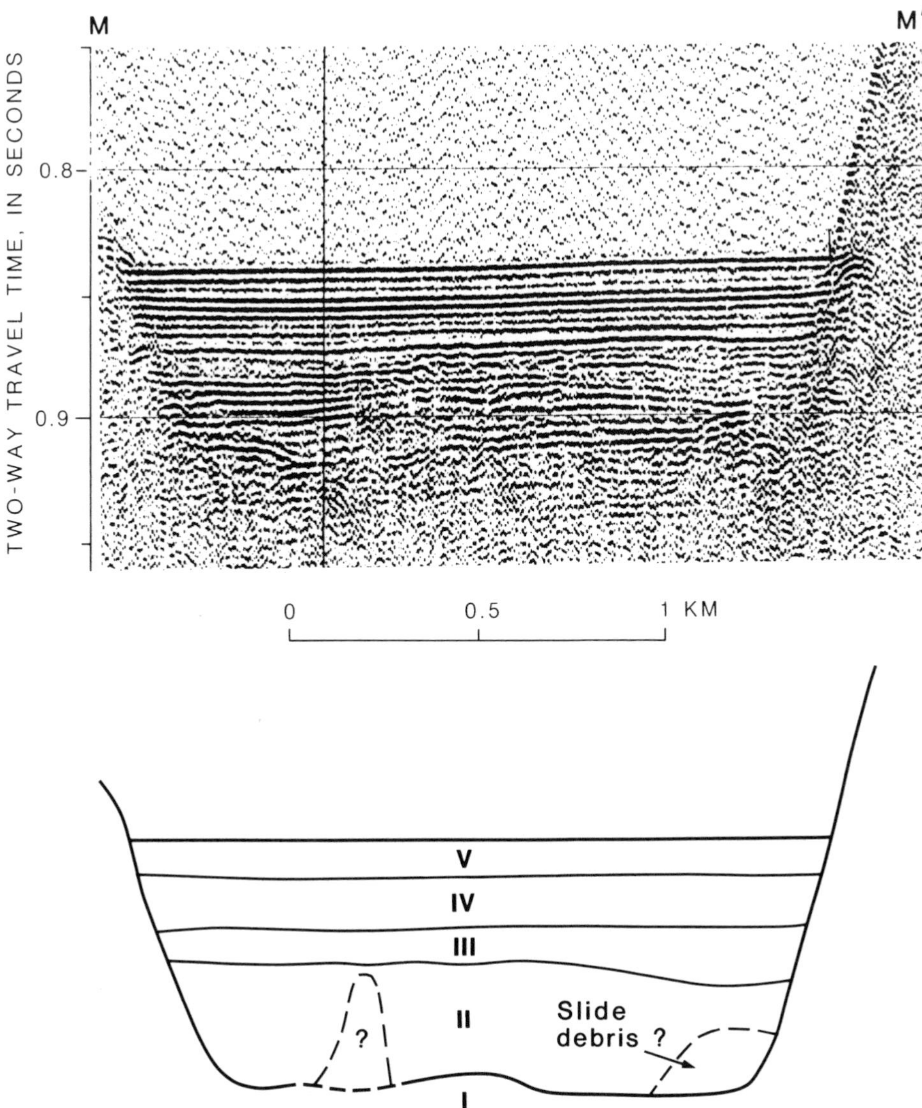

Figure 3. Airgun (1-in^3) profile oriented north-south across east sub-basin (above) and interpretive line drawing (below). Unit I, acoustic basement; unit II, bedded volcaniclastics; unit III, chaotic failure deposits; unit IV, undifferentiated lake sediment; unit V, upper lake turbidites. Vertical exaggeration varies from about 9× in the water column to about 5× in unit II (see "Methods of Study"). Location of profile shown in Figure 2.

imentary sequence of flat-lying weak reflectors that form the modern lake floor.

The uniform flat-lying beds of units IV and V cover the entire sub-basin areas (Figs. 3, 4). They are typical of lacustrine deposits composed of interbedded hemipelagic and turbidite beds like those found in near-surface sediment cores from Crater Lake (Nelson, 1967; Otis and others, 1977; Kelts, 1978; Poppe and others, 1985; Nelson and others, 1986b; Newhall and others, 1987). These upper flat-lying units, IV and V, appear to represent lake sedimentation that post-dates the older apparent subaerial sedimentation of units I through III. Subsequent discussion and the following reflector characteristics also suggest a subaerial origin for the beds of units II and III: (1) the inclined bedding, (2) the lack of continuity of reflectors, (3) the wedge-shaped geometry of many of the deposits, and (4) the general absence of uniform flat reflectors across the entire sub-basin areas (Figs. 3 and 4). Because these three older subaerial units are more closely associated with the climactic eruption of Mount Mazama, they are the main focus of this paper, and the later lake history of units IV and V will not be discussed further. These lake units, however, are described elsewhere by Nelson and others (1986b).

The complete sequence of acoustic units is best exhibited in

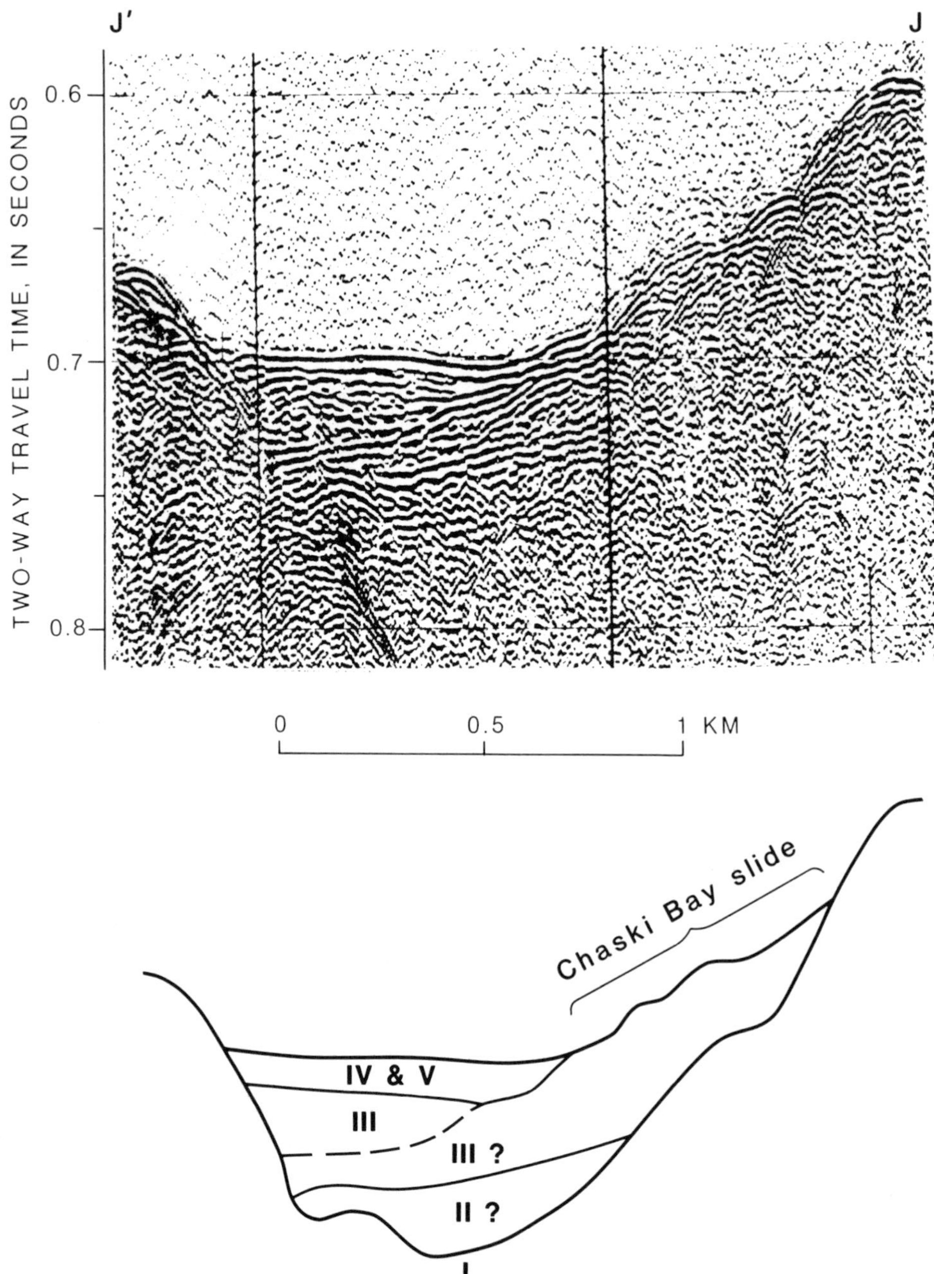

Figure 4. Airgun (1-in^3) profile oriented east-west across southwest sub-basin (above) and interpretive line drawing (below). Unit I, acoustic basement; unit II, bedded volcaniclastics; unit III, chaotic failure deposits; unit IV and V, undifferentiated lake sediment. Slope on right side of profile is over Chaski Bay slide area. Vertical exaggeration varies from about 9× in the water column to about 5x in unit II (see "Methods of Study"). Location of profile is shown in Figure 2.

the large east sub-basin (Fig. 3). In the southwestern and northwestern sub-basins units II and III cannot be clearly defined along the margins of the caldera wall, nor can unit V beds be resolved. Apparently slides and debris flows dominated and disrupted consistent development of the seismic stratigraphy in the smaller western sub-basins (Fig. 4).

Further description of units I through III is important considering there is some variation in seismic stratigraphy in different sub-basins. Acoustic basement (unit I) occurs at the surface on the steep slopes of the caldera wall or over steep volcanic edifices of the caldera floor. Here it is often delineated by overlapping hyperbolic reflectors with no acoustic penetration (Figs. 3 and 4). Numerous point-source reflectors, perhaps indicating a broken or blocky surface, characterize the acoustic basement on the caldera floor beneath sediment layers of the sub-basins. In places, the basement seems to project far into the sedimentary sequence. However, we suspect that volcanic or hydrothermal emanations may have altered the sediment and changed the seismic velocities to simulate acoustic basement (Fig. 3).

Unit II reflectors overlie acoustic unit I and occur in most of the deepest acoustic basement depressions (Figs. 5 and 6A). Unit II shows strong, relatively flat-lying reflectors (Fig. 3), but there are irregularities in the reflector continuity, particularly in the northwest and southwest sub-basins near the caldera wall (Fig. 4). This unit does not occur where the acoustic basement is high (Figs. 5 and 6A). Assuming a seismic velocity of 2,500 m/s, the maximum thickness of unit II is 50 m in the east sub-basin area and 30 m in both the northwest and southwest sub-basins (Fig. 6A).

Unit III conformably overlies the unit II reflectors, thins from the caldera walls toward the sub-basin center, and is dominated by chaotic reflectors (Figs. 3 and 6B). Although unit III generally lacks continuous flat-lying reflectors, it has discontinuous, gently dipping reflectors that wedge in from the sub-basin margins. Unlike unit II, unit III covers most of the floor of each of the three sub-basins and is not restricted to infilling of the local depressions within the sub-basins. If a velocity of 1525 m/s is assumed, maximum thickness of unit III is 20 m near the base of the caldera walls, and minimum thickness is less than 10 m near the sub-basin centers (Fig. 6B).

Collapse debris and the ring fracture; Unit I

From the caldera walls above and below the lake surface to the break in slope at the edge of the basin floor, the acoustic basement consists of the lava and pyroclastics rocks that compose Mount Mazama (Nelson, 1967; Nelson and others, 1986b). On the walls, local valleys have a cover of coarse-grained talus deposits that also appear as acoustic basement because of their restricted lateral extent and thickness. The surface of the postcaldera volcanic edifices rising above the flat lake-floor areas also appears as acoustic basement. The remaining acoustic basement signatures, consisting primarily of point-source reflectors and hummocky topography from sub-basin floors, are apparently caused by caldera-collapse rubble (Figs. 3 and 4). These areas, and those under postcaldera volcanic edifices, probably represent the top of thick intracaldera tuff and lithic breccia, such as have been described from exhumed caldera-fill deposits (Lipman, 1976, 1984; Hildebrand, 1984).

About 50 km^3 of rock was lost to caldera collapse (Bacon, 1983), but it is not known how much of this can be attributed to subsidence of a coherent block, and how much to megabreccia and lithic fragments in intracaldera tuff that may have accumulated above such a block. If the structural block was ~5 km in diameter (Fig. 5; Bacon, 1983) and all material between this block's vertical wall and the present 45° caldera walls slid into the caldera floor to contribute to the caldera deposits below the top of unit I, then as much as ~20 km^3 of caldera wall-collapse debris may have been deposited. Thus, more than 1 km of caldera wall-collapse material should be found on the caldera floor; it could be mixed with an unknown amount of juvenile intracaldera material to cause an even greater total thickness of debris above a subsided block of the mountain top. In sub-basin areas, acoustic basement apparently is the upper surface of this thick wall-collapse deposit.

A series of depressions indent the acoustic basement surface of the sub-basin floors (Fig. 5). These circular depressions may be secondary phreatic-explosion craters developed immediately after collapse when water from hydrothermal, ground-water, and/or precipitation sources interacted with hot intracaldera tuff and wall-collapse breccia around the margin of the caldera floor. Craters of 300 to 500 m in diameter and 20 to 50 m in depth on the acoustic basement (Fig. 5) resemble those observed today in active volcanic areas (Fig. 5; Fisher and Schmincke, 1984); for example, see the cover photo of *Geology,* v. 5, no. 1. This photo shows the drained lake floor of Aniakchuk caldera, which exhibits a 5-km-diameter ring of explosion craters comparable to those apparent in the acoustic basement of Crater Lake. The profiles across the Crater Lake features have complex seismic signatures, but flat-lying beds generally fill most depressions (Fig. 6A), as is typical of active phreatic-explosion craters (see Fig. 9-17 in Fisher and Schmincke, 1984). Furthermore, the scattered occurrences of probable phreatic-explosion debris on the caldera rim appear to be the final deposits observed from the climactic eruption (Bacon, 1983).

The series of depressions or possible phreatic-explosion craters in the surface of the acoustic basement appear to correlate with a ring-fracture zone along which the caldera collapsed (Fig. 5) (Bacon, 1983). An elliptical ring fracture zone 4.1 × 5.7 km in diameter can be postulated by connecting the sub-basin floor areas of small depressions in the acoustic basement with the major vents of the post-caldera cones and the areas of high heat flow on the lake floor (Fig. 5) (Williams and Von Herzen, 1983). The ring-fracture zone apparently influenced the initial sub-basin floor topography and the primary sedimentary sequence (unit II) because phreatic-explosion craters seem to have localized along the fracture contact zone with the caldera wall. These local depressions then became the site of unit II deposition (Figs. 5 and 6A).

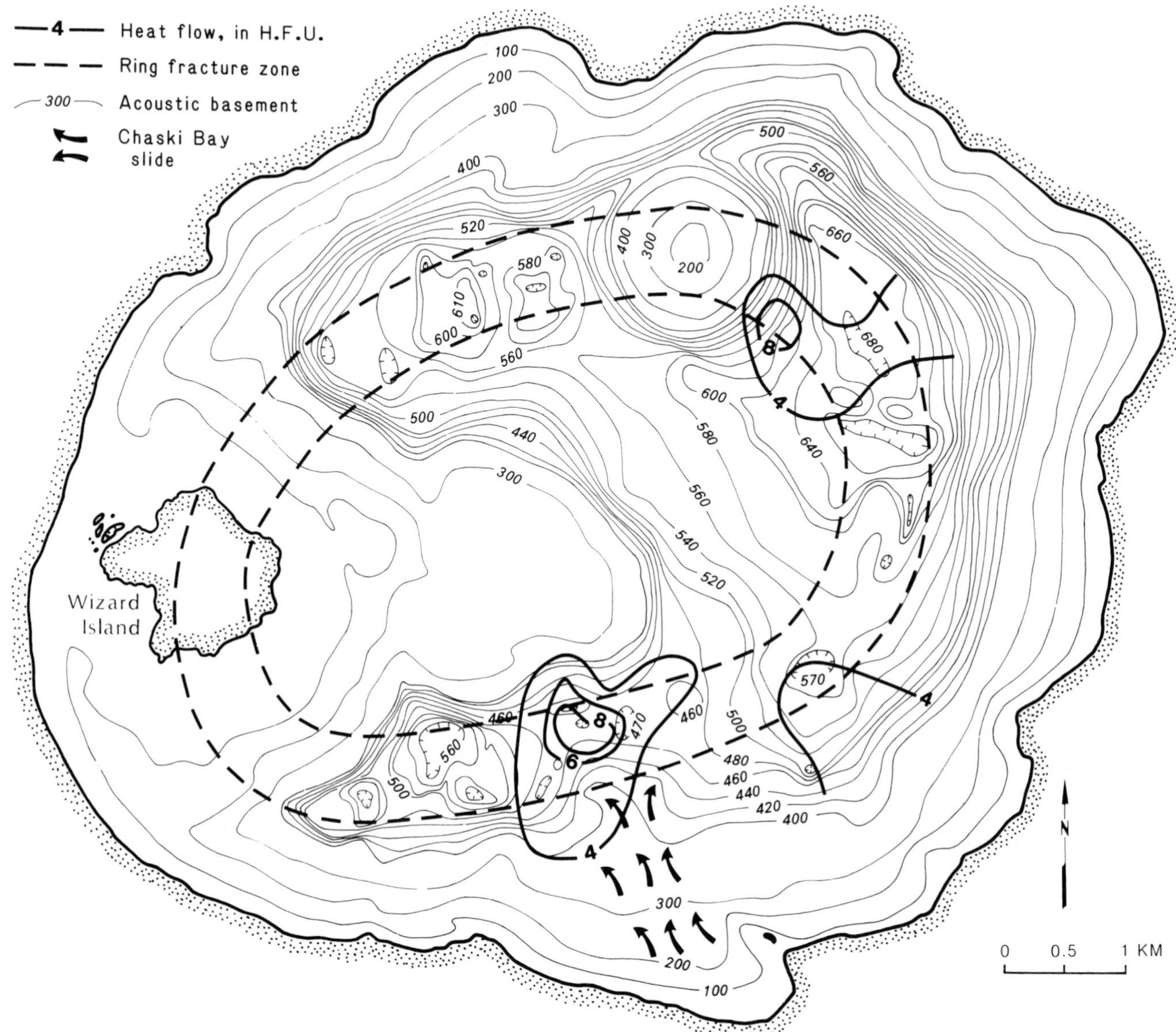

Figure 5. Structure contour map of the acoustic-basement surface (unit I), heat-flow data in H.F.U. (heat flow unit = 10^{-6} $ca^{1}/cm^{2}s$ (modified from Williams and Von Herzen, 1983) and outline of proposed ring-fracture zone. Note craters associated with fracture zone. Contour interval on acoustic basement is 100 m (above 400 m) and 20 m (below 400 m), with 10-m contours added to help define the craters.

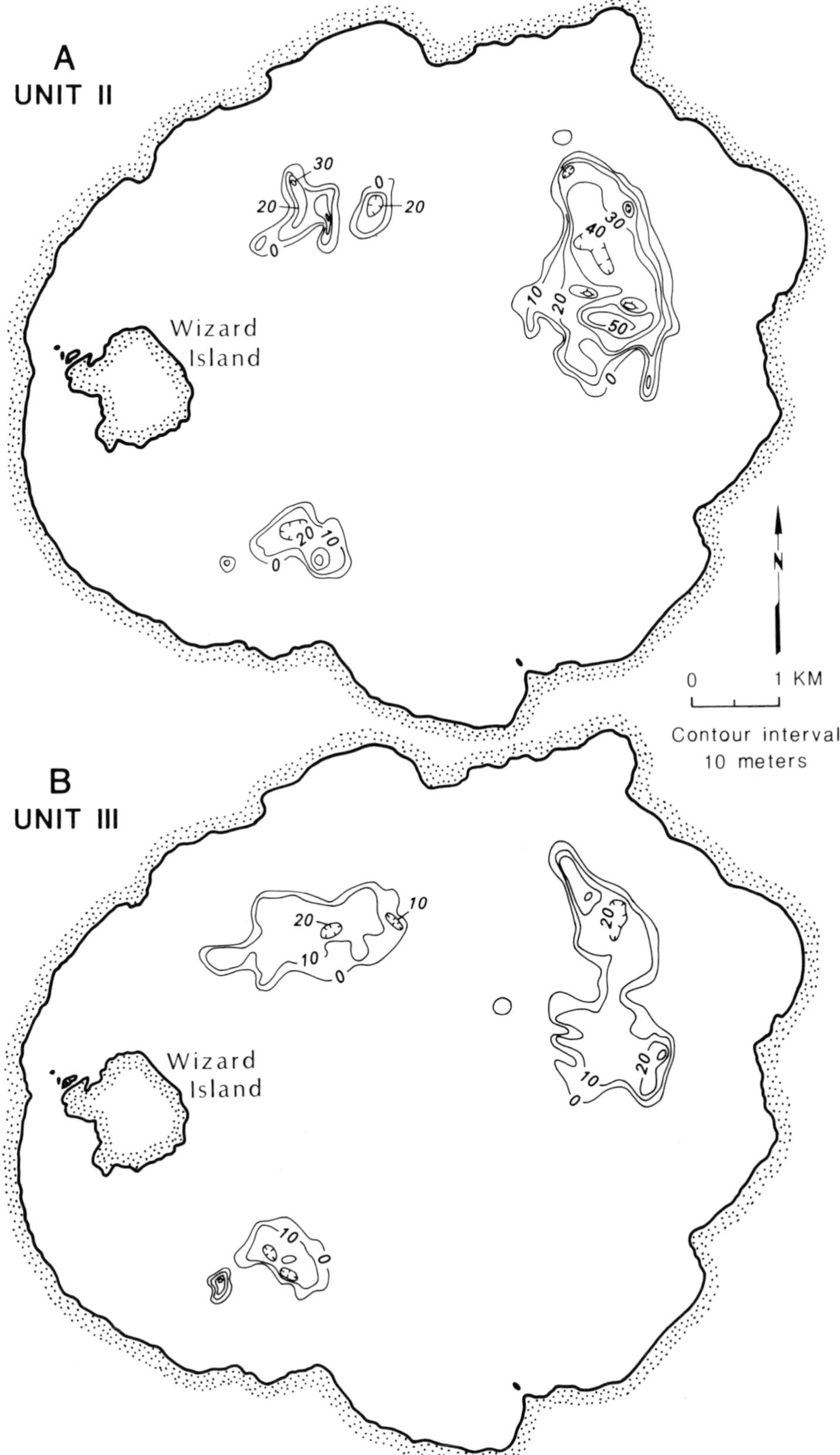

Figure 6. (A) Isopach map of unit II bedded volcaniclastics. (B) Isopach map of unit III chaotic failure deposits.

Subaerial bedded volcaniclastics; Unit II

Unit II appears to be a moderate- to well-bedded deposit that partially fills explosion craters in the collapse breccia (Fig. 6A). The bedding in the smaller northwest and southwest sub-basin areas is not as well developed and dips more steeply than the relatively well bedded and flat-lying beds of unit II in the eastern sub-basin (Figs. 3 and 4). We infer that the strong, relatively flat-lying reflectors in the eastern sub-basin area are deposits laid down by subaerial mass-wasting processes dominated by water-rich flows, such as sheetwash, hyper-concentrated flood flows (Smith, 1986), or lahar-type mudflows (Fig. 3). The interbedded wedges of debris with chaotic reflectors apparently represent rock falls and debris avalanches. These slides and avalanches must have occurred repeatedly from the initially unstable caldera walls and been relatively more prominent in the local southwest and northwest sub-basin floors because of their small floor-to-wall-ratio areas (Fig. 4).

The seismic reflection profiles suggest that the unit II sequence is much like that found in the crater of Mount St. Helens after the climactic eruption. For the first one or two years, sheetwash deposition was dominant at Mount St. Helens, but since then, mass wasting (including rock falls, snow avalanches, and lahars) has become most important (Collins and Dunne, 1986; Brantly and Power, 1985). Similarly, at Crater Lake, immediately after the climactic eruption, abundant unconsolidated sediment probably existed from fresh caldera-wall escarpments and phreatic-explosion debris in the caldera. Water for sediment mobilization and sheetwash deposition should have been available from rapid snow melt, disrupted Mount Mazama ground-water systems, and increased precipitation triggered by the climactic eruption. These comparative data between Mount Mazama and Mount St. Helens suggest that the unit II beds were deposited: (a) rapidly in the first few days or years after the Mount Mazama climactic eruption, and (b) subaerially prior to significant lake formation that required tens to hundreds of years (Nelson, 1961).

Subaerial mass movement deposits, Unit III

Compared to unit II, unit III lacks bedding continuity, flat-lying reflectors, coherent internal reflectors, and stratified beds (Figs. 3 and 4). In unit III, debris wedges cover the entire sub-basin areas and thicken toward the caldera walls, whereas in unit II, flat-lying beds characterize local crater depressions and are disrupted by small interbedded debris wedges (Figs. 3, 4, and 6). These differences in geometry and reflector characteristics suggest that the dominant processes of deposition differed. The hummocky surface, chaotic reflectors throughout, and wedge shape of unit III deposits indicate that they formed mainly by large mass movement events such as rock falls, debris avalanches, slides, slumps, and debris flows. In contrast, flat-lying and local chaotic beds of unit II imply that sheet-wash and small mass movements were the predominant depositional processes (Figs. 3 and 4).

The difference in unit II and III deposits may have resulted from triggering by earthquakes associated with volcanism (Normark and others, 1979; Lipman and Mullineaux, 1981), which caused extensive mass movements in unit III. In contrast, relative seismic quiescence and more extensive sheetwash perhaps existed immediately after collapse and during the time of unit II sedimentation. The greater seismicity and the deposition of unit III debris wedges may correlate with major post-caldera eruptions along the ring fracture. Periods of major post-caldera eruptions and quiesence apparently alternated after the climactic eruption because the petrology of post-caldera volcanic edifices varies (Williams, 1961; Bacon, 1983); unit II and III styles of sedimentation appear to reflect this alternation.

The largest of the mass-transport wedges of unit III is observed in the Chaski Bay area where a late-stage, exceptionally large wall failure is evident in subaerial morphology, lake-floor bathymetry, and seismic reflection profiles. The largest caldera-wall scallop (concave indentation), a classic signature of caldera-wall collapse (Lipman, 1976), occurs on the wall above Chaski Bay (Fig. 2). On this same subaerial slope, there is a rotated slump block of caldera wall (~1 km), the Chaski Slide, and there are steeply dipping faults that are concentric with the caldera wall and downthrown toward the lake (Fig. 2; Williams, 1942; Bacon, 1983). Below this caldera-wall area, bathymetric and acoustic basement contours bulge significantly outward onto the lake floor—another typical signature of slides (Fig. 5; Moore, 1964; Normark and others, 1979). Profiles of the lake floor in this region also show the typical hummocky and hyperbolic reflectors of a debris flow or rock avalanche surface (Fig. 4). All these data suggest that the youngest of the major wall-collapse events occurred in this area and produced intact rotated blocks on the subaerial wall and downslope slide rubble that extends out across the ring fracture and over the southwest sub-basin floor of the lake (Figs. 4 and 5).

Caldera floor history

The major slide on the caldera walls above and under Chaski Bay may have occurred during the main caldera collapse and so it is equivalent to the acoustic basement surface of unit I, or it may have taken place soon after and is correlative with the debris wedges of unit III. The lake beds of unit IV and V overlap the Chaski Bay slide debris, and it seems in the seismic reflection profiles that the deeper part of unit II was buried during this major event (Fig. 4). The apparent overlap and bulge of the Chaski Bay slide over the ring-fracture zone and lack of the phreatic-crater depressions on the slide surface also suggest that the slide could be as young as the unit III deposits (Figs. 2 and 5). On the other hand, the subaerial Chaski Slide feature and the other major scalloped headwalls of the caldera-wall collapse appear to be little modified since the time of the climactic eruption. Further proof is provided by the Cleetwood backflow of preclimactic lava that flowed down a similar scalloped area of the caldera wall during collapse and remains unchanged to the present time (Fig. 2; Bacon, 1983). In either case, the Chaski Bay slide occurred early in the caldera floor history before any extensive history of lake deposition. In this Cascade Range climate of high precipitation, the lake could have reached its present depth less than 1,000 yr after the formation of the caldera (Nelson, 1961). Thus, we conclude that deposition of both units II and III occurred as subaerial deposition immediately following the climactic eruption of Mount Mazama, because they are interbedded between the wall-collapse debris of the acoustic basement (unit I) and the extensive history of lake beds represented by units IV and V.

The subaerial units I, II, and III thus have a history mainly associated with the climactic eruption and they apparently deposited rapidly in the first few years of years prior to the nearly 7,000-yr lake history (Nelson and others, 1986a). In spite of the short depositional history, these units (I through III) make up the dominant part of the sedimentary sequence of the caldera floor. Units II and III are much thicker than units IV and V, and an extremely thick (>1 km), but unknown, quantity of mountaintop and caldera-wall collapse debris makes up the unit I acoustic basement (Figs. 3, 4, and 6). Because extensive subaerial deposits of unit I are not exposed, only thicknesses for subaerial deposits of units II and III can be calculated. Using interval velocities of 2,500 and 1,525 m/s for acoustic units II and III, respectively (see methods section), we estimate that 50 to 60 m of the total 75 to 80 m of post-collapse deposits (Nelson and others, 1986b), or about two-thirds, were deposited immediately following the climactic eruption and caldera collapse (Fig. 3). The other one-third of the acoustically mapped sediment in units IV and V took about 7,000 yr to form as lacustrine hemipelagic and turbidite deposits (Nelson, 1967; Nelson and others, 1986a).

CONTINENTAL SEDIMENTATION

The most areally extensive and complete sequence of deposits from the climactic eruption are found on the continental areas surrounding Mount Mazama. On the caldera floor, just ring-vent and final wall-collapse deposits of the climactic eruption are present, and on the ocean floor, only resedimented continental pyroclastic debris is found (Fig. 1; Nelson and others, 1968).

Single-vent phase

Climactic pumice fall and Mazama ash. The first major deposit of the climactic eruption, the climactic pumice fall, came from a convective, Plinian, eruption column (Bacon, 1983). Much of the widespread Mazama ash is the downwind equivalent of this unit (Powers and Wilcox, 1964). These deposits account for a major fraction of the total volume of pyroclastics produced in the climactic eruption. Twenty to 30 km^3 of rhyodacite magma, or well over half of the entire amount ejected from the magma chamber, was erupted during this phase. The ash was blown mainly to the north and northeast by the prevailing winds (Williams, 1942) as was the airfall from the May 18, 1980, eruption of Mount St. Helens (Figs. 1 and 7A; Sarna-Wojcicki and others, 1981). The climactic pumice fall contains <5 percent lithic fragments (Williams, 1942), pumice blocks as much as 60 cm in diameter, and lithic blocks that reach 1 m in diameter. The pumice is as much as 20 m thick at the caldera rim (Figs. 8 and 9A; Bacon, 1983), and grades to fragments <0.5 cm in diameter (Fisher, 1964) in deposits 50 cm thick, 110 km northeast of the caldera (Fig. 8); deposits are only 1 cm thick 1,200 km away in southwest Saskatchewan, Canada (Lidstrom, 1971). Mazama ash has become a distinctive stratigraphic marker for the Pacific Northwest (Fig. 9E) (Powers and Wilcox, 1964; Fryxell, 1965). Pumice and ash were deposited over at least 1,000,000 km^2 on land; some washed off the land and was redeposited in the Cascadia Basin on the Pacific Ocean floor (Nelson and others, 1968).

Wineglass Welded Tuff. The second stage of the single-vent eruption produced the Wineglass Welded Tuff (Williams, 1942), which is exposed in a prominent cliff just below the caldera rim on its north and northeast sides (Figs. 7A and 9B). The tuff is rhyodacite, identical in composition to the preceding airfall pumice, but lithic clasts up to ~1 m in diameter at the caldera rim form a minor part of the deposit (Bacon, 1983). Downslope from the rim deposits, the Wineglass Welded Tuff is overlain by pumiceous ignimbrite of the ring-vent phase, except in local valley exposures as much as 11 km from the rim (Fig. 7A). Ash flows that deposited the tuff were confined to the valleys. They were not highly mobile and were unable to surmount even minor topographic barriers. Thus they retained heat efficiently, causing the tuff to be partly to densely welded.

Collapse of the Plinian eruption column, possibly through widening of the vent, caused the change from airfall to ash-flow deposition of the Wineglass Welded Tuff (Bacon, 1983). The ash flows were derived from a high-density, relatively low eruption column. Distribution of the tuff indicates that its source vent, presumably the same vent that produced the climactic pumice fall earlier, was north of the summit of Mount Mazama so that ash flows descended from a broad glacial cirque or amphitheater (Fig. 7A). Field relations between these deposits, the ring-vent-phase deposits, and the Cleetwood lava backflow (see Fig. 2) indicate that the Wineglass Welded Tuff was still hot when the caldera collapsed (Figs. 1 and 7A; Bacon, 1983). In fact, generation of these ash flows was halted by collapse of the caldera.

Ring-vent phase

Pyroclastic flow deposits of the ring-vent phase of the climactic eruption overlie the Wineglass Welded Tuff and crop out all around the caldera rim (Figs. 7B and 8; Bacon, 1983). Because these deposits contain lithic fragments that vary systematically in composition with location, and generally correlate with local caldera-wall rock types, they are thought to have been erupted from multiple vents formed along ring fractures as the caldera collapsed (Fig. 7B). The pyroclastic-flow deposits of the ring-vent phase can be divided laterally into proximal, medial, and distal facies, and vertically into zones of silicic, mixed, and mafic flows (Bacon, 1983; Druitt and Bacon, 1986).

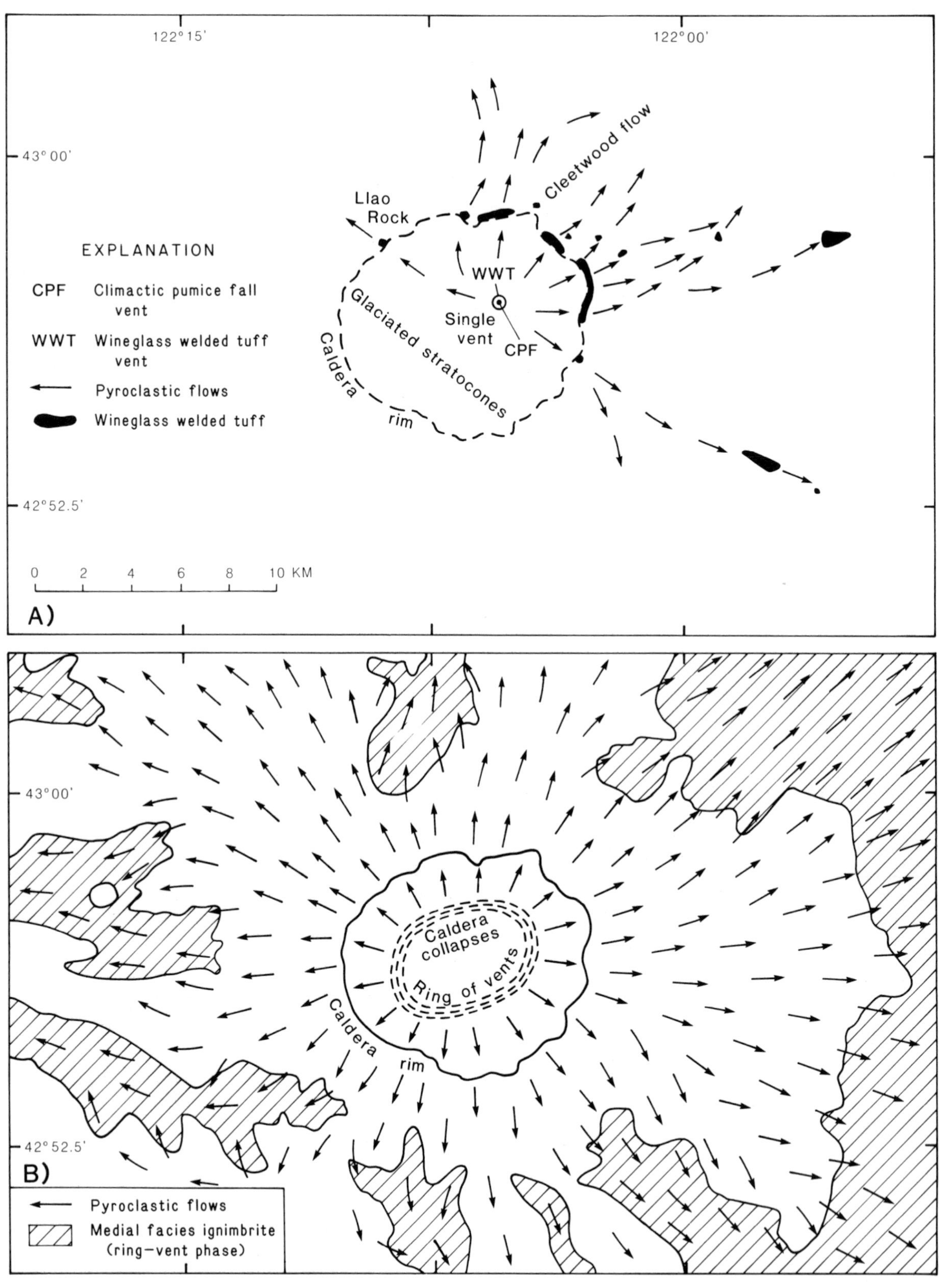

Figure 7. Distribution of deposits of the Mount Mazama climactic eruption. (A) Map showing possible approximate location of vent for single-vent phase of climactic eruption. (B) Map showing inferred paths of pyroclastic flows of the ring-vent phase of the climactic eruption.

Proximal-facies deposits of the ring-vent phase cover at least 750 km^2 and have an estimated minimum volume of 2.5 km^3 (Bacon, 1983). Relations between the Wineglass Welded Tuff and these proximal deposits on the surface of the Cleetwood backflow show that the ring-vent phase of the eruption was at least in part coincident with caldera collapse, and that the climactic eruption continued without interruption from the single- to ring-vent phase (Figs. 2 and 7A). The proximal facies is dominant within ~4 km of the caldera and locally extends up to 14 km from the rim. The deposits are as much as 20 m thick and consist of clast-supported lithic fragments up to 7 m in diameter in a fines-depleted matrix (Fig. 9C); small juvenile clasts form a minor component (Bacon, 1983; Druitt and Bacon, 1986). A thin ignimbrite veneer overlies proximal-facies deposits and contains juvenile scoria clasts to a maximum of 2 m in diameter.

The proximal flows surmounted all topographic obstacles within 14 km of the vents. They commonly cut channels into earlier deposits of the climactic eruption and locally formed longitudinal and tranverse bedforms (Figs. 7B and 8; Bacon, 1983; Druitt and Bacon, 1986). The latter are most common upstream and in the lee of topographic obstructions where they have amplitudes of as much as 10 m and wavelengths of as much as 100 m. Longitudinal bedforms are most common in areas where ash flows have accelerated and attained high velocities.

Downslope from the proximal ring-vent deposits, medial deposits as much as 110 m thick are present in valleys between 4 and ~20 km from the caldera (Fig. 8; Bacon, 1983; Druitt and Bacon, 1986). Generally, the lower half of thick sections consists of silicic ignimbrite, with >80 percent rhyodacite pumice. As much as the uppermost one-third of the section, however, is made up of mafic ignimbrite with <20 percent rhyodacite pumice. The intervening material is mixed ignimbrite with 20 to 80 percent rhyodacite pumice. About 1 m of fine ash caps typical sections and represents the ash-cloud deposit associated with the ring-vent ignimbrite. Silicic and mixed ignimbrite are commonly partly welded or indurated (Fig. 9D). Vapor-phase alteration may be localized in pipe and sheet structures, and the resulting cementation leads to differential erosion into pinnacle forms along stream valleys (Fig. 9D). Medial-facies ignimbrite was apparently deposited continuously by a highly mobile, continuous pyroclastic-flow stream.

Distal-facies ignimbrite extends as far as 60 km from the caldera (Fig. 8; Bacon, 1983). These deposits are gradational with the medial facies. They consist entirely of nonwelded silicic ignimbrite, an overlying ash-cloud layer, and reworked material. Because emplacement temperatures were lower at greater distances from the caldera, carbonized wood is locally abundant, showing that extensive forests must have been destroyed (Williams, 1942). Secondary explosion craters with attendant surge deposits and related fumarolic alteration occur in the region where ash flows covered the Rogue River drainage (Fig. 8; Bacon, 1983).

Deposits of proximal-facies material on topographic highs near the caldera and the erosive capability of the pyroclastic flows

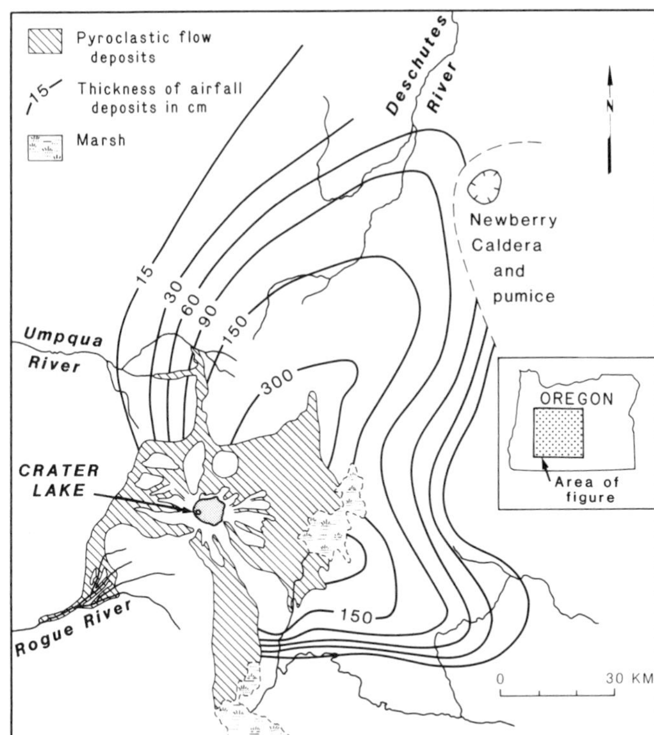

Figure 8. Isopach map for Mount Mazama climactic air-fall pumice and ash (modified from Williams, 1956). Approximate limits of pumiceous ignimbrite of ring-vent phase also shown.

indicate that flows of the ring-vent phase were highly mobile, much more so than those that deposited the Wineglass Welded Tuff. The limited welding of the medial-facies ignimbrite indicates that the eruption columns that fed the flows were comparatively high and that mixing with cool air was relatively efficient. Transport of many large lithic fragments from the vent walls and local surfaces also emphasizes the violence of this phase of the eruption.

Pyroclastic flows of the ring-vent phase were capable of eroding their substrate (Bacon, 1983; Druitt and Bacon, 1986). Consequently, climactic pumice-fall deposits have been partly stripped from the flanks of the large Llao Rock lava flow (Figs. 2 and 7). Not as much has been stripped from its upper slopes, where proximal ignimbrite fills channels scoured into the climactic pumice fall. Similar erosional effects can be seen on the Cleetwood flow (Fig. 7A), which was still hot when the climactic eruption began. Proximal deposits also fill depressions scoured into the upper part of the Wineglass Welded Tuff where poorly welded pumice and ash were removed down to the more resistant welded zone (Bacon, 1983).

Post-collapse deposits of the caldera rim

Beds of unconsolidated crystal- and lithic-rich material that are a few meters thick and contain minor rounded pumice lapilli

Figure 9. Photographs of continental deposits. For locations, see Figures 1 and 2. (A) Climactic pumice deposit 8 m thick at Pumice Point. Most beds are well-sorted coarse rhyodacite pumice. Finer-grained ashy beds are nonwelded pyroclastic-flow deposits present locally near the caldera. Llao Rock in background. (B) Wineglass Welded Tuff overlying climactic pumice fall near Skell Head. (C) Proximal-facies, ring-vent-phase, ash-flow deposits lying in glaciated dacite in roadcut south of The Watchman. (D) View looking north up Annie Creek to Munson Creek showing medial-facies ignimbrite. Exposed thickness 90 m. Silicic, mixed, and mafic ignimbrite approximately correspond to color zones from bottom to top of exposure. Light-colored band at top of exposure is zone of fumarolic alteration below ash-cloud deposit. Pinnacles result from differential erosion of poorly indurated ignimbrite between resistant spines and sheets cemented by vapor-phase crystallization. (E) Mazama ash layer, approximately 5 cm thick (see arrow), in Thompson Canyon, British Columbia (photo by M. H. Bovis; see Fig. 1 for location). (F) Barren area blanketed by ring-vent-phase ignimbrite (pumice desert) shown in lower left (lighter coloring). View is to the south. Crater Lake is in the right half of the photo.

are widespread on the caldera rim (Bacon, 1983, p. 94). In some places, single dunes of rounded pumice lapilli parallel to the caldera rim possibly represent airfall pumice redeposited by wind. The thin-bedded deposits, however, are more likely to have been deposited by airfall and pyroclastic-surge mechanisms. We believe that phreatic explosions ejected this material from within the caldera at the close of, and immediately after, collapse. The explosions could have been caused by the flow of pore water from within the remaining edifice of Mount Mazama into the caldera. This water, perhaps already quite hot, could flash to steam upon contact with hot deposits on the caldera floor. The circular depressions in the acoustic basement of the caldera floor may be the craters created by such phreatic explosions (Fig. 5). They coincide with the outer limits of a ring-fracture system, because this is where pore water would first encounter hot caldera fill. Sheetwash of wall-mantling surge and air-fall debris equivalent to the thin-bedded caldera-rim deposits from phreatic explosions, in part, may have caused the flat-bedded deposits of unit II.

FLUVIAL AND MARINE SEDIMENTATION

Mazama fluvial deposits

Fluvial and marine processes associated with channel transport have reworked and deposited distinct Mazama ash layers for distances almost equal, but in an opposite direction, to those encountered for direct air-fall deposits (Fig. 1). These processes carried Mazama debris through the Columbia River drainage system more than 600 to 700 km from the Mount Mazama source to the Columbia River mouth and then another 600 km to the ends of Cascadia and Astoria Channels. Although there have been several systematic studies of air-fall deposits and another for the Pacific sea-floor occurrences, no systematic study has been made of reworked fluvial and alluvial layers with Mazama debris (Fryxell, 1965; Kittleman, 1973; Sarna-Wojcicki and others, 1983; Nelson and others, 1968).

Based on very few reported occurrences as far as 75 km from Crater Lake, the fluvial deposits with distinct Mazama ash layers remain prominent; then they may disappear rapidly depending on the type of fluvial system. The mapped occurrence of Mazama event debris flows can be traced 55 km down the Rogue River drainage (Fig. 8), but the last remnants of the flow or reworked Mazama fluvial debris occurs 75 km from the Mt. Mazama source (Smith and others, 1982). No further Mazama deposits have been noted in the additional 235 km to the Rouge River mouth; however, Mazama tuffaceous turbidites on the sea floor in Blanco Saddle contain Klamath Mountain minerals, in part from this river drainage (Fig. 1; Duncan, 1968). Apparently, the Rogue River, which is basically a bedrock conduit, has flushed out all traces of the Mazama debris from the lower two-thirds of its valley. The Umpqua River, again draining to the west of Mount Mazama, also has prominent deposits as much as several meters thick in tributary valleys at least 75 km downstream from the Mazama source (Fig. 8; J. G. Smith, oral communication, 1987). It has some flood plains developed in its lower reaches, and thin ash horizons are present locally in overbank beds (Janda, oral communication, 1987).

The same pattern of limited fluvial deposits of Mazama ash is found downstream in the Deschutes and Columbia Rivers that drain the main pathway of Mazama air-fall deposits (Figs. 1 and 8). In the middle part of the Deschutes River, about 300 km from the Mount Mazama source, overbank deposits of 8 cm or less are encountered (Janda, oral communication, 1987). River terrace deposits of a few centimeters are observed along the Columbia River farther downstream from the Deschutes site at a location 460 km from Mount Mazama and 170 km from the river mouth (Janda, oral communication, 1987).

In contrast to the apparent poor record of fluvial deposits, prominent alluvial fan deposits of Mazama ash may occur at great distances from the source. In north-central Washington, over 700 km from Mount Mazama, ash deposits on alluvial fans range from a few cm to 2 m thick (Taber and others, 1982). Similar deposits remain several centimeters thick another 100 km north in British Columbia (Fig. 9E).

Mazama marine deposits

Ash-rich and silt layers are common in the mid-Holocene and later deposits of Cascadia Basin off the Oregon and Washington coasts (Figs. 1 and 10; Nelson and others, 1968). These beds are prominent in Willapa and Astoria Canyons, throughout the Astoria Fan, south from the fan into Blanco Saddle, along 600 km of Cascadia Channel, and continuing onto Tufts Abyssal Plain (Fig. 1).

The composition, age, and stratigraphic position of the ash-rich beds in late Holocene sediments correlate with the Mount Mazama eruption (Figs. 10 and 11; Nelson and others, 1968). Radiocarbon dates and estimated hemipelagic sedimentation rates from several canyon and channel locations show that the first occurrence of sand and silt beds containing Mazama ash immediately follows the time of the climactic eruption (Fig. 10). The two best-developed beds, 12 to 20 cm thick, are found throughout the Astoria Channel only in mid-Holocene deposits (Fig. 11). Numerous thin, ash-rich silt layers first occur in the mid-Holocene sediment of the Cascadia Channel and Astoria Canyon floors, but they are present throughout the late Holocene deposits (Fig. 10).

The sand and silt layers containing Mazama ash occur as one compositional type of the ubiquitous turbidite deposits that are interbedded with the Pleistocene and Holocene hemipelagic mud of Cascadia Basin (Fig. 10; Carlson and Nelson, 1969; Griggs and Kulm, 1970). The sand fraction of the ash-bearing layers is classified as vitric ash (>75 percent glass), tuffaceous sand (25 to 75 percent glass), and terrigenous sand or silt (<25 percent glass) after the method used by Ryan and others (1965, p. 1270–1271). The tuffaceous beds are common in all environments, whereas vitric ash occurs only in interchannel

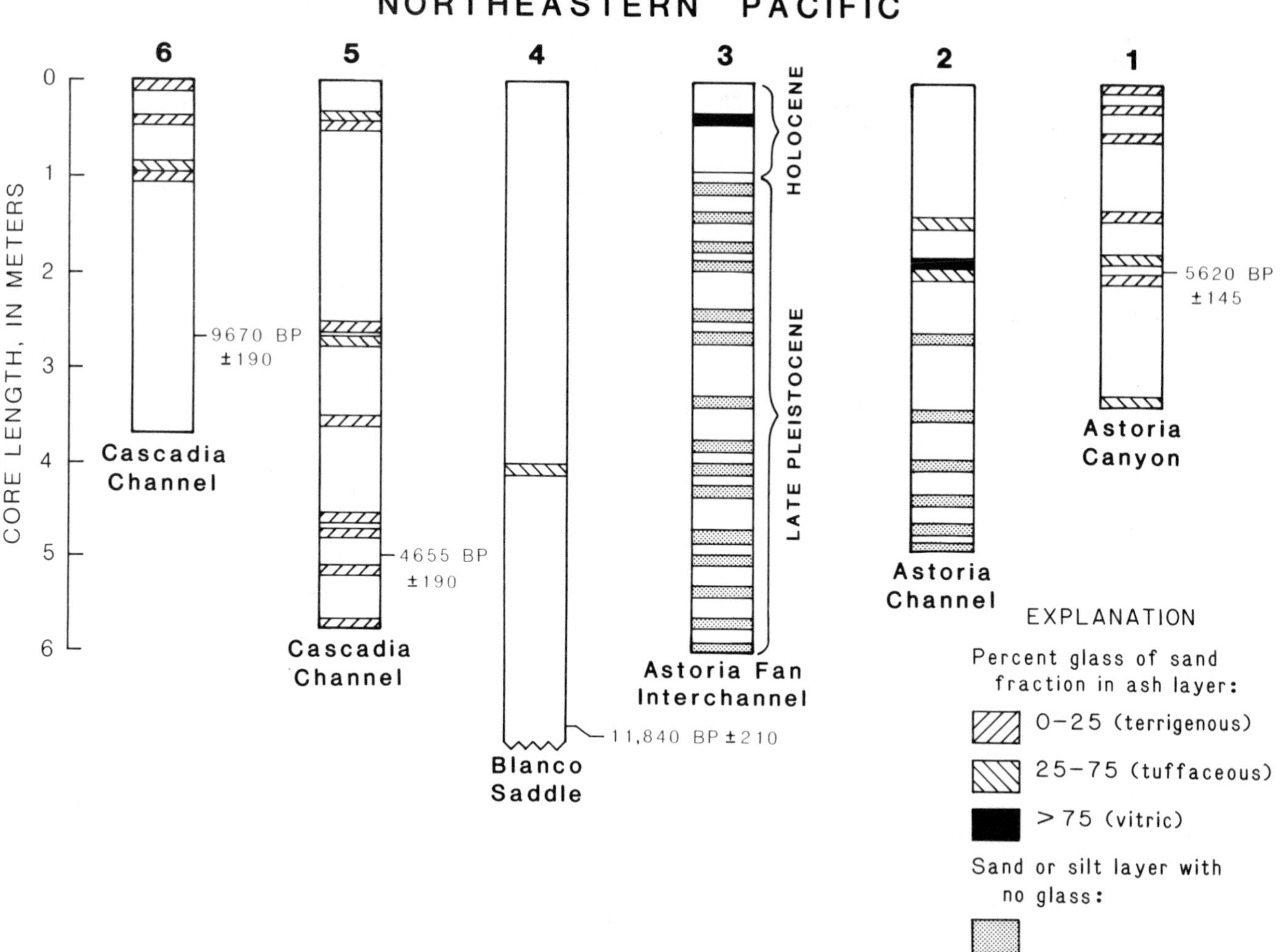

Figure 10. Vertical distribution and age correlation of Mazama ash in selected cores from the northeastern Pacific (modified from Nelson and others, 1968). Core locations shown in Figure 1.

areas and the uppermost parts of graded sand beds in channels (Fig. 11).

The two thickest tuffaceous layers, A and B, occur throughout Astoria Channel, and each becomes finer grained and richer in volcanic glass toward the top of every bed (Fig. 11). In addition to being vertically graded in composition and texture, the A and B beds exhibit proximal to distal grading in thickness, grain size, and ash content as they become thinner, finer grained, and more glass-rich downfan (Figs. 11 and 12). In the proximal Astoria Channel, the A and B beds are 20-cm-thick gravelly sands that contain pebble-sized fragments of pumice (Nelson, 1976); the correlative distal channel deposits are tuffaceous silts 10 to 12 cm thick (Fig. 11). Although the A and B beds extend the 200-km length of the Astoria Channel, the correlative, thin, glass-rich silt beds of the interchannel areas only spread 60 km to the middle of the fan (Figs. 1 and 11).

Marine transport processes of Mazama ash

The following evidence shows that the Mazama ash is redeposited from river and continental shelf sources and is not laid down directly onto the sea floor from air-fall through the water column (Fig. 12). First, the air-fall distribution pattern is northeastward and does not even reach the coast to the west (Fig. 1). Second, the main distribution of sea-floor tuffaceous beds is associated with rivers and submarine canyon-valley systems (Figs. 1, 11, and 12). Third, the Mazama pumice and volcanic glass is mixed with light and heavy minerals of Columbia River sand, except in the Blanco Saddle area where it is mixed with minerals from the Klamath Mountain metamorphic rocks (Figs. 1 and 12; Duncan, 1968; Carlson, 1967; Nelson, 1968). Fourth, transport of Mazama ash to sea-floor sites by turbidity currents is indicated by mixing with other river sediment, by the vertical grading in

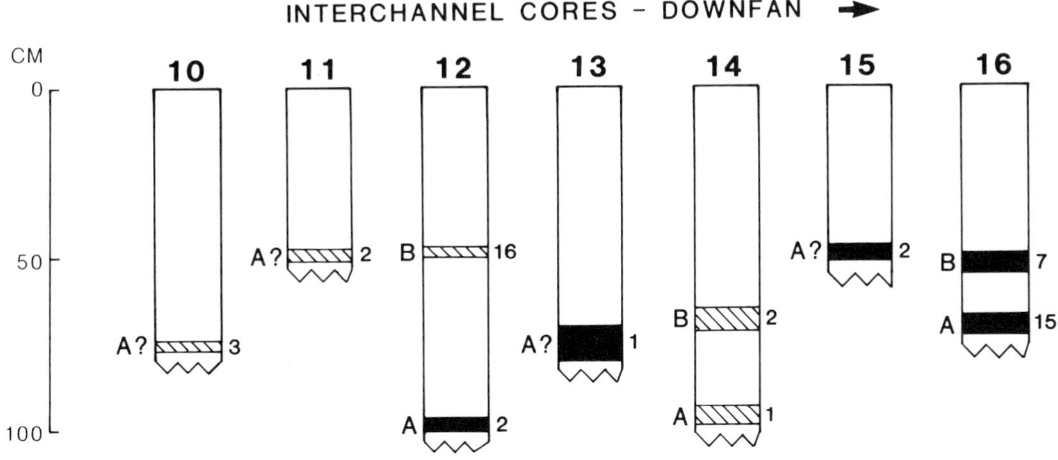

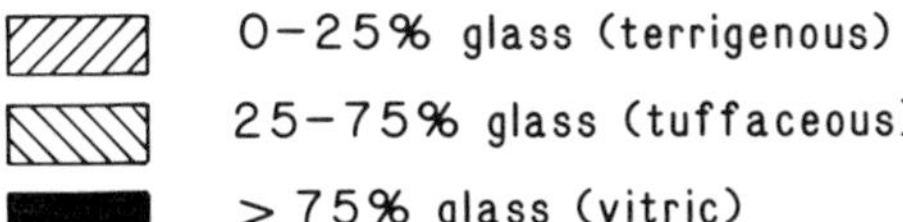

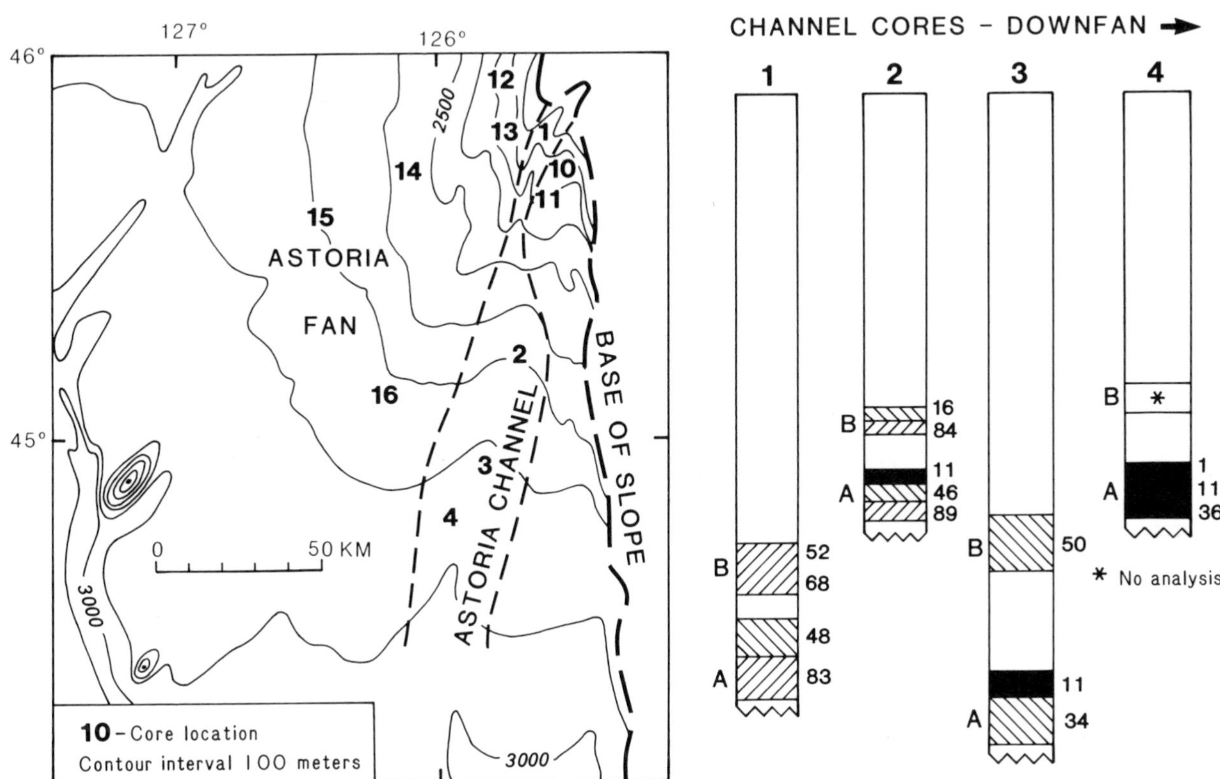

Figure 11. Distribution and composition of mid-Holocene tuffaceous turbidite layers of Astoria Fan. (A) First tuffaceous turbidite after Mount Mazama eruption. (B) Second tuffaceous turbidite. Note increased ash content in A layers relative to younger B layers, and downchannel in both layers (bathymetry from Connard and others, 1984; and Chase and others, 1981).

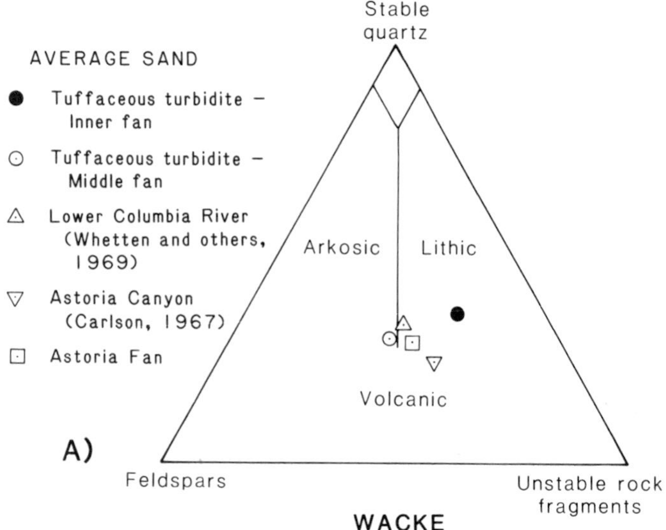

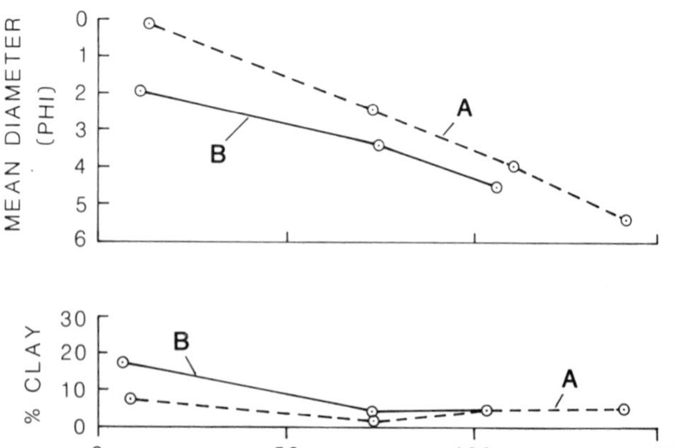

Figure 12. (A) Classification of the light-mineral sand fraction (>0.062 mm diameter) of Astoria Fan coarse-grained layers and comparison with adjacent canyon, shelf, and Columbia River sand. (B) Progressive textural gradation with distance down the fan from its apex for the basal part of correlative A and B tuffaceous turbidites (modified from Nelson, 1976).

texture and composition of tuffaceous layers (Figs. 10 and 11), by proximal to distal grading of tuffaceous layers downchannel (Figs. 11 and 12), and by the presence of displaced benthic foraminifera and occasional shallow-water macrofossils in the tuffaceous beds (Nelson, 1976).

Streams in the Columbia River drainage (Fig. 1) must have become clogged with Mazama debris immediately after the climactic eruption because the first occurrence of marine tuffaceous layers containing Columbia River sand is synchronous with the mid-Holocene age of the Mazama event (Fig. 10). A similar phenomenon was observed when the lower Columbia River changed from a channel depth of 12 m to a depth of 4 m within 24 hr after the Mount St. Helens climactic event because debris surged into the Columbia River from Mount St. Helens drainages (Schuster, 1981).

The Mazama debris probably first collected on the shelf near the head of Astoria Canyon and then was flushed through the canyon and Astoria Channel by turbidity currents. This scenario is corroborated by the presence of numerous tuffaceous layers in Astoria Canyon and its tributaries (Fig. 1). Compared to the rest of Cascadia Basin, the mid-Holocene occurrence of the two thickest tuffaceous layers (A and B), the largest pumice fragments, and the highest ash content indicate that the majority of the ash traversing the canyon was transported to Astoria Channel and Astoria Fan (Fig. 11).

Two main turbidity-current events occurred soon after the climactic eruption; this is shown by the A and B tuffaceous layers that can be correlated throughout the fan in mid-Holocene deposits (Fig. 10). The presence of the finest and purest ash at the top of the A and B graded beds in the channel and throughout equivalent interchannel layers shows that the coarsest Mazama debris funneled through Astoria Channel at the head of the flow (Fig. 11). The finest ash-rich suspension debris of the turbidity current tail lagged behind and spread laterally from the channel to interchannel areas. The flows were thick enough to overtop the 200-m-deep Astoria Channel, but the overbank flows were restricted to the middle-fan areas (Fig. 11) (Nelson, 1976). The channelized flows traveled the entire length of Astoria Channel, but became finer grained and compositionally enriched in Mazama ash as they moved down the channel (Figs. 11 and 12).

No more thick, tuffaceous turbidites were deposited on Astoria Fan after the two main turbidity currents of A and B (Figs. 10 and 11). Similarly, one main turbidity current immediately following the eruption supplied Mazama ash to the Blanco Saddle area (Fig. 10; Duncan, 1968). However, the mineralogy of the associated sand indicates that the source of this deposit was from local rivers in the Klamath Mountain region. In contrast, numerous, thin, tuffaceous silt beds are still being deposited in Cascadia Channel and Astoria Canyon. Low-density turbidity currents have apparently mobilized and redeposited Mazama ash in these conduits, and they have been active transport pathways during the late Holocene (Fig. 10).

GEOLOGIC SIGNIFICANCE

Our new data has made it possible to reconstruct a more complete scenario of the sedimentary events associated with the Mount Mazama climactic eruption. The well-known first stage of the eruption—the Plinian column and ash fall—covered the Pacific Northwest with the main deposit created by the catastrophic event (Fig. 1). Subsequent collapse of the Plinian column created the local Wineglass Welded Tuff on the northeast flank of Mount

Mazama (Fig. 7A). Since more than one-half of the magma chamber was emptied by these events, the mountain top began to founder and the eruption proceeded along a ring fracture. These eruptions filled stream drainages with ash flow deposits around the entire mountain as much as 60 km down the flanks (Fig. 8). By the completion of the ring-vent stage, the mountain top apparently had subsided like a piston, and surrounding wall debris had collapsed on top to form the acoustic basement surface of the caldera floor (Figs. 3, 4, and 5). The final eruptive phase apparently consisted of phreatic explosions associated with the ring fracture that left craters on the caldera floor and small amounts of debris on the caldera rim (Fig. 5). Sheetwash erosion down the caldera walls rapidly filled in the local craters (Figs. 4 and 5).

Erosion of Mazama ash into the Columbia River drainage resulted in turbidity-current transport and deposition offshore in the Cascadia Basin (Fig. 1). The river and submarine-channel transport, however, displaced less than 0.01 percent of continental ash over a sea-floor area of 70,000 km^2 (Figs. 1 and 11; Nelson and others, 1968; Bacon, 1983).

Post-caldera volcanic activity along the ring fracture, probably within a few years of the collapse, resulted in large edifices that cover much of the caldera floor (Fig. 2). Seismicity associated with this volcanism apparently caused many local debris flows and possibly Chaski Slide (including Chaski Bay slide now under water), the last major failure of the caldera wall (Figs. 2, 4, and 5). Sometime after the slide, lake sedimentation began, and interbedded hemipelagic and turbidite beds have been deposited for nearly 7,000 yr since (Figs. 4 and 5; Nelson and others, 1986a, Nelson and others, 1986b).

The climactic eruption of Mount Mazama immediately affected sedimentation over 1,000,000 km^2 of North America, and soon after it influenced deposition over the Cascadia Basin offshore (Fig. 1). Both the continental and marine deposits provide an excellent mid-Holocene stratigraphic marker throughout the Pacific Northwest. Only the first occurrence of the marine or lake-sediment index beds is a valid marker, however, because reworking has recycled Mazama ash to deposit ash-rich layers up until the present time (Fig. 10).

The Mazama event resulted in a wide variety of deposits, some of which have similar textural gradations but were formed by different processes and in different environments. The continental-ash layers exhibit vertical and lateral gradation of texture and composition; however, they were deposited northeast of the caldera by the prevailing winds. The marine tuffaceous layers are graded in a similar manner, but by turbidity currents that carried reworked Mazama ash westward to the deep-sea floor in a direction nearly opposite from air-fall debris. The turbidity-current deposition in submarine channels, however, is quite similar to the ash-flow deposition in stream valleys. In both instances, the coarsest-grained debris is deposited on the valley floors; then a final suspension cloud lays down a finer-grained vitric-glass-rich deposit in the upper portion of the layers on valley floors, and as thin layers on valley flanks and interchannel areas (Figs. 9C, 9D, and 11). There are differences, however, in the dispersal distance and thickness of the Mazama turbidites and ignimbrites. The turbidites extend hundreds of kilometers down submarine channels and are tens of centimeters thick. In contrast, ignimbrites extend only 60 km from the mountain, but are as much as 100 m thick.

Comparison of the ignimbrites and turbidites illustrates the important differences in the geologic preservation potential of the various Mazama deposits. Collapse debris of the caldera floor may be well preserved since such deposits are well known even from the Precambrian (Hildebrand, 1984). Sheetwash, debris flow, and lake deposits from caldera floors have not been described from the geologic record, but they are known from active caldera sites that have had several episodes of caldera formation, such as at Lake Atitlan in Central America (C. G. Newhall, oral communication, 1987). Mazama air-fall deposits on land are well preserved for the short term, but over millions of years will be eroded as the continent is denuded. Fluvial deposits of Mazama ash appear to have an even shorter residence time because rivers rapidly flush out great quantities of unconsolidated pyroclastic debris. This is well exemplified by the general lack of Mazama ash more than 75 km downstream from Crater Lake at the present time. Thus, although significant deposits now exist near the mountain and in alluvial fans, lakes, and deep-sea channels for more than 1,000 km in any direction, the correlative fluvial beds are generally missing. This great displacement of depocenters is important in reconstructing ancient events similar to the Mount Mazama episode. The long-term preservation potential for pyroclastic debris of major caldera eruptions is best in distal lake and marine basin depocenters.

The sedimentary history of the Mount Mazama event indicates that Cascade volcanoes are potentially a major geologic hazard in the Pacific Northwest. This hazard is mitigated in part because air-fall debris generally is carried to the northeast by prevailing winds, away from the main population centers of Puget Sound and the Willamette Valley (Fig. 1). Fortunately, the extremely destructive fiery-avalanche eruptions that take place down the mountain sides also appear to stay near the mountain in sparsely populated regions.

At Crater Lake, several factors suggest that present hazards may be minimal. The last of the major caldera wall failures, Chaski Slide, appears to have occurred shortly after the climactic eruption. Most of the intracaldera volcanic eruptive activity also appears to have taken place soon after the collapse, and none is evident in the sedimentation record of the past 4,000 yr (Fig. 3; Nelson, 1986a). Because none of the flat-lying reflectors of the nearly 7,000-yr-old lake beds are bowed upward, typical formation of a resurgent dome and evidence for imminent renewed post-caldera eruptions are not present (Figs. 3 and 4).

High heat-flow values in sediment, slightly elevated water temperatures, and chemical enrichment of bottom sediment plus lake water, all associated with the ring-fracture zone, indicate that some magmatic activity continues to affect the bottom water of the lake (Fig. 5; Williams and Von Herzen, 1983; Thompson and White, 1985; Jack Dymond, oral communication, 1987). There is no present evidence, however, of significant volcanic activity or hazardous gas accumulations like those recently encountered in African caldera lakes (Katzoff, 1986). In seismic reflection profiles, no acoustic anomalies associated with gas-charged sediment are noted (Figs. 3 and 4; Nelson and others, 1986b). None of the cores collected within the past 27 yr has exhibited any gas expansion from water or sediment. Furthermore, oxygen saturation and mixing of many chemical constituents throughout the water column suggests that water circulation is good (Nelson, 1961; Jack Dymond, oral communication, 1987). Thus, there is no present evidence that a sudden overturn of Crater Lake water will release poisonous gases like the tragic Lake Nyos incident in Africa (Katzoff, 1986).

The catastrophic eruption of Mount Mazama has left its mark on the sedimentary record of Crater Lake and on a large area of North America and the adjacent Pacific Ocean Basin. This eruption will continue to influence the morphology and ecology of the nearby area for thousands of years to come (Fig. 9F). However, future hazardous episodes associated with a significant volcanic event at Mount Mazama do not appear likely in the next several thousand years.

ACKNOWLEDGMENTS

We have benefited substantially from discussions with Christopher Newhall, David Williams, Timothy Druitt, and Michael Field. Field assistance in this logistically difficult area was ably provided by Matthew Larsen, Jack Lee, Audrey W. Meyer, Devin Thor, and numerous National Park Service personnel, especially Mark Forbes. Invaluable technical assistance with data compilation was furnished by John Barber Jr., Debbie Schwartz, Bradley Larsen, Clint Steele, Carolyn Degnan, and Becky Larsen. The manuscript was improved by the reviews of James Moore and Gary Greene. Any use of trade names is for descriptive purposes only and does not imply endorsement by the U.S. Geological Survey.

REFERENCES CITED

Bacon, C. R., 1983, Eruptive history of Mount Mazama and Crater Lake Caldera, Cascade Range, U.S.A.: Journal of Volcanology and Geothermal Research, v. 18, p. 57–115.

Brantly, S., and Power, J., 1985, Reports from the U.S. Geological Survey's Cascades Volcano Observatory at Vancouver, Washington: U.S. Geological Survey Earthquake Information Bulletin, v. 17, p. 20–32.

Byrne, J. V., 1962, Bathymetry of Crater Lake, Oregon: Oregon Department of Geology and Mineral Industries, The Ore Bin, v. 24, p. 161–164.

Carlson, P. R., 1967, Marine geology of Astoria Submarine Canyon [Ph.D. thesis]: Corvallis, Oregon State University, 259 p.

Carlson, P. R., and Nelson, C. H., 1969, Sediments and sedimentary structures of the Astoria submarine canyon-fan system, northeast Pacific: Journal of Sedimentary Petrology, v. 39, p. 1269–1282.

Chase, T. E., Wilde, P., Normark, W. E., Miller, C. P., Seekins, B. A., and Young, J. A., 1981, Offshore topography of the western United States between 32° and 49° north latitudes [map]: U.S. Geological Survey Open-File Report 81–0443, plate 2, scale 1:776,073.

Collins, B. D., and Dunne, T., 1986, Erosion of tephra from the 1980 eruption of Mount St. Helens: Geological Society of America Bulletin, v. 97, p. 896–905.

Connard, G., Couch, R., Pitts, G. S., and Troseth, S., compilers, 1984, Bathymetry-topography, *in* Kulm, L. D., and others, eds., Western North America continental margin and adjacent ocean floor off Oregon and Washington, Atlas 1, Ocean Margin Drilling Program, Regional Atlas Series: Woods Hole, Massachusetts, Marine Science International, sheet 1.

Druitt, T. H., and Bacon, C. R., 1986, Lithic breccia and ignimbrite erupted during the collapse of Crater Lake Caldera, Oregon: Journal of Volcanology and Geothermal Research, v. 29, p. 1–32.

Duncan, J. R., 1968, Postglacial and Late Pleistocene sedimentation and stratigraphy of deep sea environments off Oregon [Ph.D. thesis]: Corvallis, Oregon State University, 179 p.

Fisher, R. V., 1964, Maximum size, median diameter, and sorting of tephra: Journal of Geophysical Research, v. 69, p. 341–355.

Fisher, R. V., and Schmincke, H.-U., 1984, Pyroclastic rocks: New York, Springer-Verlag, 472 p.

Fryxell, R., 1965, Mazama and Glacier Peak volcanic ash layers; Relative ages: Science, v. 147, p. 1288–1290.

Griggs, G. B., and Kulm, L. D., 1970, Sedimentation in Cascadia deep-sea channel: Geological Society of America Bulletin, v. 81, p. 1361–1384.

Hildebrand, R. S., 1984, Folded cauldrons of the early Proterozoic Labine Group, northwestern Canadian Shield: Journal of Geophysical Research, v. 89, p. 8429–8440.

Hill, D. P., Kissling, E., Luetgert, J. H., and Kradolfer, U., 1985, Constraints on the upper crustal structure of the Long Valley–Mono Craters volcanic complex, eastern California, from seismic refraction measurements: Journal of Geophysical Research, v. 90, p. 11135–11150.

Hodgeman, C. S., Weast, R. C., Shankland, R. S., and Selby, S. M., 1961, Handbook of chemistry and physics, 43rd ed.: Cleveland, Ohio, The Chemical Rubber Publishing Company, 3515 p.

Katzoff, J. A., 1986, Teams release first reports on Cameroon Lake: EOS Transactions of the American Geophysical, v. 67, p. 689.

Kelts, K., 1978, Geological and sedimentary evolution of Lakes Zurich and Zug, Switerland [Ph.D. thesis]: Zurich, Switzerland, Eidgenoessischen Technischen Hochschule, 250 p.

Kittleman, L. R., 1973, Mineralogy, correlation, and grain-size distributions of Mazama tephra and other postglacial pyroclastic layers, Pacific Northwest: Geological Society of America Bulletin, v. 84, p. 2957–2980.

Lidstrom, J. W., Jr., 1971, A new model for the formation of Crater Lake caldera, Oregon [Ph.D. thesis]: Corvallis, Oregon State University, 85 p.

Lipman, P. W., 1976, Caldera-collapse breccias in the western San Juan Mountains, Colorado: Geological Society of America Bulletin, v. 87, p. 1397–1410.

—— , 1984, The roots of ash flow calderas in western North America; Windows into the tops of granitic batholiths: Journal of Geophysical Research, v. 89, p. 8801–8841.

Lipman, P. W., and Mullineaux, D. R., eds., 1981, The 1980 eruptions of Mount St. Helens, Washington: U.S. Geological Survey Professional Paper 1250, 844 p.

Moore, J. G., 1964, Giant submarine landslides on the Hawaiian Ridge: U.S. Geological Survey Professional Paper 501–D, p. D95–D98.

Mullineaux, D. R., 1974, Pumice and other pyroclastic deposits in Mt. Rainier National Park: U.S. Geological Survey Bulletin 1326, 83 p.

Nelson, C. H., 1961, Geological Limnology of Crater Lake, Oregon [M.S. thesis]: Minneapolis, University of Minnesota, 179 p.

—— , 1967, Sediments of Crater Lake, Oregon: Geological Society of America Bulletin, v. 78, p. 833–848.

—— , 1968, Marine geology of Astoria deep-sea fan [Ph.D. thesis]: Corvallis, Oregon State University, 287 p.

——, 1976, Late Pleistocene and Holocene depositional trends, processes, and history of Astoria deep-sea fan, northeast Pacific: Marine Geology, v. 20, p. 129–173.

Nelson, C. H., Kulm, L. D., Carlson, P. R., and Duncan, J. R., 1968, Mazama ash in the northeastern Pacific: Science, v. 161, p. 47–49.

Nelson, C. H., Bacon, C. R., and Robinson, S. W., 1986a, The caldera floor sedimentary history of Crater Lake, Oregon [abs.]: International Association of Sedimentologists, 12th International Sedimentological Congress, Canberra, Australia, Abstracts Volume, p. 226.

Nelson, C. H., Meyer, A. W., Thor, D., and Larsen, M., 1986b, Crater Lake, Oregon; A restricted basin with base-of-slope aprons of nonchannelized turbidites: Geology, v. 14, p. 238–241.

Newhall, C. G., Paull, C. K., Bradbury, J. P., Higuera-Gundy, A., Poppe, L. J., Self, S., Sharpless, N. B., and Ziagos, J., 1987, Recent geological history of Lake Atitlan, a caldera lake in western Guatemala: Journal of Volcanology and Geothermal Research, v. 33, no. 1–3, p. 81–108.

Normark, W. R., Lipman, P. W., and Moore, J. G., 1979, Regional slump structures on the west flank of Mauna Loa volcano, Hawaii [abs.]: Hawaii Symposium on Intraplate Volcanism and Submarine Volcanism, Hilo, Hawaii, p. 172.

Otis, R. M., Smith, R. B., and Wold, R. J., 1977, Geophysical survey of Yellowstone Lake, Wyoming: Journal of Geophysical Research, v. 82, p. 3705–3717.

Poppe, L. J., Paull, C. K., Newhall, C. G., Bradbury, J. P., and Ziagos, J., 1985, A geophysical and geological study of Laguna de Ayarza, a Guatemalan caldera lake: Journal of Volcanology and Geothermal Research, v. 25, p. 125–144.

Powers, H. A., and Wilcox, R. E., 1964, Volcanic ash from Mount Mazama (Crater Lake) and from Glacier Peak: Science, v. 144, p. 1334–1336.

Ryan, W.B.F., Workum, F., Jr., and Hersey, J. B., 1965, Sediments on the Tyrrhenian Abyssal Plain: Geological Society of America Bulletin, v. 76, p. 1261–1282.

Sarna-Wojcicki, A. M., Champion, D. E., and Davis, J. O., 1983, Holocene volcanism in the conterminous United States and the role of silicic volcanic ash layers in the correlation of latest Pleistocene and Holocene deposits, *in* Wright, H. E., Jr., ed., the Holocene; v. 2, Late Quaternary environments of the United States: Minneapolis, University of Minnesota Press, p. 52–77.

Sarna-Wojcicki, A. M., Shipley, S., Waitt, R. B., Jr., Dzurisin, D., and Wood, S. H., 1981, Areal distribution, thickness, mass, volume, and grain size of air-fall ash from the six major eruptions of 1980, *in* Lipman, P. W., and Mullineaux, D. R., eds., The 1980 eruptions of Mount St. Helens, Washington: U.S. Geological Survey Professional Paper 1250, p. 577–600.

Schuster, R. H., 1981, Effects of the eruption on civil works and operations in the Pacific Northwest, *in* Lipman, P. W., and Mullineaux, D. R., eds., The 1980 eruptions of Mount St. Helens, Washington: U.S. Geological Survey Professional Paper 1250, p. 701–718.

Smith, G. A., 1986, Coarse-grained nonmarine volcaniclastic sediment; Terminology and depositional process: Geological Society of America Bulletin, v. 97, p. 1–10.

Smith, J. G., Page, N. J., Johnson, M. G., Moring, B. C., and Gray, F., 1982, Preliminary geologic map of the Medford 1° by 2° Quadrangle, Oregon and California: U.S. Geological Survey Open-File Report 82–955, 1 sheet, scale 1:250,000.

Taber, R. W., Waitt, R. B., Frizzel, V. A., Jr., Swanson, D. A., Byerly, G. R., and Bentley, R. D., 1982, Geologic map of the Wenatchee Quadrangle, central Washington: U.S. Geological Survey Miscellaneous Investigations Map I–1311, scale 1:100,000.

Thompson, J. M., and White, L. D., 1985, Rock water interacton: Seattle, Washington, National Park Service Pacific Northwest Region 1985 Annual Science Report, p. 9–10.

Whetten, J. T., Kelley, J. C., and Hanson, L. G., 1969, Characteristics of Columbia River sediment and sediment transport: Journal of Sedimentary Petrology, v. 39, p. 1149–1166.

Williams, D. L., and Von Herzen, R. P., 1983, On the terrestrial heat flow and physical limnology of Crater Lake, Oregon: Journal of Geophysical Research, v. 88, p. 1094–1104.

Williams, H., 1942, Geology of Crater Lake National Park, Oregon: Carnegie Institute of Washington Publication 540, 162 p.

——, 1956, Crater Lake: U.S. Geological Survey topographic map Crater Lake National Park and Vicinity, scale 1:62,500.

——, 1961, The floor of Crater Lake: American Journal of Science, v. 259, p. 81–83.

Manuscript Accepted by the Society April 4, 1988

Printed in U.S.A.

Geological Society of America
Special Paper 229
1988

Sonar images of the path of recent failure events on the continental margin off Nice, France

Alberto Malinverno and William B. F. Ryan, *Lamont-Doherty Geological Observatory and Department of Geological Sciences of Columbia University, Palisades, New York 10964*
Gerard Auffret and Guy Pautot, *Institut Français de Recherche pour l'Exploitation de la Mer, Brest, France*

ABSTRACT

When examined in plan view with side-looking sonar, the path of recent sub-sea failure events on the continental margin off Nice displays morphologic features indicative of the passage of sediment flows that interacted with the sea-bed substrate. The path begins as chutes incised on a steep prodelta slope. These chutes feed into the floor of a submarine canyon, which is covered with trains of bedforms oriented at right angles to the flow direction. The bedforms have wavelengths between 35 and 100 m and heights estimated at less than 5 m. The waves are made of coarse material ranging from sand to boulders. Giant bedforms of similar character and scale in the Channeled Scabland (eastern Washington) were produced by a sudden catastrophic flood. Near the base of the slope, where the canyon floor widens and the thalweg gradient decreases, the path becomes a scoured surface with depressions up to 70 m deep produced by the erosion of bedded sediments. As the bottom slope decreases, the width of the area scoured and the cross section of the flows increase. Along the 80 km of the path that was imaged, which was entirely upslope from the location where submarine cables were disrupted by a failure event in 1979, there is little sign of deposition, but extensive evidence of substrate reworking. Our plan-view study of a present-day surface and its bedforms offers an instantaneous picture of an unconformity in development that contrasts with the classical study of similar unconformities in cross section where the emphasis is placed on the time dimension.

THE 1979 EVENT AND PREVIOUS STUDIES

On October 16, 1979, the coastline of Nice, France, was submerged for a length of about 100 km by a tsunami of several meters in amplitude. At about the same time it was found that part of the airport, which was being enlarged by landfill on the delta of the Var River, disappeared into the sea. A few hours later, two submarine telephone cables at distances of about 80 and 110 km from Nice were interrupted. Meanwhile, no earthquake was registered by the nearby Monaco observatory. This event was attributed, therefore, to a submarine slide that was not triggered by an earthquake, and that generated a turbidity current in the Var Canyon. The volume of sediment involved has been estimated at 0.4 km^3, and the maximum velocity of the current at 40 km/h (Genessaux and others, 1980).

Earlier bathymetric studies (Bourcart and others, 1960) and SeaBeam surveys carried out after the 1979 event (Pautot, 1981; Pautot and others, 1984) show that the continental slope off Nice is very steep (depths of 1,500 m are reached just 15 km off the coast). The continental margin is dissected by a number of submarine valleys and lacks a continental shelf. The physiography of the Nice continental margin is depicted in Figure 1. The Var Prodelta corresponds to a fan-shaped unit that lies between the coastline and the 1,000-m isobath, with a radius of about 5 km (Pautot, 1981). The Var Prodelta and the continental slope are incised by the Upper Var and Paillon Canyons, which are separated by a saddle that contains a flat-floored, broad (up to 2 to 3 km) valley. This feature is here named the "Median Valley." The Upper Var and Paillon Canyons join in the broader Lower Var Canyon, which in turn bends around to the east. About 60 km from this confluence, the Lower Var Canyon turns to the south and feeds into a small deep-sea fan (Genessaux and others, 1980).

A photographic study before the 1979 event carried out in the floors of the Upper Var and Paillon Canyons using a deep-

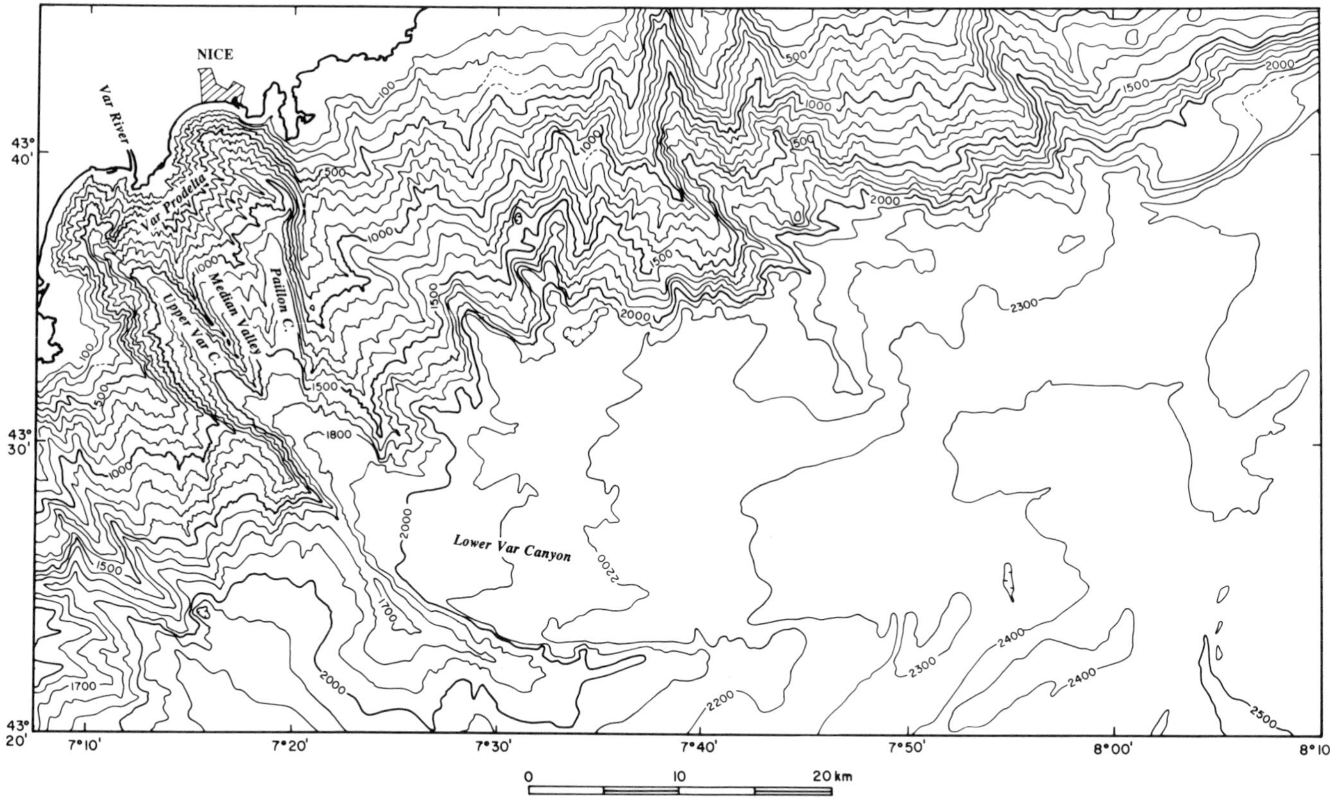

Figure 1. Bathymetry of the continental slope near Nice, France, after Pautot and others (1984). Contour interval 100 m.

towed camera sled showed that cobbles and blocks of up to 80 cm in diameter were ubiquitous and were covered by a thin dusting of sediment. These coarse clasts were associated with fresh vegetal debris, suggesting recent transport of some material from the coast. In some cases the cobbles were arranged in rows, forming ridges (Genessaux, 1966). In addition, during submersible dives carried out after the 1979 event in the Upper Var and Paillon Canyons, accumulations of pebbles and blocks were observed in the canyon floors. The canyon walls appear to be extensively eroded in a fashion that recalls a badland topography. The walls have steep step-like scarps at their base, which in effect leave side tributary gullies as hanging valleys (Auffret and others, 1982).

The Var and Paillon Canyons appear to have eroded deep into a cover of marls and clays whose maximum thickness reaches 1 km. Seismic reflection profiles show that this cover overlays an acoustic basement. Pautot (1981) interprets this basement as the Messinian (latest Miocene) erosional surface, incised in the continental slope during the desiccation of the Mediterranean Basin (Ryan and Cita, 1978). Pautot (1981) summarized the evolution of the continental slope off Nice as follows: (1) an erosional phase during the Messinian; (2) the build-up of a thick deep-sea fan fed by the sediments derived from the erosion of the Alpine chain and transported by the Var River during the Pliocene and part of the Quaternary; and (3) a progressive decrease in the sediment supply and a phase of erosion by extensive submarine sliding that started in the Quaternary and continues to the present time.

THE SeaMARC I SURVEY

From April 11 to April 14, 1984, the sonar system SeaMARC I, installed on board the research vessel *Robert D. Conrad* of Lamont-Doherty Geological Observatory, was used for the exploration of the Nice Canyon (the area imaged is shown in Fig. 2). SeaMARC I is a portable, deep-sea acoustic imaging system, consisting of a sonar vehicle and shipboard electronics, connected by a coaxial cable. The vehicle is towed 100 to 300 m above the sea floor at speeds between 1.5 and 3 knots (about 3 to 6 km/h). Options are available to insonify a swath of 1, 2, or 5 km at frequencies of 27 (port) and 30 (starboard) kHz (Chayes, 1983; Fig. 3). The side-looking sonar data are processed aboard in real time to correct for the varying altitude of the vehicle above the sea floor. The signals from each transducer are digitized in 1,024 pixels (so that the across-track resolution is about 2.5 m for a 5 km swath), printed on a facsimile graphic recorder (as the "processed side-scan" in Fig. 3), and recorded on magnetic tape. Subbottom data obtained from a 4.5-kHz, wide-beam, downward-looking sonar are also corrected for the varying depth of the vehicle (as measured by a pressure gauge), printed (as the "processed subbottom" in Fig. 3), and recorded on tape.

The next step in the analysis and interpretation is to recon-

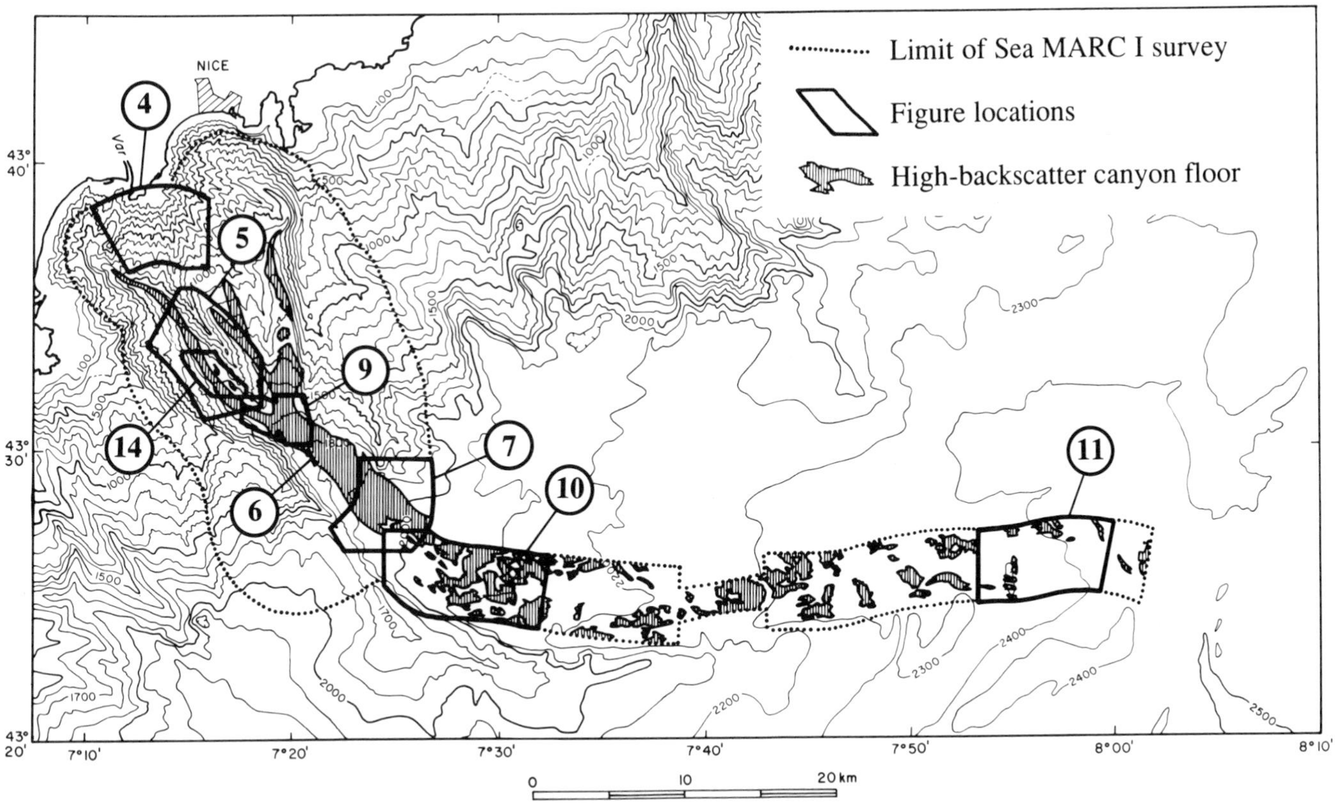

Figure 2. Area explored in the SeaMARC I survey in April 1984; distribution of high-backscatter canyon floor and location of the sonographs shown in Figures 4–7, 9–11, and 14.

struct the navigation of the sonar vehicle. This is done by plotting the ship navigation (done in this case using Loran-C calibrated by transit satellite fixes) and by swinging arcs with a radius equal to the horizontal range between the sonar vehicle and the ship, and centered on the position of the ship. The vehicle has to be on these arcs. The best position is found by matching the features on the sonar image with the bathymetry (that in this case was well known from the previous Seabeam studies). Final uncertainties in the position of the vehicle are of the order of a few tens of meters.

Successive positions of the sonar vehicle estimated every 15 min are joined using a cubic spline, to obtain a smooth and continuous vehicle path. The side-looking sonar data are then projected perpendicularly to the vehicle path, so that the final display is corrected for changes in velocity and direction of the vehicle. The final result is somewhat analogous to an aerial photograph, with the sonar vehicle being the source of illumination. Areas of high acoustic backscatter correspond to portions of the sea floor that are oriented to face the sound source, or that are rough at a scale comparable to the wavelength of the acoustic signal (5 cm in this case). Areas of low backscatter are either smooth or correspond to acoustic shadows. In all the side-looking sonar records shown in this paper, areas of high backscatter are white, and areas of low backscatter are dark.

DESCRIPTION OF THE SONAR IMAGES

Chutes in the Var Prodelta

The Var Prodelta is cut by a number of straight incisions that radiate outward from the area of the airport to feed the Upper Var and Paillon Canyons and that are easily identifiable on the bathymetric map (Fig. 1). The sonar images show that these incisions are sharply cut in the substrate, are separated by spurs that are also very sharp, and have walls that are featureless or contain only a few side tributaries. Therefore, they will be termed "chutes" (see Fig. 4; the location of this and of the following sonographs is illustrated in Fig. 2). These chutes appear to be bright on the sonar images, i.e., a large amount of acoustic energy is backscattered. This suggests that the chutes are characterized by a rough sea floor.

Drainage pattern on the canyon walls

The flanks of the Upper Var and Paillon Canyons are steep (up to about 20°), and the bathymetric maps suggest that they are dissected by a series of downslope-directed spurs and gullies. The sonar images (Fig. 5) show that the canyon walls are entirely

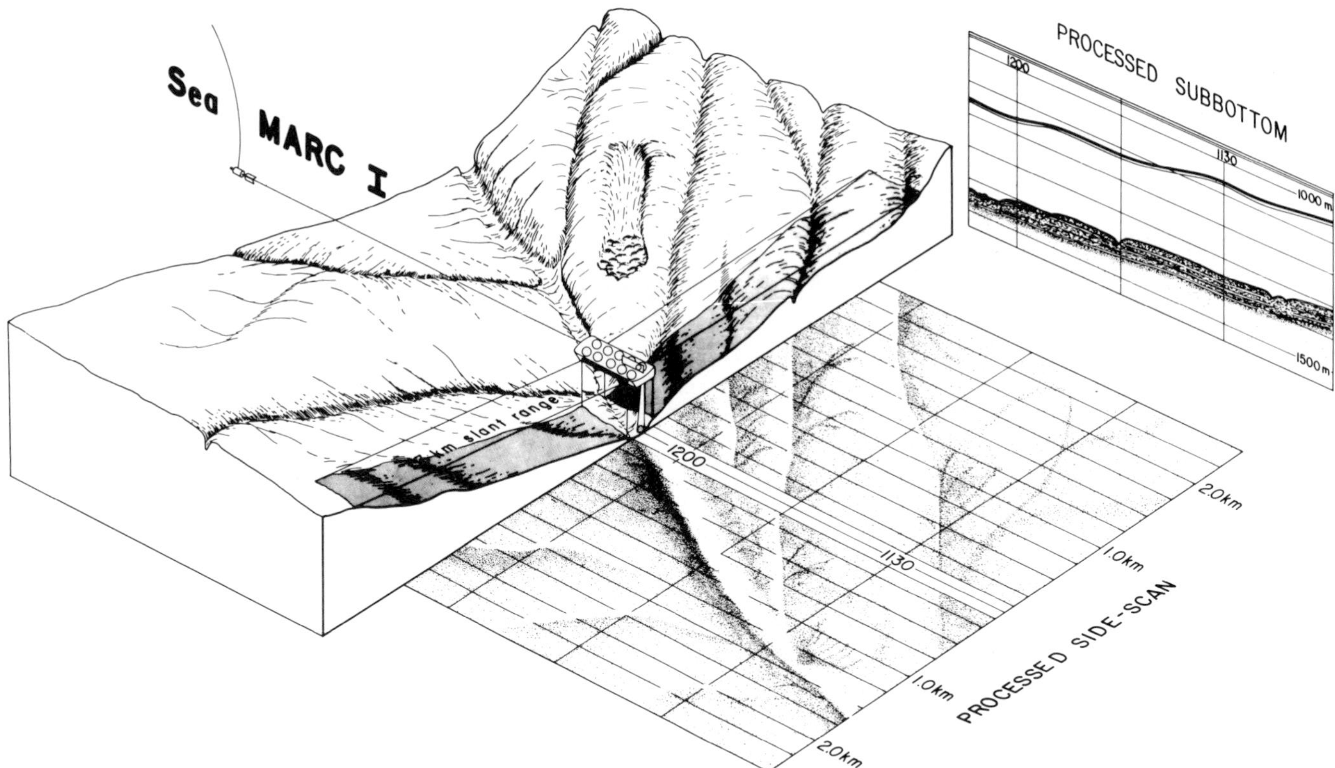

Figure 3. Pictorial representation of the SeaMARC I vehicle being towed up a canyon. The processed side-scan and processed subbottom records are output in real time and are corrected for changes in altitude of the vehicle, so that a plan view and a vertical section of the sea floor are generated.

covered by a fine-scale drainage pattern with first- and second-order tributaries, and that the spurs and gullies divide the canyon walls into a number of separate small hydrographic basins. This "herring-bone" drainage pattern, or "badland" topography, was previously observed on the canyon walls from a submersible (Auffret and others, 1982). It can be noticed that the drainage pattern reaches the spur that separates the tributaries of the Upper Var Canyon and those of the Median Valley, and that this spur is very sharp. This suggests that the erosion of the canyon walls is accomplished by dispersed mass-wasting processes, and not by abrasion of the substrate by sediment suspension traveling from the shelf to the abyssal plain (Farre and others, 1983).

High-backscatter canyon floors and bedforms

The floors of the Nice canyons down to a depth of about 2,000 m are almost entirely covered by a continuous high-backscatter acoustic unit (Fig. 5). The high-backscatter unit covers most of the floor of the Var Canyon, whereas it is present only in part of the Paillon Canyon and in the lower course of the Median Valley (Fig. 2). High-backscatter canyon floors have also been observed in Sea MARC I images from the U.S. middle Atlantic continental margin (Farre and others, 1983). The uniform high backscatter is most likely due to small-scale surface roughness, in contrast to the low-backscatter portions of the canyon floor, which are smooth. The boundary between the high- and low-backscatter canyon floor has been crossed on the western side of the uppermost Lower Var Canyon and imaged in a subbottom profile (Fig. 6). This boundary is a 30-m-high sharp step separating the shallower low-backscatter canyon floor, which contains a number of subbottom reflectors, from the high-backscatter canyon floor, into which there is no acoustic penetration. The low-backscatter canyon floor corresponds in this location to a terrace that extends to the base of the canyon wall.

The high-backscatter canyon floor is characterized also by a pattern of elongated lineaments that trend perpendicular to the canyon axis. Where this terrain is crossed along a path parallel to the strike of these structures, one can observe a series of narrow shadows in the downslope half of the sonographs and a series of narrow light stripes in the upslope half (Figs. 7, 8). This pattern can be explained if the sea floor is here incised by a train of large, asymmetric bedforms. It can be noticed that the train of bedforms in the Var Canyon appears to overlay the train in the Paillon Canyon (Fig. 9).

Patches of high-backscatter sea floor

The Lower Var Canyon at depths of more than about 2,000 m has been explored with a single swath (Fig. 2). Here the high-backscatter canyon floor that was continuous in the Upper

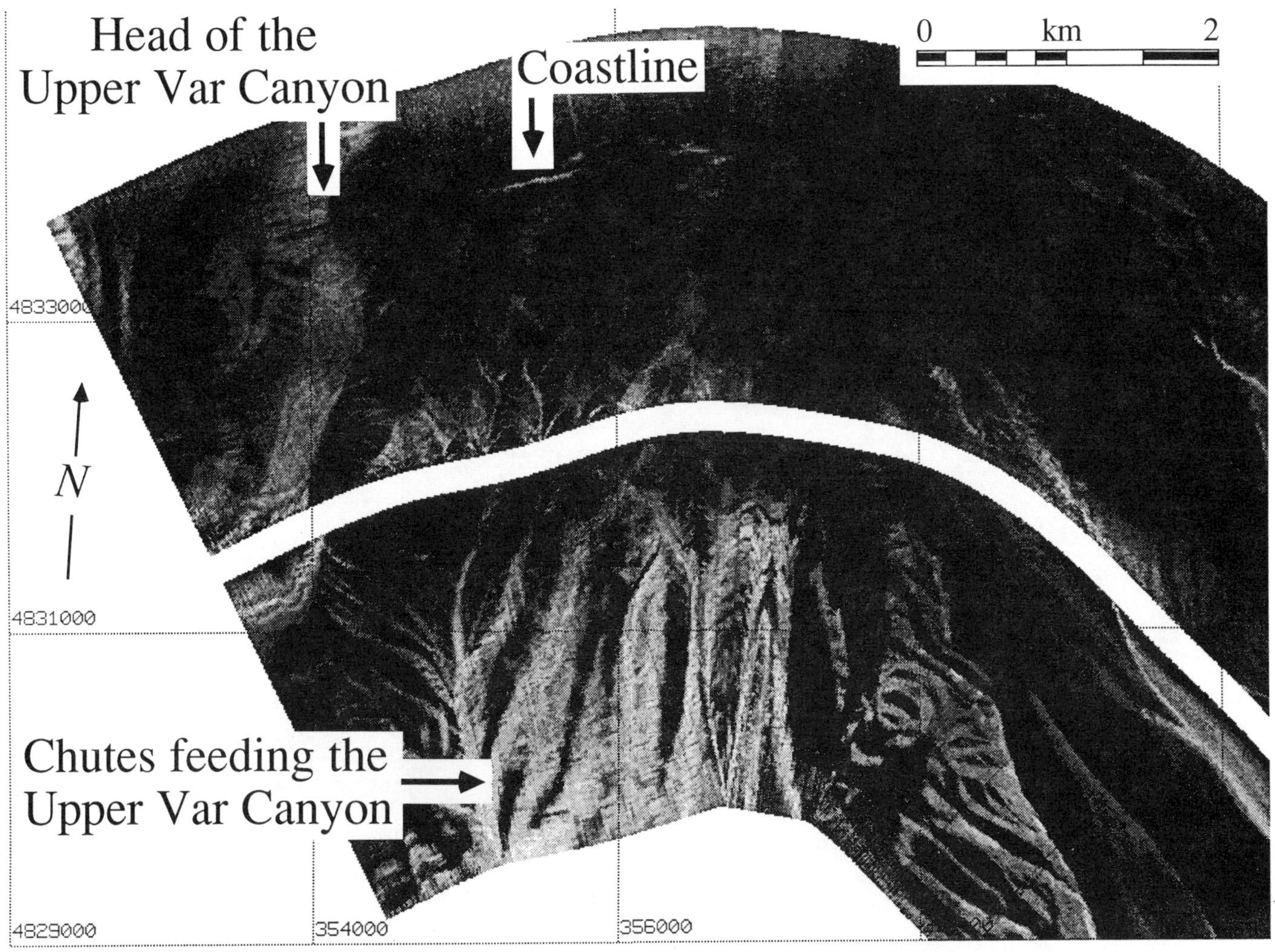

Figure 4. Side-looking sonograph of the Var Prodelta (location in Fig. 2). The image shows the head of the Upper Var Canyon, in the vicinity of the mouth of the Var River, and a number of straight high-backscatter chutes that feed into the Upper Var Canyon.

Var and Paillon Canyons is broken into a series of high-backscatter patches (Fig. 10). These patches are irregular in size (their length varies between 100 m and 4 km), and their boundaries tend to be aligned in the downslope direction. No bedforms have been observed in these patches. Here the canyon floor also displays areas of intermediate backscatter, in contrast with the black-and-white landscape of the Upper Var and Paillon Canyons. The ratio between high- and low-backscatter sea floor appears to decrease moving downslope. The bottom is nearly flat in the Lower Var Canyon, but where irregularities exist, lows correlate with high-backscatter sea floor, and highs with areas of intermediate to low backscatter.

Erosional scars

In the deepest transect explored, at depths between 2,350 and 2,450 m, the sonar vehicle crossed a series of highs and lows. The height of the oscillations is about 100 m, and the high-to-low distance 2 to 3 km. On the sonographs, the troughs display a higher backscatter (Figs. 11, 12). Two patterns of lineations etch the upslope walls of these troughs, one trending downslope and one parallel to the contours. The two patterns probably correspond to downslope-trending grooves and to across-slope outcrops of horizontal beds, respectively. These orthogonal lineations are particularly evident in a large (1.5 km in diameter and 70 m in relief) spoon-shaped scour that has been crossed by the vehicle. Fields of narrowly spaced, downslope-trending grooves can also be recognized in areas of intermediate backscatter (Fig. 11). A similar spoon-shaped scour, about 1 km wide, and a number of erosional grooves have been observed on deep-tow side-looking sonar records from the cutoff of a meander in the Reynisdjup Channel, south of Iceland (Lonsdale and Hollister, 1979). Lonsdale and Hollister (1979) termed this giant scour mark a "mega-flute."

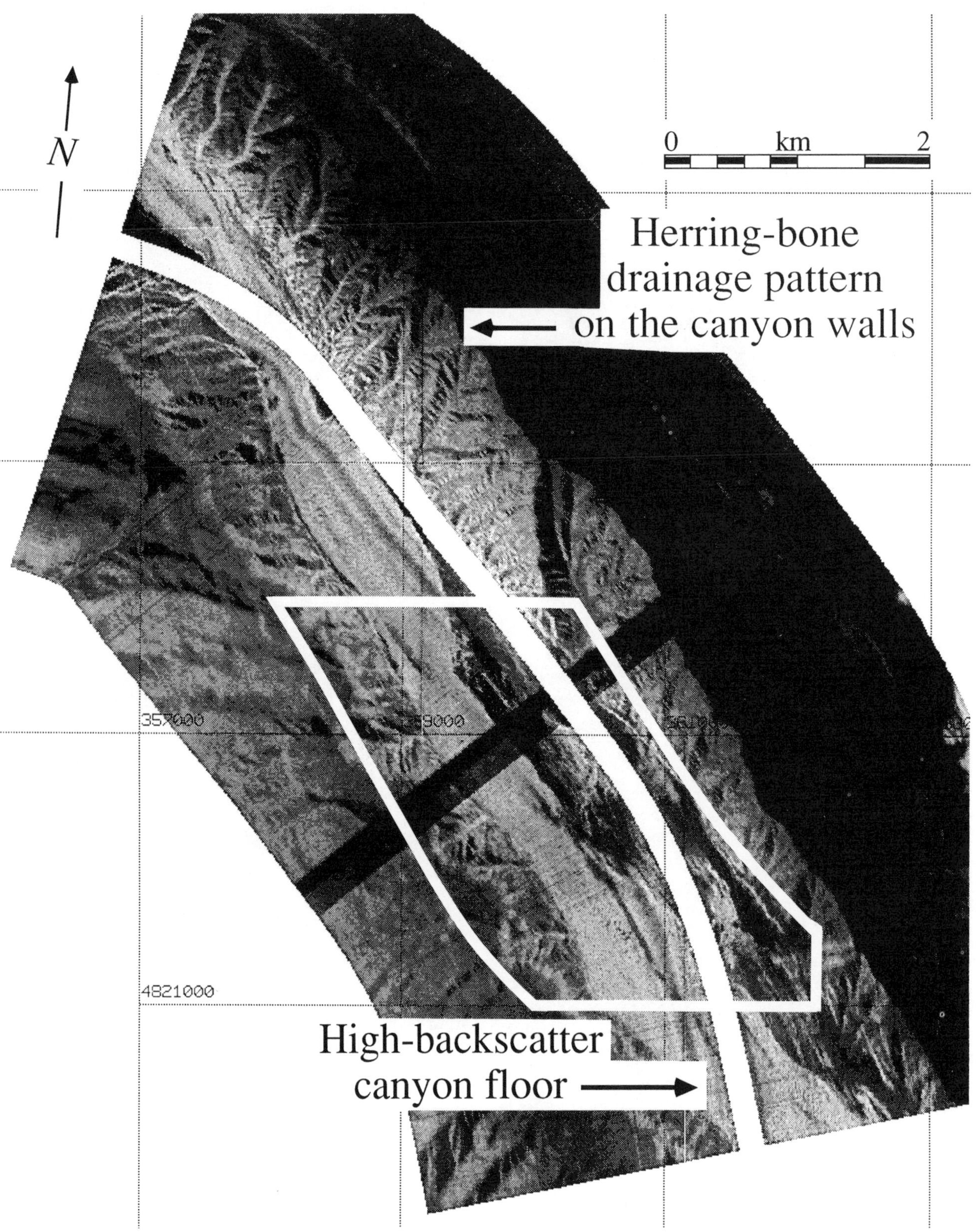

Figure 5. Side-looking sonograph of the Upper Var Canyon (location in Fig. 2). Note the bright, high-backscatter canyon floor and the fine-scale drainage pattern of the canyon walls, with first- and second-order tributaries. The white outline shows the location of Figure 14.

BEDFORMS IN THE CANYON FLOORS

Wavelength and height of bedforms in the Var Canyon

A detailed image of the canyon floor bedforms in the Lower Var Canyon (Fig. 8) shows that the bedform wavelength ranges between about 50 and 100 m.

Bedform heights are more difficult to estimate from the SeaMARC I records. On the subbottom profile, these bedforms cannot be distinguished, and thus their height probably does not exceed a few meters. The height of the bedforms can also be estimated from the length of the shadows on the side-looking records, as if one were to compute the height of a vertical pole from the length of its shadow, knowing the grazing angle of the sun rays. Then the height of the object is the shadow length times the tangent of the grazing angle. This method assumes that the grazing angle of the incident ray is less than the slope of the downstream side of the bedform, which is probably not always the case. In fact, this method predicts unreasonably large heights (up to 80 m) in the vicinity of the vehicle path. It can also be noticed that, if the black stripes on the records are shadows and if the bedform height remains constant, the shadow length should progressively increase for increasing distances from the vehicle path. The shadow length instead seems to remain more or less constant (see Fig. 8). This suggests that the black stripes are not real shadows, but that the sea floor here dips away from the vehicle at an angle smaller than the grazing angle, and that the incident rays are mostly reflected away so that very little acoustic energy is returned to the sonar vehicle. If this is the case, one can compute some maximum height for the bedforms, which will be the length of the shadow times the tangent of the grazing angle at the largest distance from the vehicle path where these shadows are observed. We can consider for this computation a distance bedform crest/vehicle path of 2.5 km and a vehicle altitude of 400 m, which altogether give a grazing angle of 9°. For a typical shadow length of 30 m, this translates into a maximum height of about 5 m, which is compatible with the fact that the bedforms are not observable on the subbottom records.

In a first submersible study, asymmetric ridges of pebbles were noticed in the canyon floors (Auffret and others, 1982). A careful re-examination of the depth data for a dive carried out in the Upper Var Canyon at depths around 1,600 m has shown that, in agreement with the sonographs, there are asymmetric oscillations in the surface of the sea floor. The average spacing of these oscillations is 35 m, and the amplitude 1 to 2 m (see insert in Fig. 9). Accumulations of pebbles are found on the crests of the oscillations.

Deep-sea sediment waves

In reviewing the available data, Normark and others (1980) found that deep-sea sediment waves have been observed mostly on the lower continental rise; their wavelengths typically vary between 0.5 and 6 km, their amplitudes vary between 10 and 40 m, and they are almost invariably composed of fine-grained sediments (lutite). The origin of deep-sea sediment waves has been attributed either to geostrophic contour currents (e.g., Hollister and others, 1974; Shepard and others, 1976) or to downslope turbidity currents (e.g., Bouma and Treadwell, 1975; Damuth, 1979). Wave crests tend to be aligned transverse to the current direction.

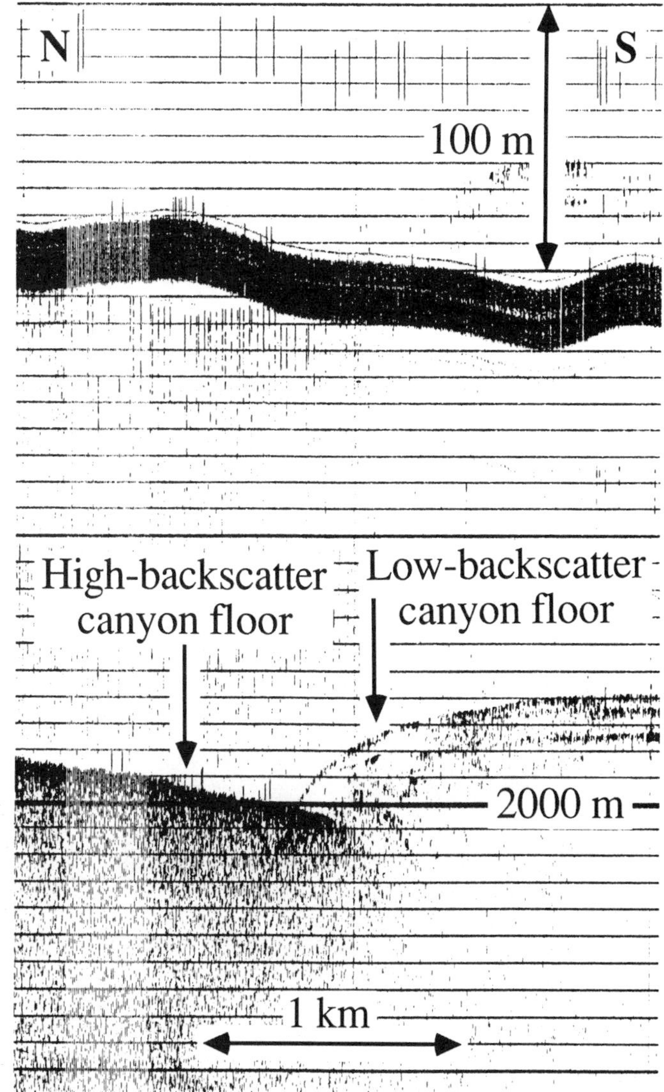

Figure 6. SeaMARC I subbottom record from the Lower Var Canyon (location in Fig. 2). Note the 30-m sharp step that separates the high- and low-backscatter canyon floor. The black stripe on the subbottom record at about 1,800 m depth is an artifact of the processing and corresponds to the depth of the sonar vehicle beneath the sea surface. The discrepancy between the absolute depth in this record and in the bathymetric map (Fig. 2) is due to an incorrect calibration of the depth sensor on the sonar vehicle.

As discussed above, the bedforms on the floor of the Var Canyon are composed of coarse material, from sand to boulders, as shown by photographic and submersible observations and by

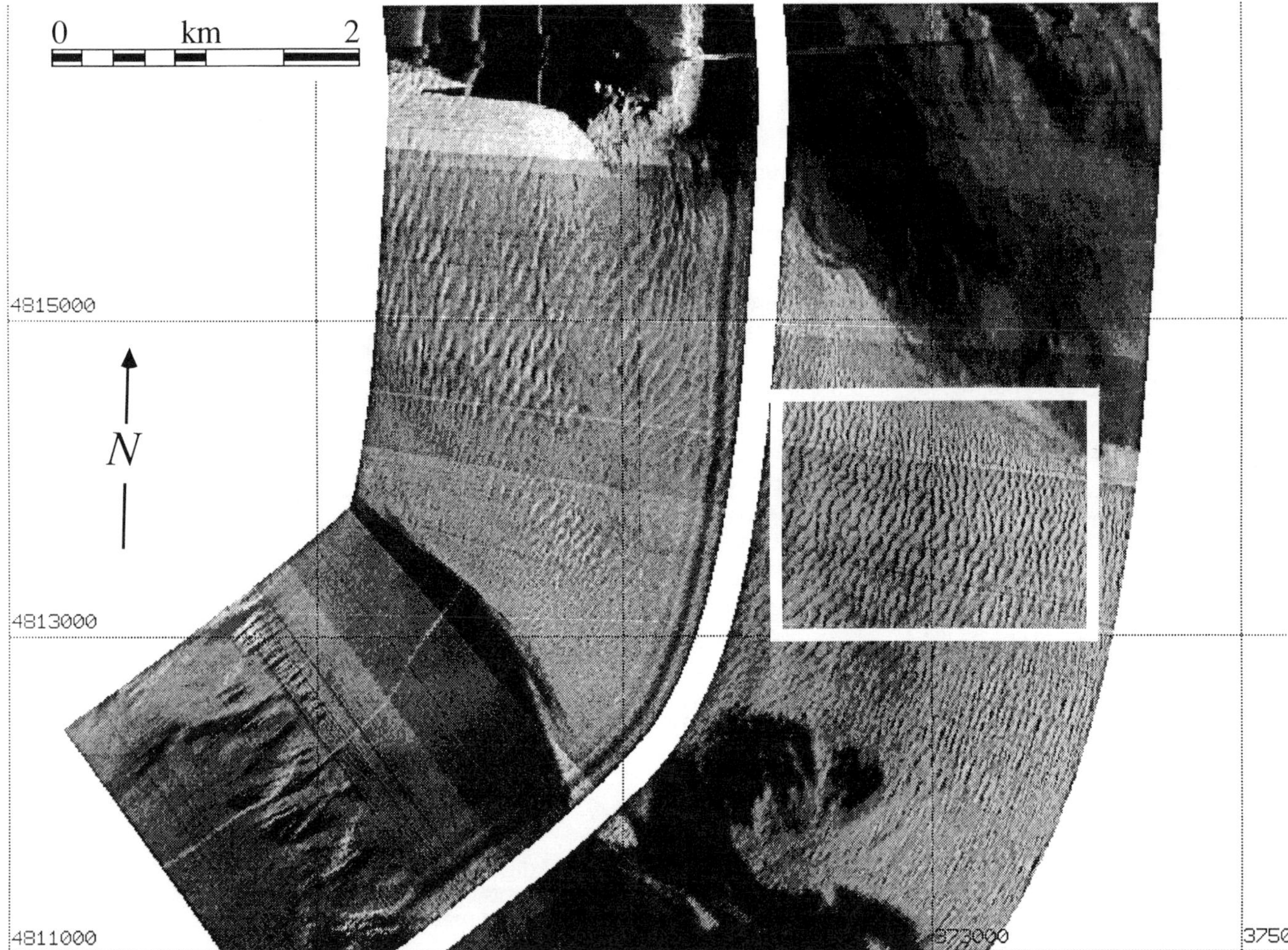

Figure 7. Side-looking sonograph of the Lower Var Canyon (location in Fig. 2). Note the occurrence in the high-backscatter canyon floor of narrow lineaments that are white in the upslope half of the record (on the left) and black in the downslope half (on the right). This pattern is created by a train of asymmetric bedforms, whose downslope-facing flank is steeper than the upslope-facing flank. The white outline shows the location of Figure 8.

the high backscatter on the sonographs. The wavelength of these bedforms ranges between a few tens of meters and 100 m, and their height does not exceed 5 m. Therefore, it does not seem justified to directly compare the Var Canyon bedforms to the "sediment waves" described by Normark and others (1980) because of the differences in composition and size.

Lonsdale and Spiess (1977) proposed a working classification of deep-sea bedforms on the basis of size and constituent material. They distinguished ripples, sand waves or dunes, mud waves, and furrows. The "mud waves," as defined by Lonsdale and Spiess (1977), seem to be the analogue of the "sediment waves" of Normark and others (1980). Sand waves or dunes are described by Lonsdale and Spiess (1977) as bedforms with typical dimensions between 10 and 100 m, aligned transverse to the flow. Because of their small wavelength and height, these bedforms cannot be recognized on conventional shipboard subbottom profiles, and can be detected only on side-looking sonar records. Their size accounts for the fact that they have been observed only rarely in the deep sea (e.g., Lonsdale and Malfait, 1974). Bedforms of size and composition similar to those described in the present study have been observed by Piper and others (1985) on SeaMARC I side-looking records from the Laurentian Fan, in the vicinity of the epicenter of the 1929 Grand Banks earthquake. Piper and others (1985) named these features "gravel waves." Since the bedforms observed in the Var Canyon floor seem also to contain large clasts besides sand and gravel, we will use the generic term "waves."

The Channeled Scabland

A very interesting terrestrial analogue is represented by the region of the Columbia Plateau in eastern Washington. Bretz (1923) found in this area a network of anastomosing channels

Figure 8. Detailed image of the train of bedforms in the high-backscatter floor of the Lower Var Canyon, at a depth of about 2,000 m (location in Fig. 7). The wavelength (crest-to-crest distance) of the bedforms ranges between 50 and 100 m.

and other topographic features such as cataracts, rock basins, and gorges that incised the basalts of the Columbia Plateau and its Pleistocene cover of lacustrine and alluvial sediments and loess. Bretz (1923) first called this region the Channeled Scabland and proposed that this unique landscape was gouged by a sudden flood whose depth reached hundreds of meters, as shown by the height above the valley floors of flood-eroded crossings of former stream divides. It is interesting, considering the topic of this volume, to note that this hypothesis was strongly challenged, since it involved a catastrophic event and, therefore, was non-uniformitarian (for a complete history of this debate see Baker, 1981). Only 30 years after the first investigations of the Channeled Scabland, a number of trains of "giant current ripples" were identified from aerial photographs of the area (Bretz and others, 1956). They were not observed even in detailed land surveys because of their large wavelength and small height. Sixty ripple trains were studied in detail by Baker (1973). The ripple wavelengths range between about 20 and 130 m, and their heights between 0.5 and 7 m. The dimensions of the ripples are found to correlate with hydraulic parameters such as the depth of the flood and the mean flow velocity.

Figure 13 shows the height of giant ripples as measured in the Channeled Scabland versus their wavelength (Baker, 1973). Ranges in the estimates of these two parameters are plotted from submersible and side-looking sonar observations (A) for the Var Canyon waves and from side-looking sonar observations done by Piper and others (1985) on "gravel waves" in the Laurentian Fan (B). Baker (1973, p. 51) argues that the regression line shown in Figure 13 is steeper than the regression line previously computed by Allen (1968) for giant sand ripples because the Scabland ripples are composed of gravel and sand instead of sand only. The waves in the Nice and Laurentian canyon floors shown in Figure 13 seem to follow the Scabland regression line, which agrees with the observation that they are composed of coarse material.

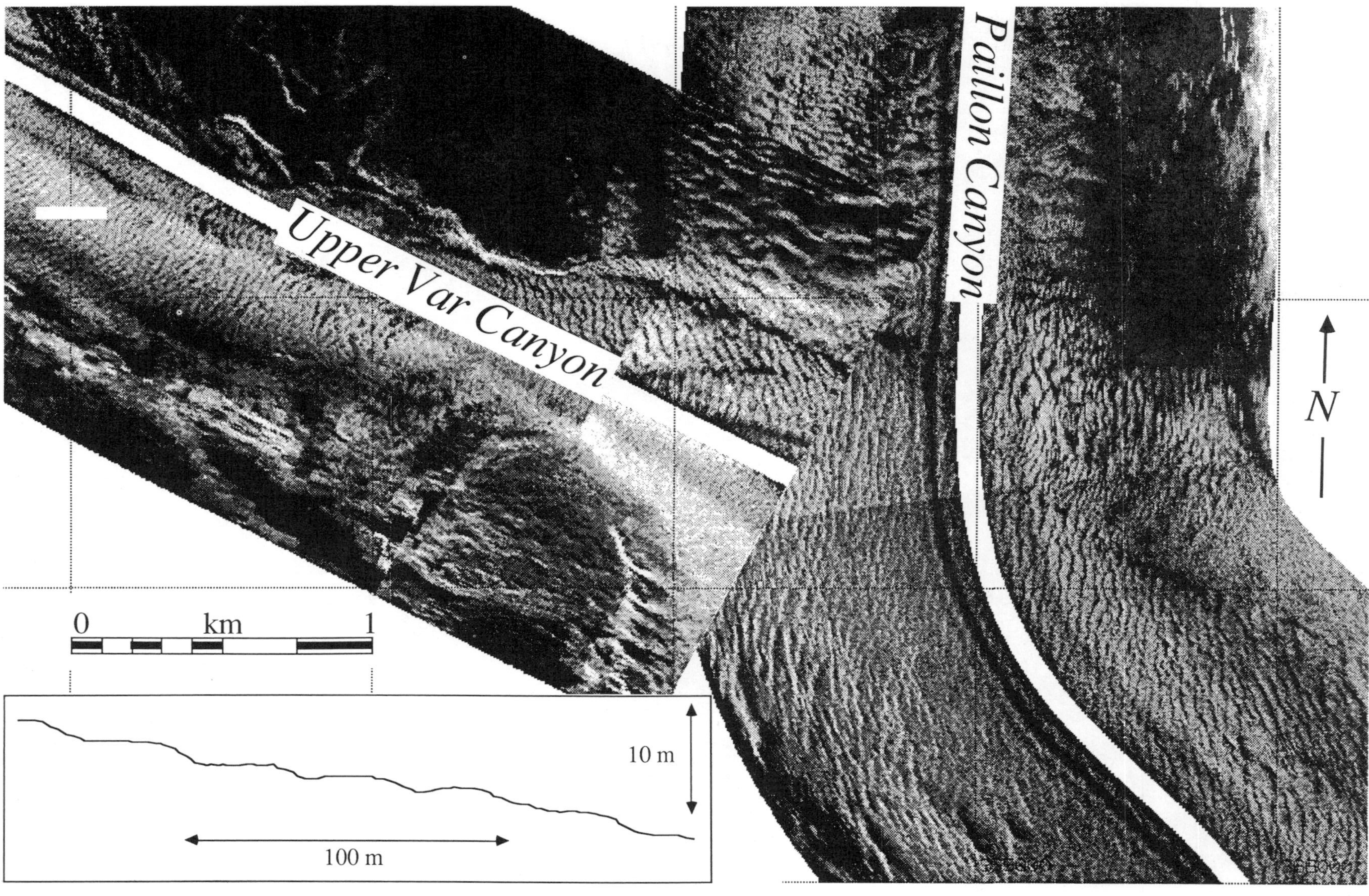

Figure 9. The train of bedforms in the Upper Var Canyon overlays an analogous set in the Paillon Canyon. The insert shows a detailed bathymetric profile in the Upper Var Canyon from submersible data. The location of the profile is shown as a white segment in the upper left of the sonograph.

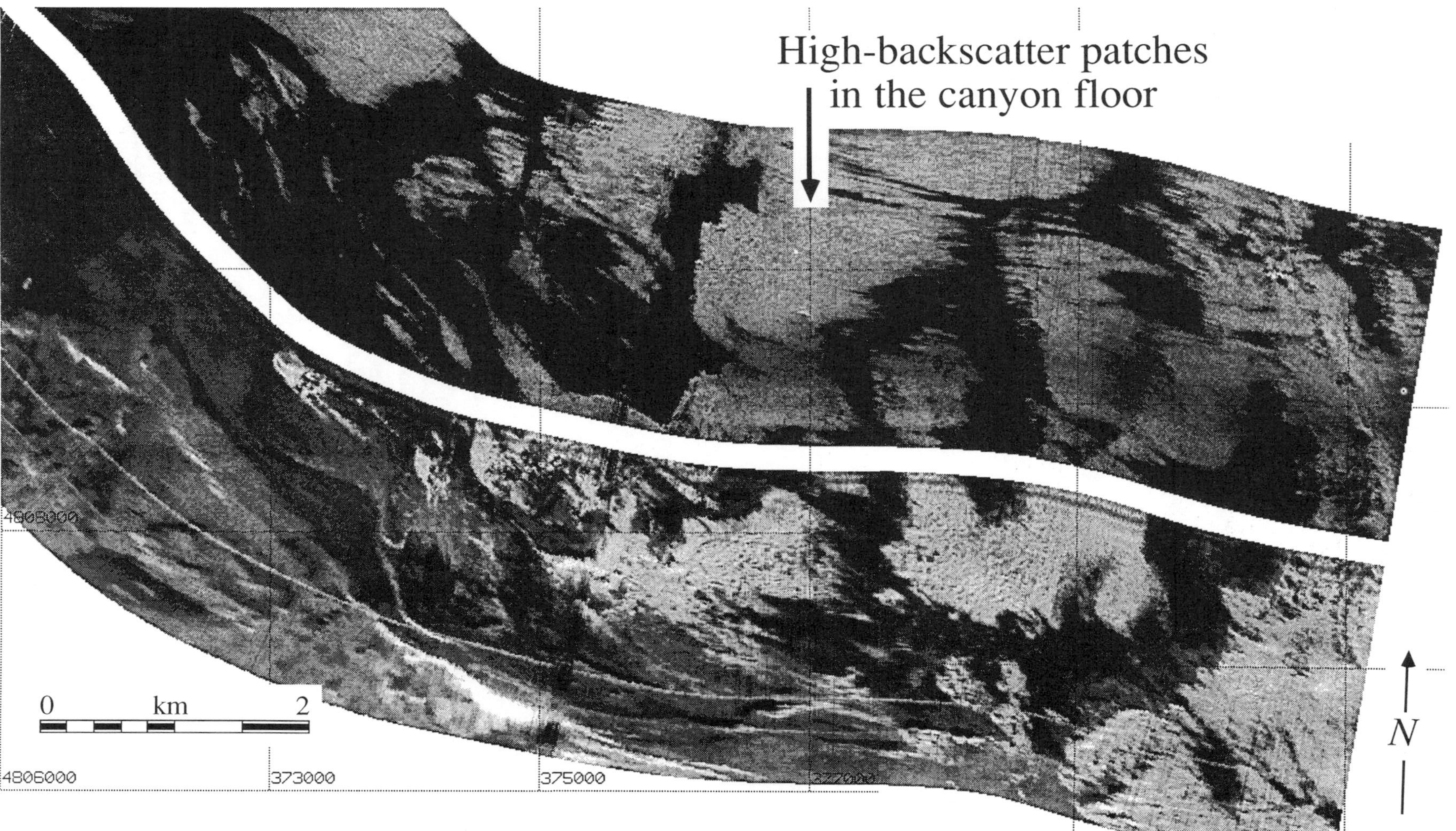

Figure 10. Side-looking sonograph of the Lower Var Canyon (location in Fig. 2). The high-backscatter canyon floor that was continuous in the shallower portion of the Lower Var Canyon is here broken into a number of high-backscatter patches.

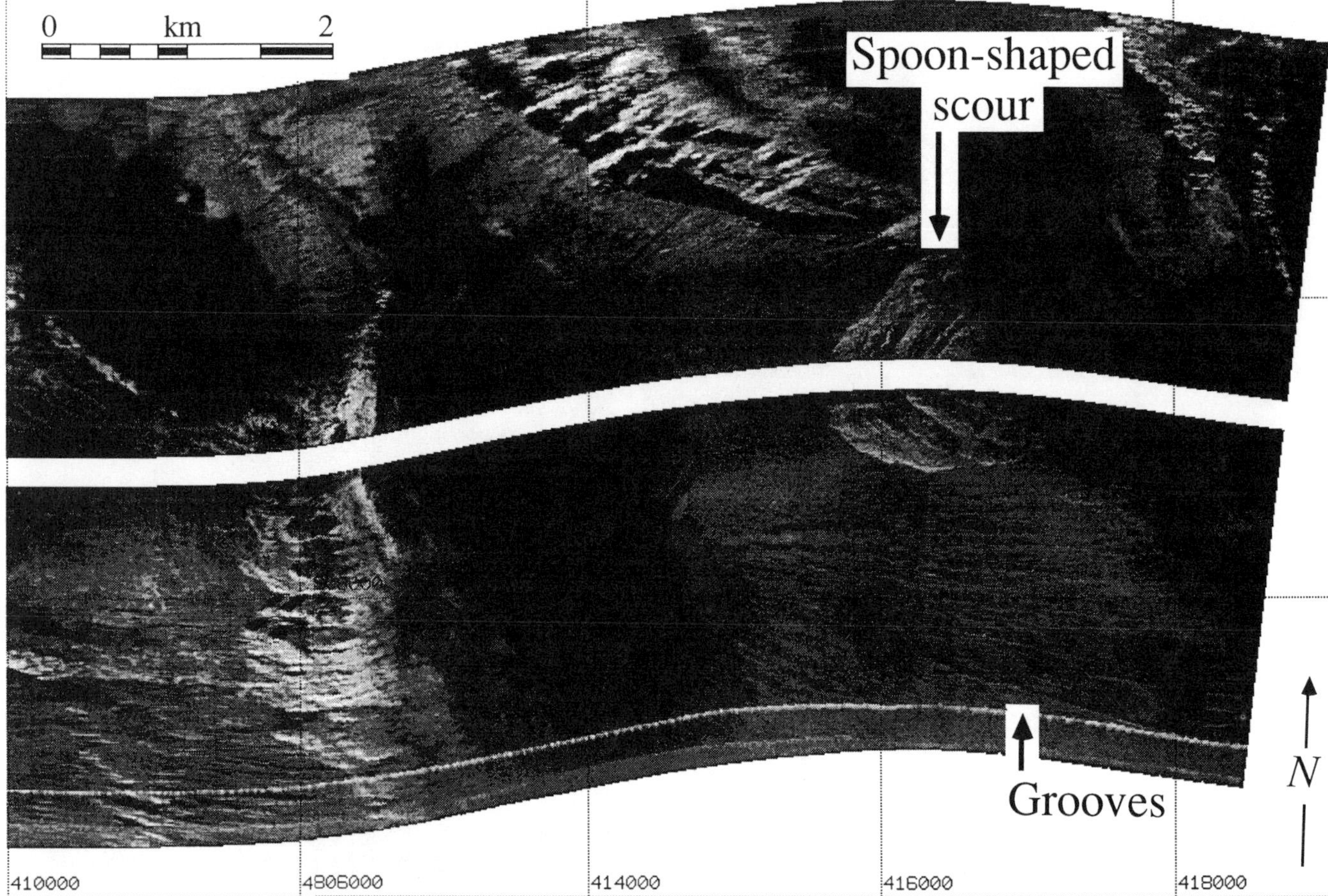

Figure 11. Side-looking sonograph of the Lower Var Canyon (location in Fig. 2, detailed bathymetry in Fig. 12). Note the occurrence of downslope-trending grooves and a spoon-shaped scour, with outcrops of eroded layers.

DISCUSSION

Failure events in the Upper and Lower Var Canyon

In this discussion, we concentrate on failure events that originate in the Var Prodelta and travel down the Upper and Lower Var Canyon. Although the sonar images cannot conclusively demonstrate this, it is likely that the most recent event traveled along this path. In particular, this inference is supported by: (1) small-scale surface roughness of the chutes radiating from the Var Prodelta and feeding the Upper Var Canyon, (2) truncation of an analogous wave field in the Paillon Canyon at its junction with the Upper Var Canyon, and (3) consistent orientation of the wave field from the Upper Var Canyon to the Lower Var Canyon. Observation (1) above, the absence of a mature drainage system on most of the Var Prodelta, and the fact that part of the Nice airport collapsed into the sea in the 1979 event, suggest that failures occurred on the steep slopes of the prodelta quite near the coastline. We will refer mostly to multiple failure events, since it does not seem possible to discriminate between the effects of several events that followed the same path.

Cross section of the flows as deduced from the sonar images

Sediment flows traveling down the Upper Var Canyon not only created the wave fields described above, but also left signs of scouring in the low-backscatter floor of the Upper Var Canyon (Fig. 14). The flows apparently never reached the canyon walls, since the mature drainage pattern on the walls is preserved. A reasonable estimate would be to limit the height of the flows in the Upper Var Canyon to the height of the scarps at the base of the canyon walls. The height of these scarps can be estimated from the subbottom profiles as about 30 m. Sediment flows in the Upper Var Canyon then passed through a section that was never wider than 1 km or higher than 30 m.

As already observed, a sharp step about 30 m high separates the high- and low-backscatter canyon floors of the uppermost Lower Var Canyon. Here, at depths of less than 2,000 m, the high-backscatter canyon floor is continuous and contains waves, and there are no signs of scouring outside the high-backscatter canyon floor (Fig. 7). Apparently, the flows passed through a section that was about as high as in the Upper Var Canyon, but about 4 km wide. No definite statements about the thickness of

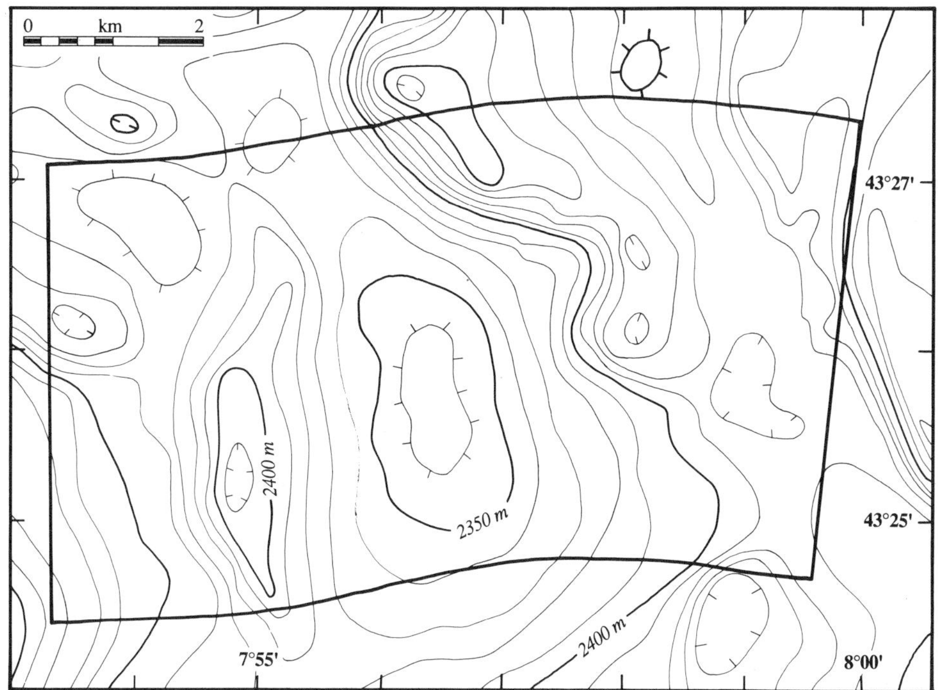

Figure 12. Detailed bathymetry of the area imaged in the sonograph of Figure 11 (bathymetry after Pautot and others, 1984; contour interval 10 m). Highs correspond to areas of low backscatter in the sonograph, and lows to areas of high-backscatter and scouring.

the flows can be made from the sonar observations for the rest of the Lower Var Canyon, at depths of more than 2,000 m. However, it can be observed that scouring extends throughout the whole swath, i.e., for a width of more than 5 km.

Although we have no indication of a change in thickness, we conclude that flows traveling in the Var Canyon passed through a wider cross section as the bottom slope decreased from about 5 percent in the Upper Var Canyon to about 0.5 percent in the deepest part of the Lower Var Canyon explored during the SeaMARC I survey.

Origin of coarse material in the Var Canyon

Coarse material could have originated from deposits in the Var Prodelta, or could have been transported into the sea as the bed load of the Var River. Another possibility is that these coarse materials were excavated from the substrate. The Messinian erosional surface is an attractive candidate because it was a subaerial surface exposed to alluvial processes during the desiccation of the Mediterranean. The Messinian erosional surface has been recognized throughout the margins of the Mediterranean on seismic reflection records (Ryan and Cita, 1978) and has been penetrated by drilling in the Gulf of Lion (Cravatte and others, 1974), the Ebro margin of Spain (Maldonado and Riba, 1974), and the western margin of Sardinia (Ryan and others, 1973). At the latter location (DSDP site 133), drilling penetrated a thickness of 150 m into a deposit of fluvial silts and sands with cobbles of Paleozoic schists. On seismic reflection profiles that cross the Upper Var Canyon and the Paillon Canyon, the reflector interpreted by Pautot (1981) as the Messinian erosional surface appears to converge with the canyon floors. A viable possibility is that the Messinian erosional surface has been exposed during the development of the Var Canyon, and its coarse clasts have been reworked into the canyon floor waves.

Agents that shaped the Var Canyon bedforms

The only reasonable candidates for the generation of the canyon floor bedforms are sediment gravity flows, or in short, sediment flows (Middleton and Hampton, 1973; Lowe, 1982). In the following discussions, we will refer to sediment flows as a generic term encompassing flows supported by matrix strength (debris flows) and turbulent flows (turbidity currents). It has been shown from hydraulic considerations that turbidity currents can transport coarse material both as bed load and in suspension (Johnson, 1964). This conclusion is also supported by the presence of gravels and cobbles in turbidite deposits (Komar, 1970). Sediment flows of higher densities than those normally accepted for a turbidity current would be capable of transporting coarse clasts (Kuenen, 1950). One could conclude that sediment flows derived from slides originating in the vicinity of the Nice coastline were in some way responsible for the formation of the canyon floor waves.

As an alternative to mass wasting of the slope, sediment flow could be generated by river flooding. The head of the Upper Var Canyon reaches the mouth of the Var River; there is no continen-

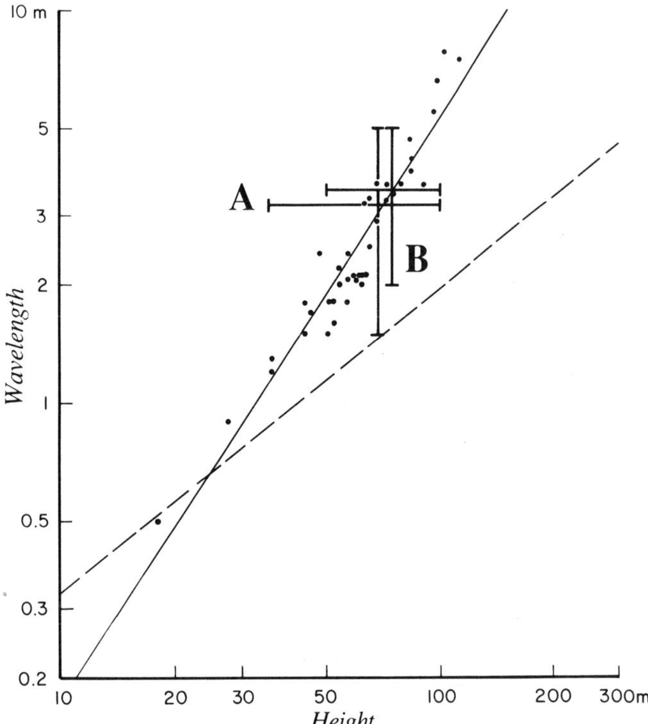

Figure 13. Logarithmic regression of bedform wavelength versus height. The black dots represent the averages of 40 trains of giant current ripples measured in the Channeled Scabland as reported by Baker (1973). The straight line represents the best-fit regression. The dashed line instead is the best-fit regression computed from observations of sand dunes by Allen (1968). The two crosses represent ranges of estimates from side-looking sonar observations for the Nice Canyon (A; 35 to 100-m wavelength, 1.5 to 5-m height) and for the Laurentian Fan (B; 50 to 100-m wavelength, 2 to 5-m height; after Piper and others, 1985).

tal shelf, and no sign of any accumulation of sediment in the vicinity of the estuary (see Fig. 1). A similar case is represented by the Congo Submarine Canyon, whose head penetrates inland in the estuary as much as 25 km (Shepard and Emery, 1973). Heezen and others (1964) reviewed the history of cable breaks in the Congo Canyon and found that the ruptures took place during periods of high river discharge. They proposed that the suspended load of the river was transported at the head of the canyon during times of peak discharge. This material would settle into the underlying sea water and could then generate turbidity flow. Heezen and others (1964) also proposed that the river bed load might separate at the head of the canyon and simply continue as a turbidity flow. In fact, it is possible, although unlikely, that during flood periods the concentration of sediment in the fresh water of a river is so high that the density exceeds that of deep-sea water, so that the river could flow as a density current in the canyon floor (Stetson and Smith, 1938; Menard and Ludwick, 1951). These possibilities cannot be discounted for the Var Canyon.

Genessaux and others (1971) report current measurements carried out for about one month on the floor of the Upper Var Canyon, at a depth of 850 m. They found sudden bursts of down-canyon currents that reached velocities of about 1 m/sec (3.6 km/hr). In particular, these currents were observed during a one-day period of flooding of the Var River, and were interpreted as turbidity currents (Genessaux and others, 1971). Flows originated by the sediment transported by the Var River are probably not as important as the 1979 event; on the other hand, these flows could periodically flush the canyon floor, reworking the substrate and eroding fine material.

Speculations on the characteristics of the 1979 event

If we consider the 1979 event as the last major mass-wasting event, we can draw some interesting speculations on the characteristics of the associated sediment flow in the Var Canyon. We do not know how much the 1979 event contributed to the wave train, either in terms of clasts and matrix now incorporated in the waves or in terms of its ability to shape the waves. Furthermore, we do not know if the wave train is a composite deposit of numerous events, both greater and smaller in magnitude than the 1979 event. The herring-bone drainage pattern on the canyon walls requires innumerable mass-wasting events, each of which might contribute material to the waves and energy to shape them. Flooding events from the Var River could do likewise. However, it seems likely that gravel, cobbles, and boulders were not transported in 1979 beyond the continuous high-backscatter floor of the Var Canyon, where the last occurrence of the waves is observed (Fig. 7). Beyond this limit, high-backscatter patches, scours, and grooves have been observed (Figs. 10 and 11). Here, the latest event seems to have had only a scouring effect, and our imagery and profiles do not indicate an identifiable deposit that can be attributed to it ("identifiable" here means more than about 2 m thick, since this is the thinnest layer that can be realistically detected in the SeaMARC I subbottom records). If this is the case, then the flow passed through the Lower Var Canyon as a turbidity current, and the turbulent suspension was the agent that produced the observed high-backscatter patches, scoured depressions, and grooves. The coarse material that should be exposed in the high-backscatter patches was not transported there in 1979, but was simply excavated. Thus, there have been events in the past greater in magnitude and capable of transporting coarse bedload farther than the 1979 event.

Komar (1973) has noted that if the turbidity flow is continuous (i.e., there is no water entrainment, no erosion, and no deposition), a decrease in velocity caused by a decrease in bottom slope will be matched by an increase in the flow cross section. The widening of the cross section of flow in the Lower Var Canyon might be related to a decreasing flow velocity. One can speculate that the transition from the continuous high-backscatter canyon floor to the patches is related to the inferred decrease in flow velocity. The flow velocity possibly became too small to move coarse material, so that the effect of the flows beyond this point was mainly scouring the substrate. Scouring phenomena are particularly evident in the deepest part of the Lower Var Canyon that was imaged (Fig. 11). The sonographs show elevations char-

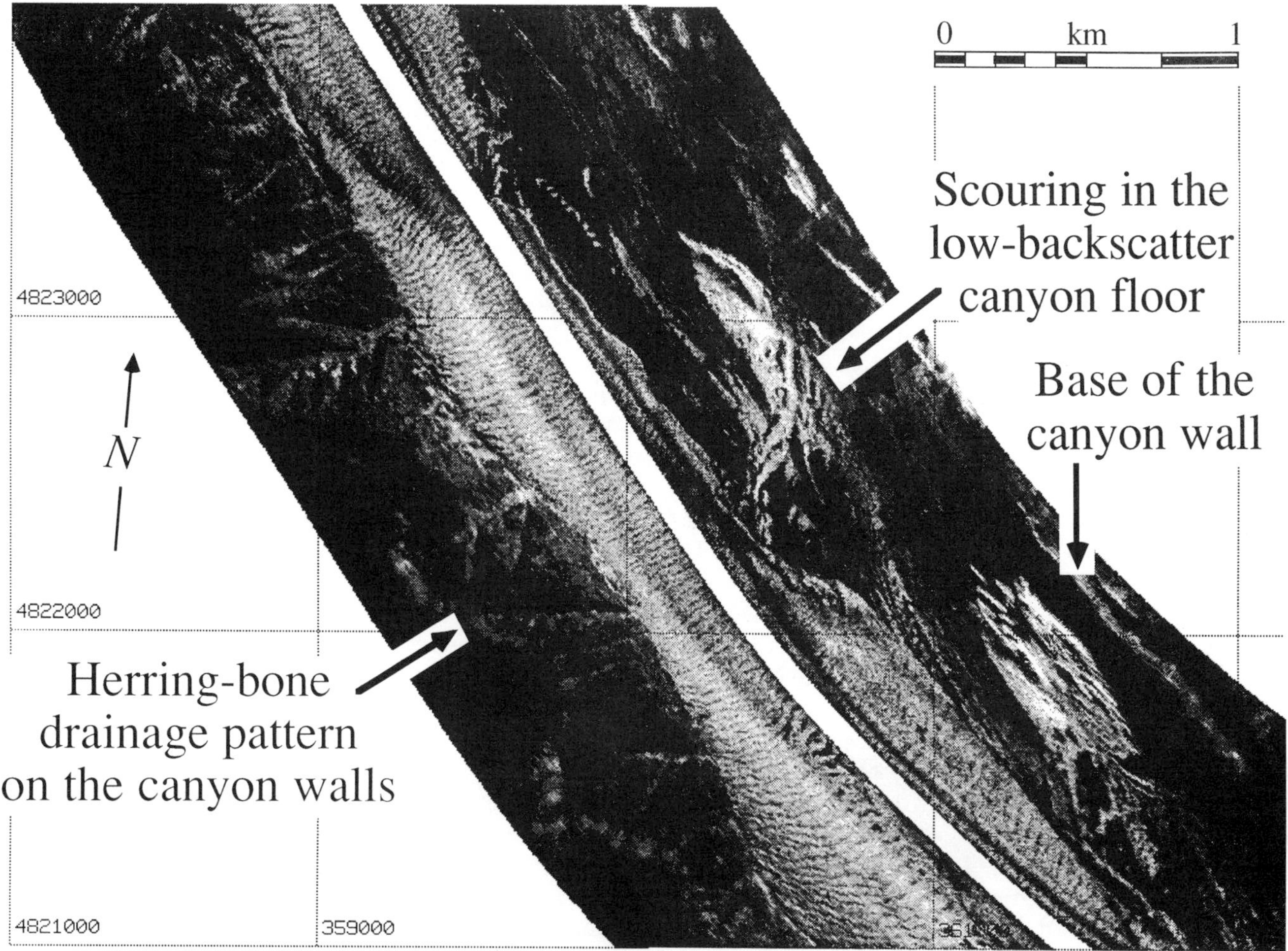

Figure 14. Side-looking sonograph of the Upper Var Canyon (location in Figs. 2, 5). Note signs of scouring in the low-backscatter portion of the canyon floor, on the right. The base of the canyon wall corresponds to a sharp step, imaged as a discontinuous bright line in the extreme right portion of the swath. The mature drainage pattern on the canyon wall seems to be untouched by the flows that flushed the canyon floor. The waves in the high-backscatter canyon floor have wavelengths of about 30 m, shorter than those in the Lower Var Canyon (Figs. 7, 8).

acterized by very low backscatter, i.e., very smooth, and depressions with grooves and scours. The alternation of smooth and scoured terrain suggests that the flow here intermittently touched the sea floor, jumping over highs and eroding depressions. The ultimate deposition of the fine material held in suspension occurred seaward of the area imaged, when the flows arrived on the abyssal plain.

Our most plausible scenario for the 1979 event is as follows: (1) the event originated as a submarine slide on the slope of the Var Prodelta; (2) the slide traveled as a sediment flow down the Upper Var Canyon and the uppermost Lower Var Canyon; (3) the sediment flow gave rise to a turbidity current by water entrainment before reaching the area where the high-backscatter patches are found today; (4) the turbidity current thus generated intermittently eroded the substrate in its path, until it broke the cables and turned into a depositional event in the abyssal plain.

The characteristics of the flow in the Upper Var Canyon and the uppermost Lower Var Canyon, where the wave fields were formed, are most uncertain. The 1979 failure had to both involve a large volume of sediments and take place suddenly in order to generate the observed tidal wave (Genessaux and others, 1980). If this is the case, then the flow had to travel at a high velocity to pass in a short time through the small cross section described above for the Upper Var Canyon. If one requires the high flow velocities to be associated with a turbulent regime, then the flow had to be a turbidity current. This turbidity current would have shaped the coarse material of the canyon floor into the wave fields observed. Another possibility is that the coarse material in the canyon floor is a deposit from a high-velocity debris flow. While this debris flow was traveling down the Var Canyon, fine material would have been brought in suspension to eventually generate a turbidity current. In this case, however, it is not clear how the waves could form.

These speculations cannot be accurately tested, since we

cannot precisely constrain the nature and amount of sediment that slid in the Upper Var Canyon, and the velocity of the flow in its uppermost course for the 1979 event. Nevertheless, we suggest that further studies of this kind will cast some light on the transition from a submarine slide to a turbidity current, which at present is poorly known (Kennett, 1982).

A plan view perspective

The entire continental slope off Nice is in a process of destruction through mass wasting, and the path of any failure event is an ephemeral feature. Present scars will be obliterated by future ones, and the coarse material in the canyon floors will continue to be reworked into new orientations. The geologic record as a vertical section of strata will preserve at best a tiny trace of any single event as an unconformity, and this trace may never reveal the magnitude and dynamics of the event or collection of events that produced the unconformity. The preservation of the 1979 event took place on the abyssal plain beyond the broken telephone cables. In a hypothetical future study of the slope unconformity in the geologic record, one might postulate catastrophic events to wipe away a kilometer or so of the Plio-Quaternary cover. On the other hand, cores from the abyssal plain may create the impression of continuous rhythmic sedimentation. In the geologic record, the unconformity corresponding to the erosion in the Nice canyons may then remain unlinked with deposits in the Balearic Abyssal Plain. In other words, an event that fits uniformitarianism to the classical geologist who studies turbidite sequences in time is viewed as catastrophic by the planetary geologist who images plan-view surfaces in space. This conclusion should not be interpreted as a criticism of either of these two approaches to the study of geologic events. Rather, we want to emphasize that it is not completely straightforward to apply the results of our plan view study of a landscape in an instant of time to the investigation of the geologic record, where the spatial relationships are difficult to resolve.

CONCLUSIONS

(1) Recent failure events followed a path that today possesses erosional chutes in the Var Prodelta, large bedforms in the Upper Var Canyon and uppermost Lower Var Canyon, and intermittent mega-scour marks in the rest of the Lower Var Canyon. The changes along the path correspond to a decrease in the downslope gradient and a corresponding increase in the flow cross section.

(2) The canyon floor bedforms are asymmetric waves composed of coarse material. The coarse material was reworked into bedforms by sediment flows. The wave trains have analogues in the Channeled Scabland of eastern Washington, which has been the site of catastrophic floods. There is a quantitative similarity in bedform sizes.

(3) Coarse material in the floor of the Var Canyon probably slid from the Var Prodelta during failure events. There is also the possibility that these coarse clasts may be reworked alluvial materials originally deposited on the margin of the Mediterranean during the time of its late Miocene desiccation.

(4) Sediment flows in the floor of the Var Canyon most likely originated from failures of the Var Prodelta. Sediment flows could also have generated by the failure of accumulations of sediments transported by the Var River into the head of the Var Canyon during flood periods.

(5) Sediment flows left abundant signs of their occurrence in bedforms that can be attributed to reworking or erosion of existing deposits, but there is no clear sign of deposition in the area imaged.

(6) The plan view images presented here are a glimpse of a landscape in an instant of time. They give a detailed view of a present-day unconformity. When viewed in the vertical geologic record, the morphology will be very difficult to reconstruct, and it will be almost impossible to resolve the sequence of events that shaped the landscape. A look at the geologic column leads thus to a more uniformitarian view than an examination of the sea-floor surface at a given time.

ACKNOWLEDGMENTS

We thank the officers and crew of R/V *Robert D. Conrad* for their assistance at sea. Dan Chayes, Dale Chayes, John Di Bernardo, John Freitag, Bernie Gallagher, and Jim Smith operated the SeaMARC I and SeaBeam sonar systems. Kim Kastens, Suzanne O'Connell, and Hans Nelson helped in directing the survey operations. Michelle Henrion assisted in data reduction and Mary Ann Luckman in drafting. Nick Christie-Blick, Roger Flood, Ken Hunkins, Neil Kenyon, and Barry Raleigh reviewed the manuscript and provided a number of suggestions. This work was supported by IFREMER, France, under contract number 83/7313/Y and by the Submarine Geology and Geophysics Program of the National Science Foundation, U.S.A., with grant OCE 81-18069. This is Lamont-Doherty Geological Observatory contribution number 4325.

REFERENCES

Allen, J.R.L., 1968, Current ripples; Their relation to patterns of water and sediment motion: Amsterdam, North-Holland Publishing Co., 433 p.

Auffret, G. A., Auzende, J. M., Genessaux, M., Monti, S., Pastouret, L., Pautot, G., and Vanney, J. R., 1982, Recent mass wasting processes on the Provençal margin (western Mediterranean), *in* Saxov, S., and Nieuwenhuis, J. K., eds., Marine slides and other mass movements: New York, Plenum Publishing Co., p. 53–58.

Baker, V. R., 1973, Paleohydrology and sedimentology of Lake Missoula flooding in eastern Washington: Geological Society of America Special Paper 144, 79 p.

—— , ed., 1981, Catastrophic flooding; The origin of the Channeled Scabland, *in* Benchmark papers in Geology, v. 55: Stroudsburg, Pennsylvania, Dowden, Hutchinson and Ross, Inc., 361 p.

Bouma, A. H., and Treadwell, T. K., 1975, Deep-sea dune-like features: Marine Geology, v. 19, p. M53–M59.

Bourcart, J., Genessaux, M., and Klimek, F., 1960, Ecoulements profonds des sables et de galets dans la grande vallèe sous-marine de Nice: Comptes Rendus de l'Académie des Sciences, Paris, ser. D, v. 250, p. 3761–3765.

Bretz, J. H., 1923, The Channeled Scablands of the Columbia Plateau: Journal of Geology, v. 31, p. 617–649.

Bretz, J. H., Smith, H.T.U., and Neff, G. E., 1956, Channeled Scablands of Washington; New data and interpretations: Geological Society of America Bulletin, v. 67, p. 957–1049.

Chayes, D. N., 1983, Evolution of SeaMARC I, *in* Institute of Electrical and Electronic Engineers Proceedings of the third working symposium on oceanographic data systems, p. 103–108.

Cravatte, J., Dufaure, P., Prim, M., and Rouaix, S., 1974, Les sondages du Golfe du Lion; Stratigraphie et sedimentologie: Notes Memoirs Compagnie Française Pétrol, 11 p. 209–274.

Damuth, J. E., 1979, Migrating sediment waves created by turbidity currents in the northern South China Basin: Geology, v. 7, p. 520–523.

Farre, J. A., McGregor, B., Ryan, W.B.F., and Robb, J. M., 1983, Breaching the shelfbreak: Passage from youthful to mature phase in submarine canyon evolution: Society of Economic Paleontologists and Mineralogists Special Publication 33, p. 25–39.

Genessaux, M., 1966, Prospection photographique des canyons sous-marins du Var et du Paillon (Alpes maritimes) au moyen de la troika: Revu de Géographie Physique et de Geologie Dynamique, v. 8, p. 3–38.

Genessaux, M., Guibout, P., and Lacombe, H., 1971, Enregistrement de courants de turbiditè dans la vallèe sous-marine du Var (Alpes Maritimes): Comptes Rendus de l'Académie des sciences, Paris, ser. D, v. 273, p. 2456–2459.

Genessaux, M., Mauffret, A., and Pautot, G., 1980, Les glissements sous-marins de la pente continentale niçoise et la rupture des cables en Mer Ligure (Mediterrañe occidentale): Comptes Rendus de l'Académie des Sciences, Paris, ser. D, v. 290, p. 959–962.

Heezen, B. C., Menzies, R. J., Schneider, E. D., Ewing, W. M., and Granelli, N.C.L., 1964, Congo Submarine Canyon: American Association of Petroleum Geologists Bulletin, v. 48, p. 1126–1149.

Hollister, C. D., Flood, R. D., Johnson, D. A., Lonsdale, P., and Southard, J. B., 1974, Abyssal furrows and hyperbolic echo traces on the Bahama Outer Ridge: Geology, v. 2, p. 395–400.

Johnson, M. A., 1964, Turbidity currents, Oceanography and Marine Biology Annual Review, v., 2, p. 31–43.

Kennett, J. P., 1982, Marine geology: Englewood, New Jersey, Prentice-Hall, 813 p.

Komar, P. D., 1970, The competence of turbidity current flow: Geological Society of America Bulletin, v. 81, p. 1555–1562.

—— , 1973, Continuity of turbidity current flow and systematic variations in deep-sea channel morphology: Geological Society of America Bulletin, v. 84, p. 3329–3338.

Kuenen, P. H., 1950, Turbidity currents of high density: 18th International Geological Congress, London, Reports, Part 8, The geology of sea and ocean floors, p. 44–52.

Lonsdale, P., and Hollister, C. D., 1979, Cut-offs at an abyssal meander south of Iceland: Geology, v. 7, p. 597–601.

Lonsdale, P., and Malfait, B., 1974, Abyssal dunes of foraminiferal sand on Carnegie Ridge: Geological Society of America Bulletin, v. 85, p. 1697–1712.

Lonsdale, P., and Spiess, F. N., 1977, Abyssal bedforms explored with a deep-towed instrument package: Marine Geology, v. 23, p. 57–75.

Lowe, D. R., 1982, Sediment gravity flows; Two depositional models with special reference to the deposits of higher density turbidity currents: Journal of Sedimentary Petrology, v. 52, p. 279–297.

Maldonado, A., and Riba, O., 1974, Les rapports sedimentaires du Neogene et du Quaternaire dans le plateau continental aux environs du delta de l'Ebre (Espagne): Memoir Institute of Geology Bassin Aquitaine, v. 7, p. 321–329.

Menard, H. W., and Ludwick, J. C., 1951, Applications of hydraulics to the study of marine turbidity currents: Society of Economic Paleontologists and Mineralogists Special Publication 2, p. 2–13.

Middleton, G. V., and Hampton, M. A., 1973, Sediment gravity flows; Mechanics of flow and deposition, *in* Middleton, G. V., and Bouma, A. H., eds., Turbidites and deep-water sedimentation: Society of Economic Paleontologists and Mineralogists Short Course 1, p. 1–38.

Normark, W. R., Hess, G. R., Stow, D.A.V., and Bowen, A. J., 1980, Sediment waves on the Monterey Fan levee; A preliminary physical interpretation: Marine Geology, v. 37, p. 1–18.

Pautot, G., 1981, Cadre morphologique de la Baie des Anges (Nice-Côte d'Azur); Modèle d'instabilitè de la pente continentale: Oceanological Acta, v. 4, p. 203–211.

Pautot, G., Le Cann, C., Coutelle, A., and Mart, J., 1984, Morphology and extension of the evaporitic structures of the Liguro-Provencal basin; New Sea-beam data: Marine Geology, v. 55, p. 387–409.

Piper, D. J., Shor, A. N., Farre, J. A., O'Connell, S., and Jacobi, R., 1985, Sediment slides and turbidity currents on the Laurentian Fan; Sidescan sonar investigations near the epicenter of the 1929 Grand Banks earthquake: Geology, v. 13, p. 538–541.

Ryan, W.B.F., and Cita, M. B., 1978, The nature and distribution of Messinian erosional surfaces; Indicators of a several-kilometer-deep Mediterranean in the Miocene: Marine Geology, v. 27, p. 193–230.

Ryan, W.B.F., Hsü, K. J., Cita, M. B., Dumitrica, P., Lort, J., Maync, W., Nesteroff, W. D., Pautot, G., Stradner, H., and Wezel, F. C., 1973, Initial reports of the Deep Sea Drilling Project, part 1 and 2: Washington, D.C., U.S. Government Printing Office, v. 13, 1447 p.

Shepard, F. P., and Emery, K. O., 1973, Congo Submarine Canyon and fan valley: American Association of Petroleum Geologists Bulletin, v. 57, p. 1679–1691.

Shepard, F. P., Marshall, N. F., McLoughlin, P. A., and Fisher, R. L., 1976, Sediment waves (giant ripples) transverse to the west coast of Mexico: Marine Geology, v. 20, p. 1–6.

Stetson, H. C., and Smith, J. F., 1938, Behavior of suspension currents and mud slides on the continental slope: American Journal of Science, ser. 5, v. 35, p. 1–13.

MANUSCRIPT ACCEPTED BY THE SOCIETY APRIL 4, 1988

Printed in U.S.A.

Geological Society of America
Special Paper 229
1988

The 1929 "Grand Banks" earthquake, slump, and turbidity current

David J. W. Piper, *Atlantic Geoscience Centre, Geological Survey of Canada, Bedford Institute of Oceanography, P.O. Box 1006, Dartmouth, Nova Scotia B2Y 4A2, Canada*
Alexander N. Shor, *Lamont-Doherty Geological Observatory of Columbia University, Palisades, New York 10964*
John E. Hughes Clarke, *Department of Oceanography, Dalhousie University, Halifax, Nova Scotia B3H 4J1, Canada*

ABSTRACT

The epicenter of the 1929 "Grand Banks" earthquake (M_S = 7.2) was on the continental slope above the Laurentian Fan. The zone in which cables broke instantaneously due to the earthquake is characterized by surface slumping up to 100 km from the epicenter as shown by sidescan sonographs and seismic reflection profiles. The uppermost continental slope, however, is almost undisturbed and is underlain by till deposited from grounded ice.

The Eastern Valley of the Laurentian Fan contains surficial gravels molded into large sediment waves, believed to have formed during the passage of the 1929 turbidity current. Sand sheets and ribbons overlie gravel waves in the lower reaches of Eastern Valley. Cable-break times indicate a maximum flow velocity of 67 km/hr (19 m/s). The occurrence of erosional lineations and gravel on valley walls and low intravalley ridges suggest that the turbidity current was several hundred meters thick. The current deposited at least 175 km^3 of sediment, primarily in a vast lobe on the northern Sohm Abyssal Plain where a bed more than 1 m thick contains material ranging in size from gravel to coarse silt.

There is no apparent source for so much coarse sediment on the slumped areas of the muddy continental slope. We therefore infer that there was a large volume of sand and gravel available in the upper fan valley deposits before the earthquake. This coarse sediment was discharged from sub-glacial meltwater streams when the major ice outlet through the Laurentian Channel was grounded on the upper slope during middle Wisconsinan time. This sediment liquefied during the 1929 event, and the resulting flow was augmented by slumping of proglacial silts and gas-charged Holocene mud on the slope. Although earthquakes of this magnitude probably have a recurrence interval of a few hundred years on the eastern Canadian margin, we know of no other deposits of the size of the 1929 turbidite off eastern Canada. For such convulsive events, both a large-magnitude earthquake and a sufficient accumulation of sediment are required.

GEOLOGIC SETTING

The epicenter of the "Grand Banks" earthquake of November 18, 1929, was located on the continental slope between St. Pierre Bank and the Laurentian Fan, at the extreme western margin of the Grand Banks of Newfoundland (Fig. 1). This part of the continental slope appears to be a typical passive margin developed during the late Triassic rifting of the central Atlantic Ocean (Haworth and Keen, 1979). It is more than 60 km from the northwest-to-southeast-trending transform margin along the Southwestern Grand Banks Slope that marks the northeastern extent of the Mesozoic central Atlantic Ocean. The epicenter lies south of a major, east-west, Late Paleozoic strike-slip fault zone on the outer continental shelf (Chedabucto Fault on the Scotian Shelf, Collector anomaly on the Grand Banks: Haworth and Keen, 1979). The Mesozoic–Tertiary continental margin sediment wedge beneath the continental slope near the 1929 epicenter is up to 8 km thick (Uchupi and Austin, 1979).

The surface morphology of the continental margin south

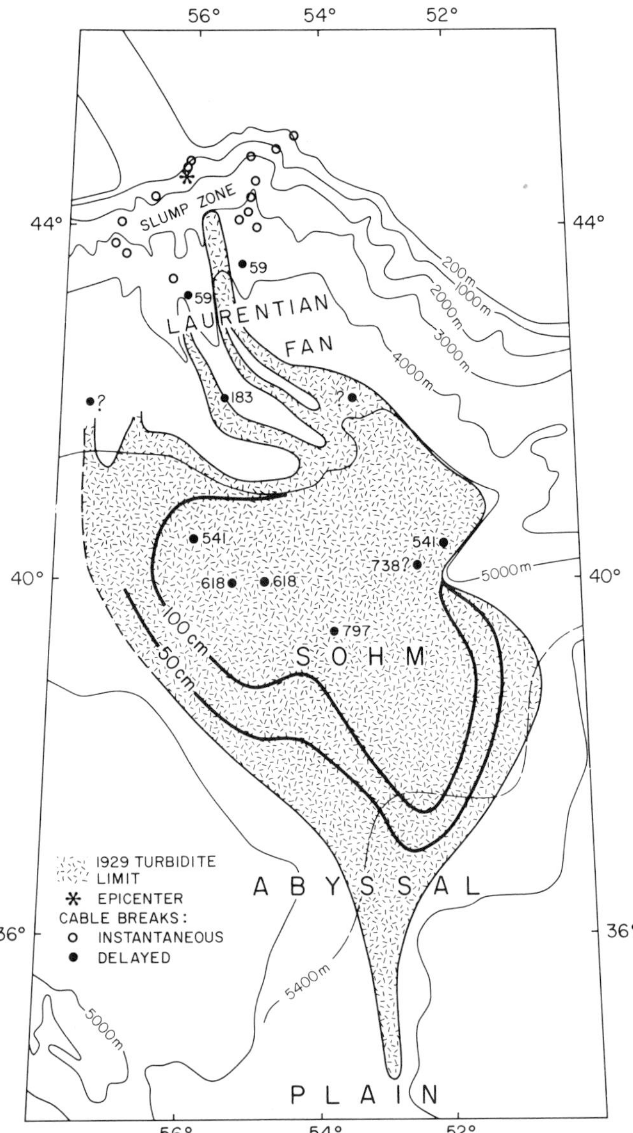

Figure 1. Map showing cable breaks and thickness of turbidite deposit resulting from the 1929 earthquake. Epicenter from Dewey and Gordon (1984). Location and time (minutes after main shock) of cable breaks from Doxsee (1948). Limit and isopachs of turbidite on the Sohm Abyssal Plain from Fruth (1965) and Piper and Aksu (1987).

and west of the Grand Banks is the result of late Tertiary and Quaternary geomorphic processes. The 400-m-deep U-shaped trough of the Laurentian Channel was glacially excavated in the Pleistocene along the locus of a late Tertiary fluvial system (King and MacLean, 1970). It was the major ice outlet for a large area bordering on the Gulf of St. Lawrence, and large quantities of glacial sediment passed through it to the continental slope above the Laurentian Fan (Stow, 1981). The continental slope seaward of the Laurentian Channel shows net late Quaternary erosion, following a phase of late Tertiary–early Quaternary progradation (Piper and Normark, 1982b). In contrast, the slope off St. Pierre Bank shows progradation throughout much of the Quaternary, broken only by minor erosional events (Meagher, 1984). High-resolution seismic profiles from the St. Pierre Slope show till tongues, indicating the presence of grounded ice, extending to water depths of 500 m (Piper and others, 1984b). Pebbly mudstones that outcrop in the upper fan valleys are probably of glaciomarine origin.

The Laurentian Fan is a major deep-water sediment accumulation seaward of the Laurentian Channel. Several tributary valleys on the continental slope lead to two main valleys (Eastern and Western) (Masson and others, 1985) that extend to water depths of about 5,000 m, where they debouch onto the Sohm Abyssal Plain (Fig. 2). Regional bathymetric and seismic data indicate that the valleys off the Grand Banks do not contribute significant amounts of sediment to the fan.

The upper Western Valley, which drains the slope south of the Laurentian Channel, and the Central Valley, which drains the region between the Eastern and Western Valleys, join near the 3,600-m isobath to form the main Western Valley (Fig. 2). Its floor is around 3 to 5 km wide, and the valley becomes sinuous below 4,000 m.

The Eastern Valley floor is typically 25 km wide, with a western levee that stands as much as 950 m above the valley floor. The tributary Grand Banks Valley (Fig. 2) drains the slope between St. Pierre Bank and Whale Bank, forming a major reentrant feature into the upper slope. The Eastern Valley splits into two branches below 4,300 m, the east and the south branches. The Eastern Valley has been the locus of much of the recent research into the effects of the 1929 event (Piper and others, 1985a; Hughes Clarke, 1986). In this chapter, the term "upper valley" refers to Eastern Valley above 2,000 m water depth, "middle valley" to 2,000 to 4,500 m, and "lower valley" to below 4,500 m.

South of 42°N, both the Eastern and Western valleys turn to the ESE (Normark and others, 1983). Beyond this region, which is termed the valley termination zone (Piper and others, 1984a), the valleys are reduced to shallow low depressions whose axis and trend are very poorly defined.

METHODS AND DATA

This paper reviews a large amount of mostly unpublished data collected between 1981 and 1985 in the area of the 1929 earthquake epicenter and turbidite, as well as drawing on earlier published work on the 1929 event and the regional studies of Stow (1981) and Piper and his co-workers (Piper and Normark, 1982b; Normark and others, 1983; Piper and others, 1984a). SeaMARC I mid-range sidescan data were collected around the epicenter of the 1929 earthquake (Piper and others, 1985a) and along much of the length of Eastern Valley (Hughes Clarke, 1986) on cruises in 1983 and 1984. About 20 new piston cores from the epicentral region and Eastern Valley were obtained on

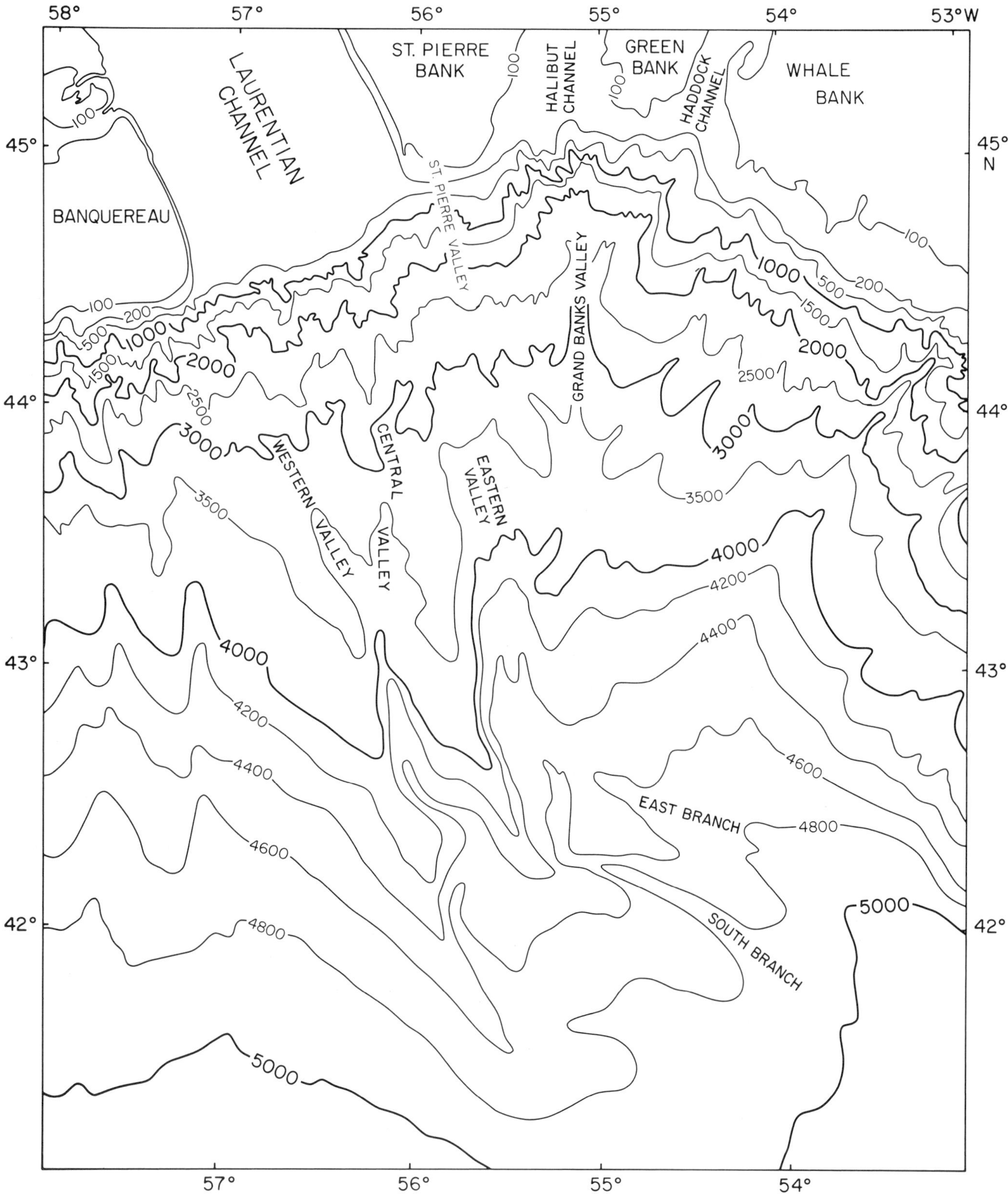

Figure 2. Bathymetry of Laurentian Fan north of 41°N. Compilation based on Canadian Hydrographic Survey charts 801 and 802, updated using C.H.S. chart Nk 21-B and echo-sounding data from Bedford Institute of Oceanography cruises 78-022 and 83-017. Contour intervals are 100 m, 200 m, then at 500 m to 4,000 m, and 200 m to 5,000 m.

three cruises between 1981 and 1985. A bottom camera system was used to obtain seabed photographs in 1985. Both cores and photographs allow detailed interpretation of acoustic facies identified in sidescan images. Four dives were made using the Pisces IV submersible in 1985 in the epicentral region and the uppermost part of Eastern Valley. Bathymetric, 3.5-kHz, and seismic profiles were obtained on most of these cruises.

THE 1929 EARTHQUAKE

Location and magnitude

The Grand Banks earthquake occurred at 2032 Z on November 18, 1929. There was a pronounced aftershock at about 2302 Z. The most recent relocation by Dewey and Gordon (1984) places the epicenter at 44°41.5′N 56°00.4′W, with $M_S = 7.2$.

Effects of the earthquake

The Grand Banks earthquake is known for the remarkable series of cable breaks on the ocean floor south of the epicenter (Doxsee, 1948). These breaks were used by Heezen and Ewing (1952) and Heezen and others (1954) to propose the occurrence of a large turbidity current. The earthquake was also responsible for 27 of the 28 recorded earthquake-related deaths in Canadian history. The summary of the earthquake effects presented here is based on Doxsee (1948). We have also used his numbering of cable breaks throughout this paper.

Severe seismic-shock effects on land were restricted to southeastern Cape Breton Island, where some chimneys fell and highways were blocked by small landslides. Four ships at sea within 250 km of the epicenter reported severe tremors (de Smitt, 1932).

The tsunami associated with the earthquake coincided with a storm high tide, accentuating its effect and making it difficult to accurately reconstruct the behavior of the tsunami itself. Most damage occurred in southwestern Placentia Bay, on the south coast of Newfoundland, particularly in places where the tsunami was funneled and concentrated by local relief. The tsunami was also detected in Halifax, Nova Scotia, and the islands of Bermuda and the Azores.

Twenty-eight breaks were reported in 12 different cables in the area around and to the south of the epicenter (Fig. 1). The time of five of these breaks is unknown, and 12 are recorded as instantaneous at the time of the earthquake. The remaining 11 cables broke at times up to 13 hr after the earthquake, approximately in sequence from north to south. The instantaneous breaks, within 100 km of the epicenter, were interpreted by Heezen and Ewing (1952) to result from sediment failure around the epicenter, whereas the sequential breaks resulted from a turbidity current. The distances between successive cable breaks imply current velocities of at least 67 km/hr on the upper fan, decreasing to 32 km/hr on the northern Sohm Abyssal Plain, 550 km south of the epicenter. (These estimates are reviewed in detail below.)

FAILURE ON THE CONTINENTAL SLOPE

The distribution and style of failure on the continental slope (Fig. 3) around the earthquake epicenter is known from SeaMARC I 5-km-swath side-scan sonar surveys (Piper and others, 1985a). High-resolution seismic reflection profiles, cores (Piper and others, 1984b), bottom photographs, and 27 hr of Pisces IV submersible observation have provided further ground truth.

The present morphology of the continental slope off the Laurentian Channel appears to be largely erosional relief in Pleistocene muds and pebbly mudstones (Fig. 4). The valleys of the Laurentian Fan are floored by gravel and sand, with a veneer of overlying mud. The uppermost part of the slope off St. Pierre Bank is underlain by glacial till, which interfingers downslope with proglacial silt and mud. Both lithofacies are overlain by an organic-rich gas-charged mud of Holocene age (typically 5 to 10 m thick) that thins and becomes more sandy above the 500-m isobath. In deeper waters, Holocene biogenic-rich mud (typically 1 to 2 m thick) overlies Late Pleistocene red mud with silt laminae in intervalley areas.

The uppermost part of the continental slope, in water depths of less than 500 to 800 m, appears smooth and undisturbed. The proglacial silts and overlying Holocene gas-charged muds in this area are truncated by a series of arcuate scarps (Fig. 3b). Both sidescan and submersible observations show large blocks of debris lying downslope from these scarps, thus indicating a probable slump origin for the scarps.

At the mouth of the Laurentian Channel is a series of steep valleys that lead from these slump scarps and coalesce near the 1,800-m isobath to form the Eastern Valley of the Laurentian Fan. Sidescan images and submersible observations suggest that these upper slope valley floors are erosional (Fig. 3c), with step-like terraces cut in stratified sediments. Blocks apparently derived from these scarps have been seen overlying a twisted cable segment near the 1,200-m isobath, indicating that the 1929 event resulted only in local erosion of the stiff mud of the valley floor in this area.

Tributary valleys west of the Eastern Valley have been eroded in older Quaternary sediments. Valley walls show a ridge and gully morphology in sidescan images (Fig. 3a). Submersible observations show that this mesotopography is highly modified by bioturbation. Comparison with the slumped areas suggests that this degree of modification requires more than 55 yr to develop, implying that there was little failure on these tributary valley walls in 1929.

Sediment failure is widespread in the soft surficial sediment of the upper St. Pierre Slope to the east of Eastern Valley, below the arcuate slump scarps. Shallow valleys lead from the slump scars (Fig. 3d) to St. Pierre Valley and Grand Banks Valley, which are larger prominent slope valleys leading to the Eastern Valley. Sidescan images and submersible observations show that

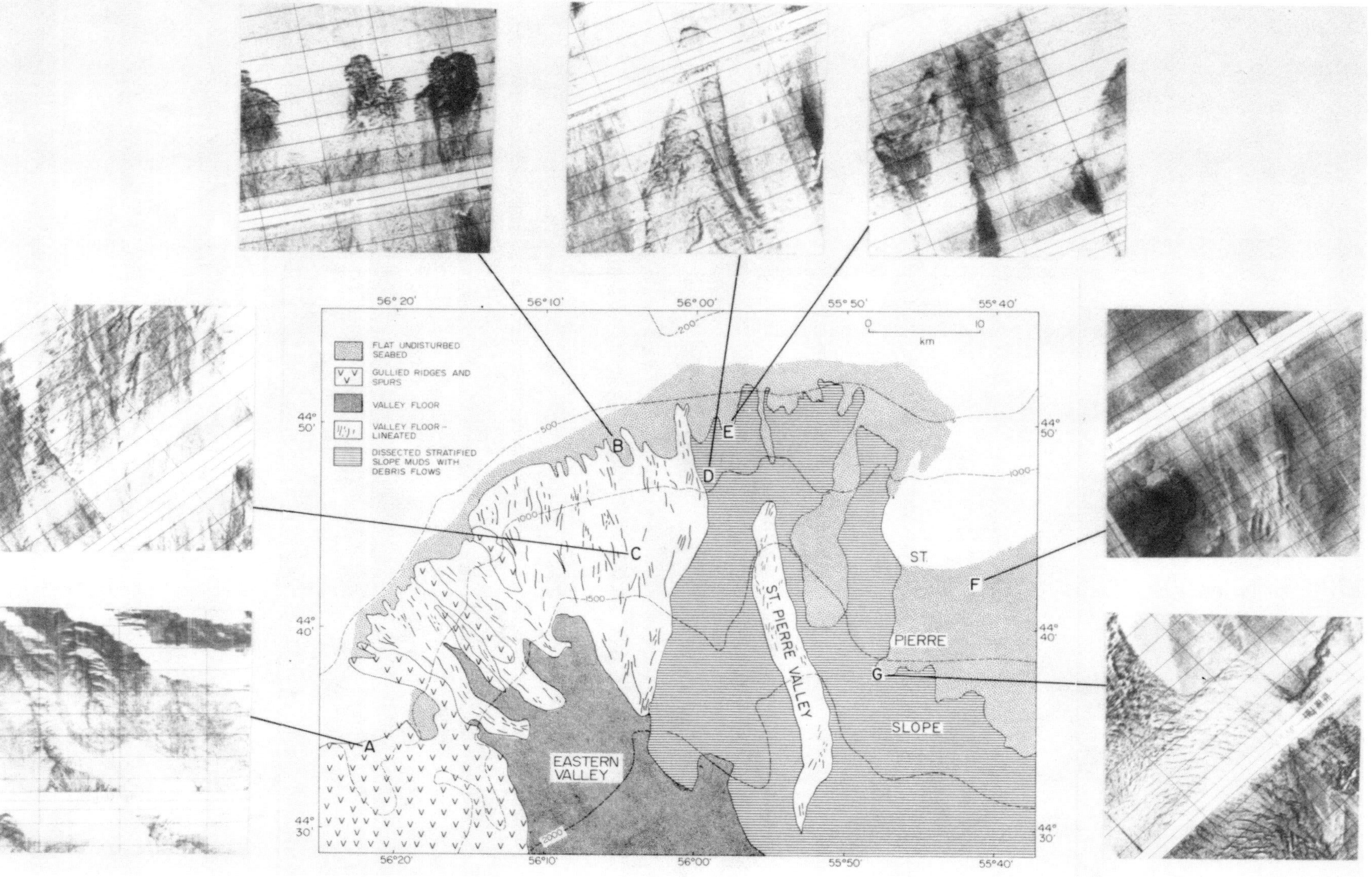

Figure 3. Morphology of the continental slope around the 1929 earthquake epicenter, illustrated by sidescan images. Each image represents 2.8 × 2.8 km^2 of sea floor: a. Ridge and gully topography. b. Arcuate slump scars on upper slope. c. Lineated erosional valley floor with pits and blocks. d. Arcuate slump scar with prominent bedding planes. e. Pockmarks (gas escape craters). f. Creased and streaked seabed. g. "Thumbprint" seabed interpreted as regressive rotational slump.

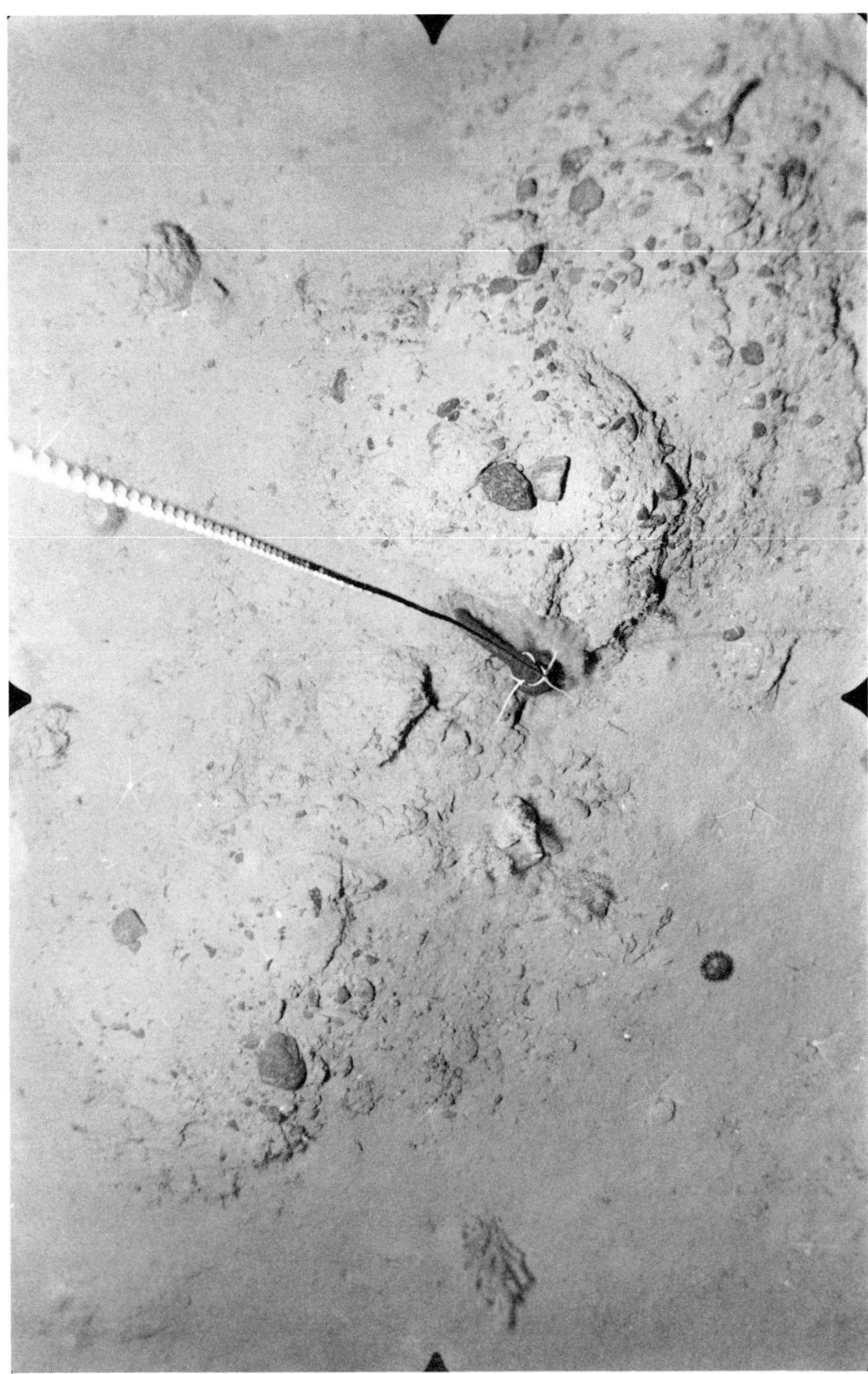

Figure 4. Photograph of Pleistocene pebbly mudstone cropping out on the floor of Eastern Valley in the epicentral region of the 1929 earthquake (1,550 m). Exhumed gravel is visible. Field of view approximately 2 by 3 m.

slumping is common on the steep valley walls. Over large areas of the slope between Eastern Valley and Grand Banks Valley, the uppermost 5 to 15 m of sediment appears to be missing on high-resolution seismic and 3.5-kHz profiles. The seabed surface commonly has a creased or streaked appearance on sidescan images (Fig. 3f), and submersible observations show a gentle undulating topography with a relief of a few meters and wavelength of tens to hundreds of meters. Where surface sediment is present, it frequently contains pockmarks, interpreted to be gas escape craters (Fig. 3e).

On the St. Pierre Slope there are large areas of seabed with prominent surface ridges a few meters high and a spacing of many tens to hundreds of meters (Fig. 3g). They occur downslope from 5 to 50-m-high scarps. Submersible observations show that these ridge crests are composed of steeply dipping stratified muds, with only limited modification by bioturbation. This "thumbprint" seabed is similar to regressive slumps on land such as the South Nation River landslip in eastern Canada (LaRochelle and others, 1970; Bentley and Smalley, 1984). Similar wrinkled topography from the Scotian Slope (Piper and others, 1985b) overlies rotational slumps, which have been sampled in several piston cores. Over much of the St. Pierre Slope, sidescan images show that the seabed is very irregular, with near-surface sediments appearing incoherent in high-resolution seismic profiles. Elsewhere, similar acoustic characteristics result from both debris flows and shallow slumps, and distinguishing between these two interpretations is not possible without more detailed local investigations.

Surface sediment failure thus appears widespread on the St. Pierre Slope. The age of this failure has not been determined directly. Seismic reflection profiles, particularly high-resolution profiles from the upper slope, show that the 30 to 100 m of sediment below the surface failures is undisturbed, suggesting that major failure occurs rarely. The distribution of surface failures seen in regional seismic reflection profiles corresponds approximately to the limit of instantaneous cable breaks in 1929. For these reasons, we consider that the surface failures that we have surveyed on the upper Laurentian Fan and St. Pierre Slope date from the 1929 earthquake, or subsequent movement of sediment made unstable during the earthquake.

SURFACE SEDIMENTS DEPOSITED BY THE 1929 EVENT

There is a distinctive suite of surface sediments in the fan valleys of the Laurentian Fan and on the Sohm Abyssal Plain that were first interpreted as the deposits of the 1929 turbidity current by Heezen and others (1954). Within the fan valleys, cores have recovered coarse sand and gravel. Some gravel at 4,700-m water depth occurs in graded, well-sorted beds as much as 5.8 m thick (Stow, 1981). These coarse sediments have been interpreted as resulting from the 1929 event because of the high velocities required to transport well-sorted gravel to these water depths. The valley floors are underlain by a discrete incoherent seismic facies, suggesting that there has been long-term accretion of coarse clastic sediments. Thus, there would have been a source of coarse sediment available on the valley floor prior to 1929 that could have been incorporated into a sufficiently competent turbidity current.

On the northern Sohm Abyssal Plain, an extensive sheet of fine sand and coarse silt exceeding 1 m thickness (Fruth, 1965) overlies pelagic sediments. A bulk sample of this pelagic material yielded a radiocarbon age of 5,130 ± 130 yr (Ericson and others, 1961), implying a late Holocene age for the thick surficial sand/silt unit. The sand sheet varies in grain size from medium to coarse sand in the valley termination zone of the Laurentian Fan to fine sand on the northern Sohm Abyssal Plain and to coarse silt near the distal limit of the sheet. The sand sheet thins from over a meter of sand to less than 10 cm of turbidite mud near its margin over a distance of less than 100 km (Piper and Aksu, 1987).

The levees of the Laurentian Fan valleys have Holocene hemipelagic sediments, overlying Pleistocene overbank turbidite silt-mud couplets (Stow and Bowen, 1980). There is no trace of a surficial turbidite mud. In the lower part of Eastern Valley, near the 4,500-m isobath, there is widespread erosion of the valley walls and an interdistributary high to about 300 m above the thalweg. These observations suggest that the 1929 turbidity current was only a few hundred meters thick and did not overtop the levees of the main fan valleys.

The volume of the deposits of the 1929 turbidity current has been estimated by Fruth (1965) as 1.4 to 1.6 × 10^{11} m^3 (i.e., about 150 km^3). At least 50 percent of the deposit is fine sand and coarse silt, and less than 10 percent is fine silt and clay.

On the Sohm Abyssal Plain, apart from the surficial sand sheet, the Holocene sequence is entirely muddy ooze. The underlying Pleistocene section is made up of thinner beds of mud, coarse silt, and sand. These are a product of the same activity that created the silt-mud couplets on the levees. Thus it appears that here has not been an event comparable to the 1929 turbidity current during at least the Holocene, which is represented by the muddy ooze.

COARSE SEDIMENTS AND BEDFORMS IN EASTERN VALLEY

Sidescan sonographs reveal a variety of flow-generated features on the Eastern Valley floor developed in sediments with a high acoustic reflectivity. We attribute these features to the passage of the 1929 turbidity current.

Eastern Valley from 1,500 m to 4,300 m is a straight, wide (more than 25 km), largely flat-floored conduit that subsequently divides into two branches (Fig. 5). The upper valley consists of a central raised floor with two marginal thalwegs cut 30 to 40 m below the main floor level (Fig. 6). In the middle reach of Eastern Valley there are two to three valley-floor channels with sections of raised floor in between. These channels appear to grow and die along the valley, but neither cut obliquely across the valley nor meander.

The Grand Banks Valley joins Eastern Valley from the

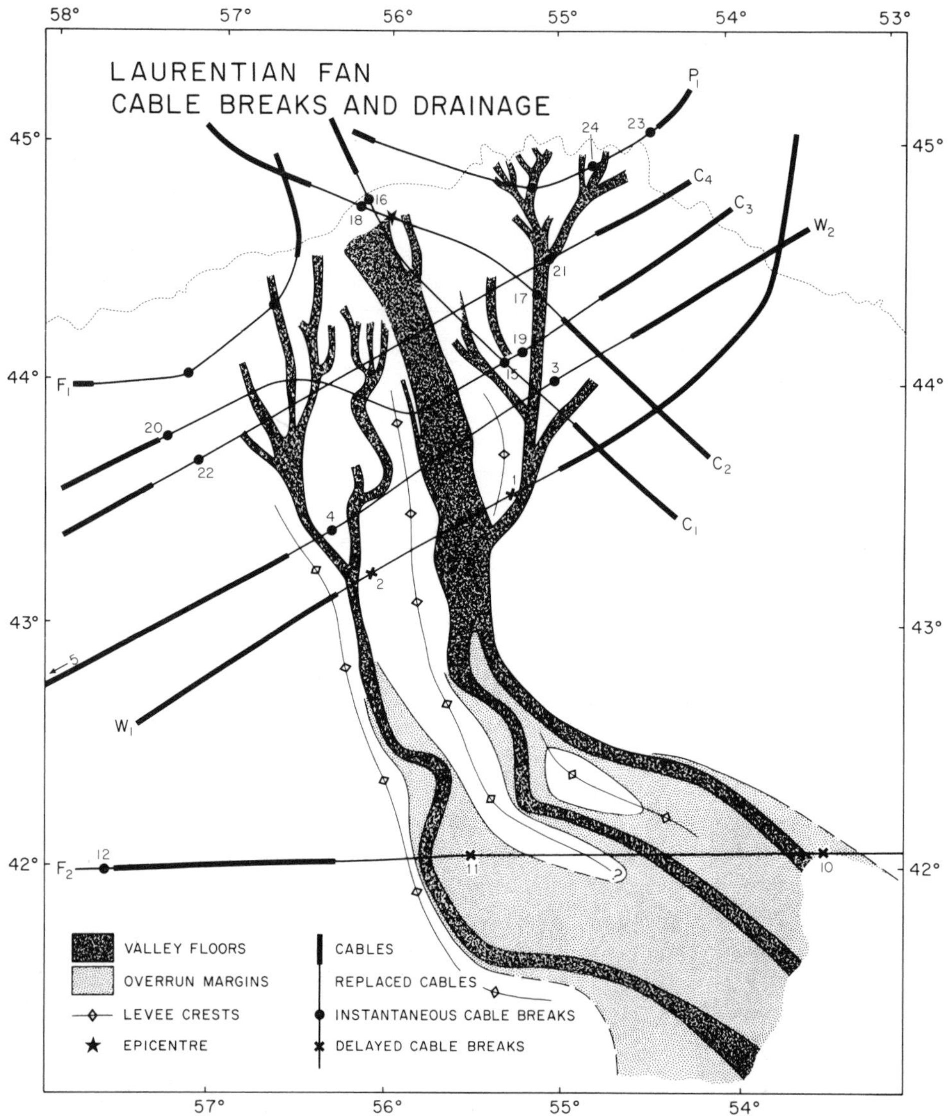

Figure 5. Diagram showing relationship of earthquake epicenter, cable locations, and cable breaks to the valley floors and margins affected by the 1929 event. Cable names and break numbers are in accordance with Doxsee (1948). Channel patterns for the upper and middle fan valleys based on Masson and others (1985).

Figure 6. Bedform succession along Eastern Valley. Nine orthorectified sidescan images are presented for the locations indicated on the valley floor. Each image represents 2.8 × 2.8 km^2 of sea floor: darker returns represent higher amplitude of acoustic backscatter. All images are oriented so that down valley is toward the bottom of the page: Images a, b, and c demonstrate the progressive evolution of the gravel waves; Images c, d, and e demonstrate the increasing importance of sand sheets; Images f and g demonstrate the periodic troughs and swells developed in places in the middle and lower valley; Image h shows the simpler morphology within the south branch of Eastern Valley; Image i shows the transverse longer-wavelength dunes developed in the valley termination zone.

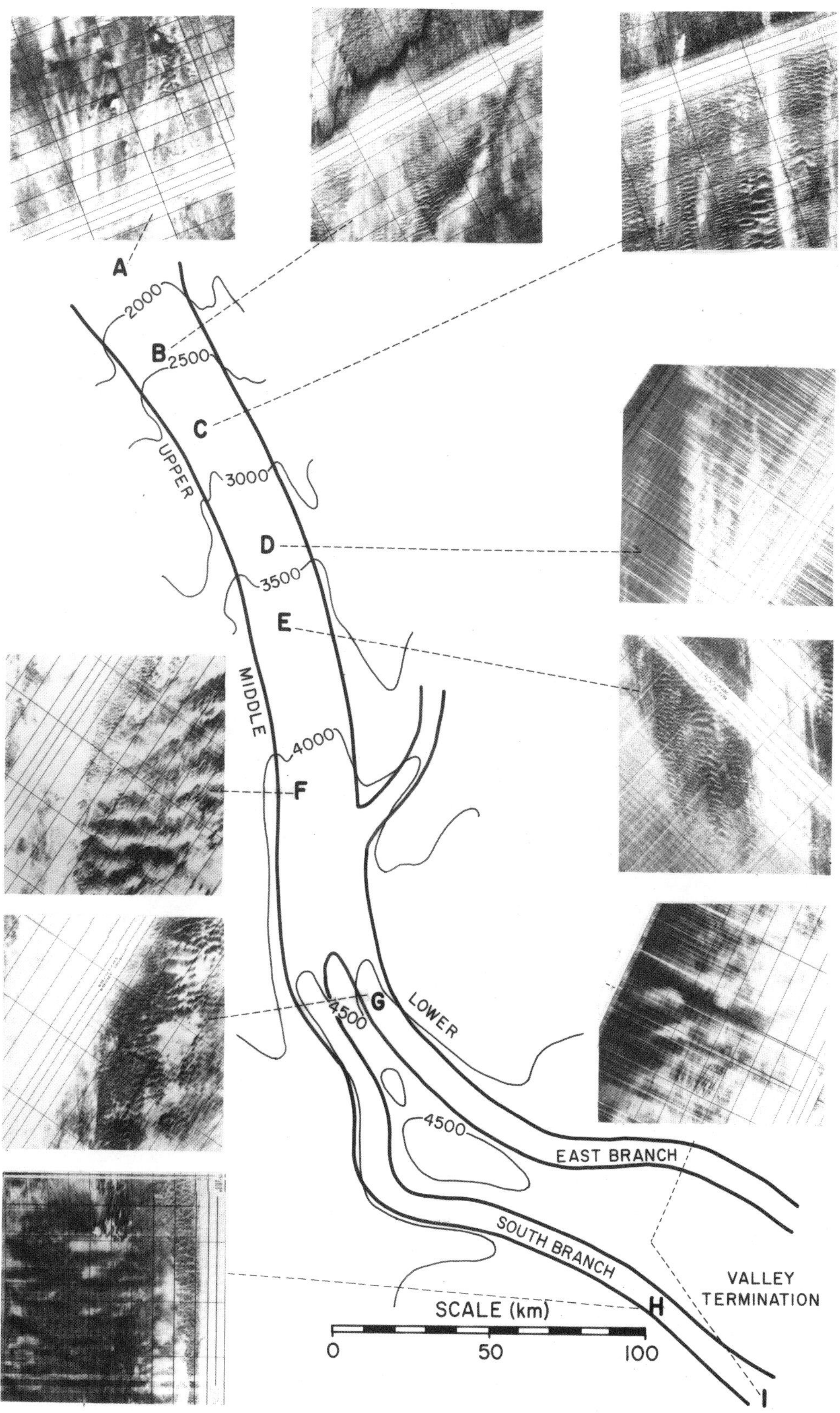

EASTERN VALLEY-BEDFORM SUCCESSION

northeast at 4,000-m water depth. The east and south branches of Eastern Valley diverge at 4,500 m. The east branch continues the trend of the main valley, curling gently to the southeast. It is the broader of the two branches. The south branch, in contrast, becomes sinuous and entrenched, and its overall morphology more closely resembles the Western Valley at this depth.

Bedforms in Eastern Valley are first observed at 1,600 m at the southern limit of the erosional lineated terrain of Piper and others (1985a). Small, poorly defined bedforms with transverse crests and a spacing of 10 to 15 m are seen in strips of higher acoustic reflectivity (darker areas in Fig. 6). By 2,300-m depth, there are well-defined higher-reflectivity regions (granule to cobble gravel) of asymmetric dunes with sinuous elongate crests transverse to inferred flow and wavelengths of 40 to 50 m that occur interspersed with smooth regions of lower-reflectivity sheets of coarse sand (Fig. 6b). At 3,000 m, the asymmetric dunes (hereafter referred to as gravel waves) are the dominant feature seen on the valley floor. They make up a wavefield broken by elongate (5 to 25 km long), thin (30 to 100 m wide) sand ribbons that are oriented downvalley (Fig. 6c). The gravel waves are best developed on elevated regions of the valley floor, whereas the sand sheets and ribbons are situated in shallow elongate depressions. The sand overlies the gravel waves and hence appears to be a late-stage deposit postdating the period of wave growth. The sand ribbons coalesce farther down the valley to form wider (600 m) composite sand sheets (Fig. 6d). By 3,500-m water depth, the smooth sand cover predominates over the gravel wavefields (Fig. 6e). From 4,000-m water depth through the east branch, to the valley termination zone, the valley is floored by sand (ribbons, sheets, and streaks) and gravel waves (wavelength 50 to 80 m), which are locally superimposed on large swells and troughs spaced 400 to 600 m apart (Fig. 6f and g). In the south branch, however, a simple axial sand belt is developed with gravel waves on either flank. Finally, in the valley termination zone, where the valley walls cease to confine larger flows, the gravel waves coexist with (Fig. 6h) and eventually are replaced by large (300-m-wavelength) transverse asymmetrical bedforms (Fig. 6i).

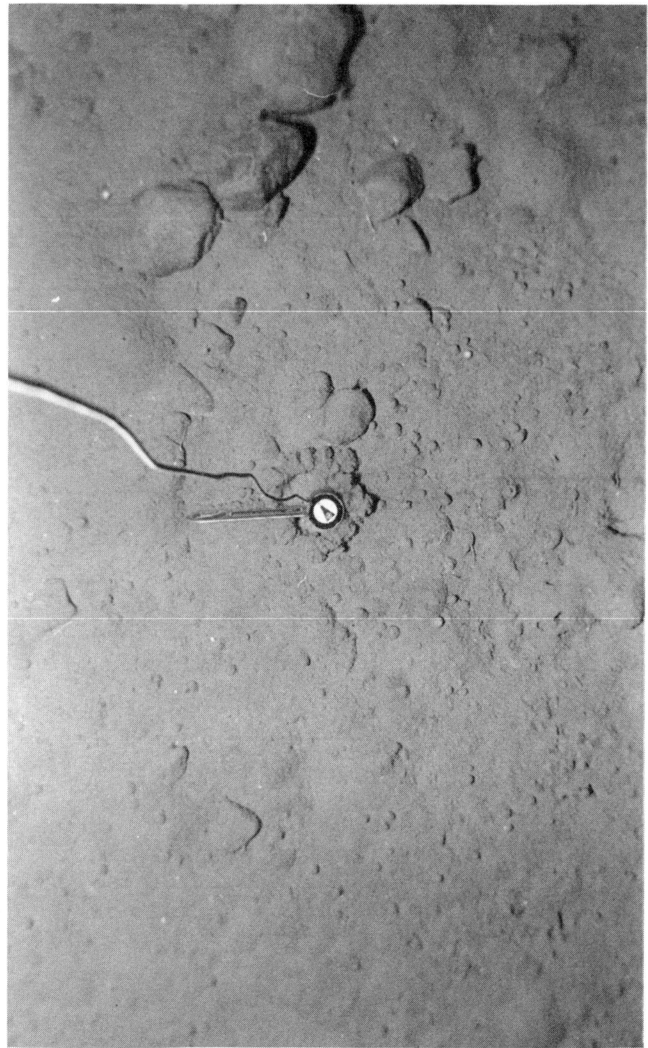

Figure 7. Photograph showing gravel pavement at 4,300 m in a gravel wave field of Eastern Valley. Cobble gravel is visible at the surface, partially covered by a thin mud drape. Field of view is 1.5 by 2.5 m.

These features on the valley floor are built of coarse clastic sediment ranging in size from coarse sand to boulders. The gravel waves in the upper and middle valley are made up of poorly sorted cobble to granule gravel. Cores have not penetrated more than 130 cm into such coarse sediment, so that there is no information on internal structure of the gravel waves. A thin mud drape mantles the surface. Our few short cores suggest that the surface of the gravel waves is variably draped with a normally graded sand–gravel sequence. In one core, a reverse-graded gravel-sand surface layer (37 cm) overlies the normally graded sequence. From bottom photography, the surface of the waves is either smooth or appears to be a coarse lag (Fig. 7). Probing with the mechanical arm of the submersible demonstrated that the bedforms are constructed of gravel even where draped by finer sediment. The boulder lags appear on the stoss sides of the waves, whereas the lee slopes are more commonly thickly draped with sand or mud. Similar gravel bedforms have now been recognized on the floor of Nice canyon (Malinverno and others, this volume). Most sidescan surveys of deep-sea channels, however, have not found similar wave fields.

The ribbons, in contrast, are made up of well-sorted coarse sand and granule gravel that is normally graded in the top 40 cm. In the lower Eastern Valley, cores up to 580 cm long reveal a normally graded sequence from well-sorted sand to pebble-sized (15-mm) clasts in the top 100 to 150 cm, overlying a complex alternation of variably sorted pebbly sand and gravel. Imbrication or good stratification within the coarse facies cannot be demonstrated due to disturbance during coring.

Pleistocene supply of glacial sediment to the upper slope (Stow, 1981) yielded a source of lithologically heterogeneous and poorly sorted clasts that has been incorporated into the valley floor facies. There is a downslope trend in Eastern Valley of increased sorting and decreasing grain size in the valley floor facies. The grain-size distribution is a result of varying degrees of sorting of an initially poorly-sorted, glacially derived population. The sand sheets represent a higher degree of maturity and may result from late-stage winnowing of the surface of the gravel waves.

The gravel waves imply bedload transport involving traction processes. The morphology of these gravel waves closely resembles that of the giant Scabland ripples of Washington (Bretz and others, 1956). Estimated flow depths and velocities during the formation of the Scabland ripples are 50 to 70 m and 10 to 20 m/s, with corresponding Froude numbers of 0.6 to 0.9 (Baker, 1973). By analogy with fluvial asymmetric dunes and the Scabland ripples, the gravel waves in Eastern Valley are considered to be bedforms deposited under subcritical flow conditions (Froude number <1).

COMPOSITION AND VOLUME OF FAILED SEDIMENT

The total area within which failure occurred on the continental slope in 1929 is about 20×10^9 m^2. Within this region, however, there are large areas of undisturbed surficial sediment, and it appears that only half of this area was involved in surface failure. Seismic reflection profiles on the slope off St. Pierre Bank indicate that the average thickness of missing sediment is about 5 m. If this thickness is applicable to the entire slope area, then the total volume of sediment lost would be about 100×10^9 m^3 (Piper and Aksu, 1987). On the St. Pierre Slope, slump scars cut mostly proglacial muddy sediment with an estimated 30 percent gravel, sand, and coarse silt.

The volume of gravel, sand, and coarse silt on the lower Laurentian Fan and Sohm Abyssal Plain is conservatively estimated at 175×10^9 m^3 (including sediment porosity) (Piper and Aksu, 1987). Less than 20 percent of this volume can be accounted for by the coarse sediment derived from muddy areas of the continental slope. The undisturbed nature of the entire uppermost slope means that there was no massive loss of till from the shelf break, although some distal till tongues may have failed on the slope above the Laurentian Fan. Therefore, it is necessary to postulate a large reservoir of sand in the uppermost parts of the fan valleys. This sand may have accumulated proglacially during middle Wisconsinan time when the major ice flow through the Laurentian Channel was grounded on the upper slope. During the 1929 event, a large volume of this coarse sediment was removed, perhaps by seismically induced liquefaction failure or in part by turbidity current erosion during the earthquake. Removal of 30 to 40 m of coarse sediment from the upper fan valleys would account for the observed volume of the 1929 turbidite (Piper and Aksu, 1987).

The observed volume of mud (fine silt and clay) in the 1929 turbidite is estimated as 10×10^9 m^3. This is several times less than the volume observed to be missing from the continental slope. This missing mud was probably transported southwest by deep-ocean circulation and settled out beyond the Laurentian Fan and Sohm Abyssal Plain.

ANALYSIS OF THE CABLE BREAKS

Previous work

Many authors have attempted to use the sequential cable breaks to estimate the velocity of the 1929 turbidity current. Heezen and Ewing (1952) assumed that the current advanced on a broad front and was initiated close to the epicenter. Most subsequent authors (Menard, 1964; Emery and others, 1970; Uchupi and Austin, 1979) assumed that the flow was channelized and did not make estimates based solely on the timing of the first delayed break. Differences between the estimates of various authors depend on assumptions made about the path of the current.

Significance of reported breaks

The time of a cable failure is recorded automatically by shore stations, and the position can be located in terms of the distance from the shore station (Keith, 1930). Subsequent breaks between the original break and shore station can be detected only as a drop in resistivity (Kullenberg, 1954). Thus, the time of the first break will always be known, but the second and subsequent ones may not be detected. Further breaks or damaged zones may be recognized during cable replacement.

Only two breaks were reported by de Smitt (1932) for most cables, delineating the east and west limit of breakage. The style of breaking, however, was a wide zone of parting or damage to the cables between the reported end breaks. Cables were buried but not broken on either side of the central tract where the cables had been broken and removed (Heezen and Hollister, 1971). As many as ten breaks over a distance of 140 mi went unreported for one cable. The entire section was replaced because the remaining cable was so badly damaged (Higgins, 1930).

A third break is reported for four cables. Two of these (breaks 5 and 12) occur west of the Laurentian Fan, separated from the main zone of breakage by unaffected cable (Higgins, 1930). These breaks are thus attributed to a synchronous but geographically separate event farther west on the Scotian Slope. This event must have been only minor, as cable W_1, which lies between these breaks, was unaffected.

The instantaneous breaks and slope failure

The instantaneous breaks occur on the walls of valleys and other areas of steep slopes. They may thus be attributed to slope failure resulting from the initial shock wave. The breaks delineate the limit of synchronous failure. They are not a result of passage of turbidity current, but rather are due to dislocations of the sediment surface producing an immediate breaking stress on the cable.

The pattern of instantaneous cable breaks indicates that an area of roughly 100 km in diameter around the epicenter experienced simultaneous local mass wasting. Thus we cannot identify a point source for the turbidity current against which the first delayed breaks (numbers 1 and 2) can be referenced. Estimates of flow velocity based on travel times from the epicenter to the first delayed breaks (Heezen and Ewing, 1952) are thus unreliable.

Delayed breaks and the turbidity current

Subsequent breaks may be attributed to downslope passage of mobile sediment in a turbidity current. Delayed breaks 1 and 2 are positioned at the confluence of the Western and Central Valleys and slightly up the Grand Banks Valley from the zone of confluence with the Eastern Valley (Fig. 5). Thus the passage of the turbidity current head down the Western or Central and Grand Banks Valleys was apparently faster than down the Eastern Valley (which lies in the middle of the broken cable segments).

The next delayed break, number 11, is situated on the margin of the Western Valley, delineating the western limit of cable damage. The time of the subsequent break delineating the eastern limit (number 10 in the east branch of Eastern Valley) is unknown. This again suggests a faster passage down the Western Valley than the Eastern Valley (the Grand Banks and Central Valleys are no longer discrete features at this latitude). A minimum velocity of 19 m/s can be calculated for the head of the turbidity current down the Western Valley between breaks 2 and 11 (measuring the straight-line distance between the two breaks). If the path of the current followed the axis of Western Valley, then a velocity of 20 m/s is obtained. These estimates are almost identical to those of Menard (1964) and Uchupi and Austin (1979), and represent the maximum demonstrable velocity for the turbidity current.

FLOW CONDITIONS IN THE TURBIDITY CURRENT

The character of turbidity current flow in Eastern Valley can be determined in some detail, because we know the floor slope, the valley cross section, and an upper limit on the velocity from cable breaks in the Western Valley. We also have limited information on the sediment transported, and a variety of observations to constrain the flow thickness. All of these parameters vary down valley. We have focused, therefore, on the middle reach of Eastern Valley at 43°N in about 4,400-m water depth, as it was sufficiently far away from the source region for the flow to be approaching a steady-state condition, yet was still constrained within the valley. Furthermore, this location is below the confluence with the Grand Banks Valley, so that its contribution is included. At this location, the slope is 5×10^{-3}.

Previous estimates in other deposits of current velocity from valley-floor facies (e.g., Komar, 1969; Walker, 1975; Winn and Dott, 1977, 1979) have used the estimated threshold velocity for bedload transport of the largest observed clast and the velocity required to transport an inferred suspended load.

In Eastern Valley, the observations on maximum clast size are limited to sparse bottom-camera coverage of the top of the clastic sequence. Submersible observations are available at 2,000-m water depth, where 2-m boulders have been observed on the valley floor. In the middle valley, around 4,000 m, a maximum size of 50 cm has been observed in bottom photographs; and in the lower valley, there do not appear to be surface boulders, as the floor is covered by a thick, surface, accretionary, graded layer of gravel to mud. In the upper valley, it is not known whether the boulders have been merely exhumed in place from outcrop (as has been observed from submersible), or whether they have been truly transported.

The thick, graded gravel in the lower valley (Stow, 1981), with a median grain size of about 15 mm at the bottom of the core, is interpreted as a suspension load deposit. Using the suspension criteria of Bagnold (1966) and Bridge (1981), this corresponds to flow velocities of the order of 13 to 16 m/s.

Estimates of bedload and suspended-load threshold criteria for such coarse clastics, however, are poorly constrained by empirical data. Bedload thresholds are hard to define for natural poorly sorted deposits and have been demonstrated to also be a function of hydraulic radius (Carling, 1983) and sorting (relative protrusion, Hammond and others, 1984), as well as the grain size, shape, and density. Empirical calibration of suspension criteria has not been made for coarse clasts under conditions of high sediment concentration. For these reasons, we do not place much reliance on estimates of flow velocity from grain size.

Flow thickness can be physically constrained in the middle reach of Eastern Valley to between 270 and 420 m, on the basis of erosional lineations and the surface distribution of coarse sediment as interpreted from sidescan sonographs and 12-kHz profiles. We can also use the presence of the gravel waves to constrain the flow thickness by two methods. First, the gravel waves are interpreted as subcritical bedforms, the internal Froude number of the flow must have been less than unity (at least for the period of bedform growth). Second, Baker (1973, Fig. 51) was able to demonstrate an empirical relationship between the depth (D)–slope (S) product and the ripple chords (wavelengths) of the giant Scabland ripples:

$$\text{chord} = 263.7 \cdot (D \cdot S)^{0.66}$$

If allowance is made for the smaller density contrast, the following empirical relationship may be used for the gravel waves:

$$\text{chord} = 263.7 \cdot [(\frac{\rho_t - \rho}{\rho_t}) \cdot D \cdot S]^{0.66} \qquad (1)$$

To use these relationships, an estimate of flow density is necessary. Kuenen (1952) suggested a density of 1,600 kg/m^3 for the 1929 turbidity current. More recent work, however, suggests that this density is far too great for natural turbidity currents. Bagnold (1954) suggested that grain-to-grain interaction dominates when the volume concentration of the sediment exceeds 9 percent (1,173 kg/m^3 for a grain density of 2,650 kg/m^3). Densities estimated for sandy turbidity currents range from 1 to 7 percent (Bowen and others, 1984, Fig. 12), in contrast with values as low as 0.1 percent for muddy turbidity currents (Stow and Bowen, 1980). The 1929 current did mobilize cobbles and boulders, but they were probably moved as bedload. In the lower valley, a mean grain size of 15 mm is found at the base of graded cores, implying deposition from suspended load. Thus, this turbidity current was exceptionally coarse, and for our calculations we have taken a relatively high value of 6 percent (1,124 kg/m^3) as an estimate for the density.

Using the first method, we estimate that at the upper limit of velocity (19 m/s), the flow depth must have exceeded 397 m (and would be 186 m for 13 m/s). Using the second method, for observed chords of 50 to 80 m, the calculated flow thicknesses are 186 to 379 m. This flow thickness would correspond to the body of the turbidity current during the period of bedform growth.

Thus, a variety of methods all indicate maximum flow thicknesses of the order of 300 to 400 m. In the analysis below, we therefore examine two cases: one for a thickness of 300 m, the other 400 m. We apply standard equations of flow (Komar, 1969) to examine the flow conditions of the 1929 turbidity current in the middle reach of Eastern Valley.

Chezy-type equation for steady-state flow can be derived from the Bagnold equation (Komar, 1969; Bagnold, 1962):

$$u = \left[\frac{(\rho_t - \rho)}{(\rho_t)} \cdot \frac{g \cdot h_b \cdot S}{(1 + \alpha)\, C_f} \right]^{1/2} \quad (2)$$

and estimates of head velocity can be made using Middleton's (1966) equation:

$$v = 0.75 \left[\frac{(\rho_t - \rho)}{(\rho_t)} \cdot g \cdot h_h \right]^{1/2} \quad (3)$$

where u = body velocity, v = head velocity, ρ_t = current density, ρ = seawater density (1027 kg/m^{-3}), h_b = body thickness, h_h = head thickness, S = slope (5×10^{-3}), α = ratio of interfacial shear to bed shear (0.43), and C_f = drag coefficient.

In previous numerical modeling of flows in turbidite channels, the appropriate values for bed roughness have been a matter of controversy (Komar, 1975; Hand, 1974). This was in part because the contribution of isolated roughness elements—like megaflutes (Lonsdale and Hollister, 1979) and other erosional features (McGregor and others, 1982)—to overall channel drag was hard to quantify. The gravel wave field, however, covers almost the whole of the upper valley floor and rarely less than half of the floor in the middle and lower valley.

If we take the gravel waves as bed roughness elements of the order of 5 m amplitude, using the Von Karman–Prandtl equation for a rough bed, the drag coefficient is between 4.5 and 4.1×10^{-3} for flow thicknesses of 300 to 400 m. We, therefore, use a value of 4.3×10^{-3}, which is higher than has been used in most other studies in the deep sea (e.g., 3.5×10^{-3}, for the Maury Channel, Lonsdale and others, 1981), reflecting the additional presence of the gravel waves.

The flow velocity of 19 m/s obtained from the Western Valley is taken as an upper limit for the head velocity. Using estimates of flow thickness of 300 and 400 m (which may be either a body or a head thickness) in equations 2 and 3, we obtain the following velocity estimates: for head thicknesses of 300 and 400 m, head velocities of 12.0 and 13.8 m/s; for body thicknesses of 300 and 400 m, body velocities of 14.4 and 16.6 m/s; a corresponding subcritical Froude number of 0.88.

Estimates of discharge and duration of the flow depend on body velocity. The above estimates of body velocity give discharges between 1.1 and 1.7×10^8 m^3/s. If we assume that two-thirds of the volume of sediment in the turbidite (with 40 percent porosity) on the Sohm Abyssal Plain passed through the middle reach of Eastern Valley, a flow concentration of 6 percent by volume implies the minimum duration of turbidity current flow was 3 to 2 hr. The actual flow duration probably exceeded this estimate, however, because both flow velocity and concentration would have decreased in the later stages of the flow.

DISCUSSION

Recurrence interval

Earthquakes of magnitude sufficient to cause widespread sediment failure occur somewhere on the eastern Canadian continental margin every few hundred years (Basham and Adams, 1983). For example, there were large earthquakes in southern Baffin Bay in 1931 and off central Labrador in the nineteenth century. Piper and others (1985b) recognized widespread slumping on the central Scotian Slope caused by a large earthquake about 12,000 yr ago. There is, however, no evidence of a very large earthquake affecting the upper Laurentian Fan for many thousands of years before the 1929 event (Piper and Normark, 1982a).

No other thick, widespread, sandy turbidite is known in the upper 10 m of sediment on the eastern Canadian continental margin. This implies that the 1929 turbidite was an unusual event resulting from the fortuitous occurrence of a large earthquake near a large glacial depocenter on the steep upper continental slope. Furthermore, the peculiar topography of the Laurentian Fan, with a large number of upper slope valleys focused into two main fan valleys with very high relict levees, allowed a confined high-density flow to maintain itself across the slope and rise. Thus, a large volume of sediment was transported to the deep sea, rather than being redeposited on the slope and upper rise.

Fan and abyssal plain growth

The 1929 event has been taken as a type example of a dense, sandy turbidity current. Its thick, sandy deposit, with distinctive proximal to distal changes in grain size and bedforms, would have a high preservation potential in the ancient record, but for its location on oceanic crust. In contrast, lower-density, thicker, muddy turbidity currents are almost certainly a more important mechanism for the long-term accretion of the fan. Sandy turbidity currents of the magnitude of the 1929 event are too infrequent to make a major volumetric contribution. Such coarse units might be thought important to the growth of abyssal plains, because they form mappable marker horizons. Volumetrically, however, they may be overemphasized because they are more easily correlatable than silt or mud units.

During Wisconsinan time, bankfull turbidity currents capable of moving silt and clay appear to be the more common type of event on the Laurentian Fan (Stow and Bowen, 1980). Triggering of these events was much more frequent; they were a response to rapid deposition on the upper slope as a result of glacially controlled outwash and iceberg melting. Under these conditions, failure may have been induced by small earthquakes, large dropstones, storm waves, or even whale impacts (Ryan, 1982).

Although turbidites like the 1929 deposit are volumetrically of minor significance on the fan, such large turbidity currents are of great importance in controlling its growth pattern. The 1929 event flushed out the fan valleys: without such a process, the valleys would tend to aggrade in the manner seen on many large passive-margin fans (such as the Bengal Fan; Emmel and Curray, 1984), resulting in channel migration by avulsion.

The erosive power of the 1929 turbidity current was related to its large load of sand and gravel. In most passive-margin deep-sea fans, gravel and coarse sand are not available in the source sediment for turbidity currents: such coarse clastics are trapped in terrestrial deposits in the hinterland. The coarse sediment supply to the 1929 turbidity current is the result of a glacial source. Fan systems supplied by fan deltas in orogenically active areas also have a significant source of coarse sediments, so that it is in these environments that ancient analogues of the 1929 event might be sought.

The volume of the 1929 turbidite is similar to that of "seismoturbidites," which Mutti and others (1984) describe as occurring intercalated within normal flysch sequences. The 1929 turbidite differs from many typical carbonate-rich seismoturbidites in lacking a basal breccia in most areas. Furthermore, the 1929 turbidites do show systematic proximal to distal changes in grain size and in sedimentary structures, as far as they can be inferred from surficial bedforms. The very well developed preexisting valley system on the Laurentian Fan may have played an important role in focusing and confining the turbidity current, compared with many seismoturbidites that are inferred to have traveled more as sheet flows across slopes with lesser bathymetric relief. The cable breaks provide evidence for sediment failures over a wide source area, as is inferred for other seismoturbidites. The sediment budget calculations, however, indicate that most of the sediment in the 1929 turbidite was derived from reservoirs of coarse sediment in the heads of the fan valleys, rather than from regional failure of continental slope sediments. The widely observed down-fan decrease in the frequency of turbidity currents (Piper and Normark, 1983, Figs. 8 and 9) suggests that fan valleys may be an important reservoir of coarse sediment to many fans. Flushing of fan valleys may thus provide a mechanism for supplying clastic sediment to extremely large turbidites.

CONCLUSIONS

The 1929 sediment failure and resulting turbidity current was an impressive convulsive event. Transatlantic communications were disrupted for six months. There was sediment failure over an area of 20,000 km^2 on the continental slope. As much as 200 km^3 of sediment was transported in the turbidity current, flowing with thicknesses of a few hundred meters and velocities up to 19 m/s. Graded beds of gravel were deposited more than 300 km from the shelf break, and a bed of sand more than 1 m thick was deposited over an area of 150,000 km^2 on the Sohm Abyssal Plain. Where the flow was confined within the fan valleys, it molded the coarse seabed sediments into a series of bedforms. Gravel wave fields in the upper valley became increasingly masked by sand sheets and ribbons downvalley, and long-wavelength, symmetrical, sandy bedforms predominate in the lower valley.

Although the 1929 turbidite is not of major volumetric importance on the fan, the behavior of the 1929 turbidity current was important for the overall growth pattern of the fan. Flushing of sediment from the upper valley system, such as occurred in the 1929 event, prevents aggradation of the valley system, a process which occurs in many large passive-margin deep-sea fans and which eventually leads to channel avulsion.

The 1929 event was an unusual occurrence, resulting from the fortuitous coincidence of a large earthquake at the site of a large accumulation of coarse Upper Pleistocene sediment on the steep continental slope. Many of the distinctive features of the turbidite on the Laurentian Fan are the result of the large amount of gravel available to the turbidity current. Similar turbidites in the ancient geologic record occur not only in glaciated terrains, but also in orogenic belts where coarse sediment is delivered to the coastline by fan deltas.

REFERENCES CITED

Bagnold, R. A., 1954, Experiments on gravity free dispersion of large solid spheres in a Newtonian fluid under stress: Proceedings of the Royal Society of London, A, v. 225, p. 49–63.

—— , 1962, Auto suspension of transported sediments; Turbidity currents: Proceedings of the Royal Society of London, A, v. 265, p. 315–319.

—— , 1966, An approach to the sediment transport problem from general physics: U.S. Geological Survey Professional Paper 422–I, 37 p.

Baker, V. R., 1973, Paleohydrology and sedimentology of Lake Missoula flooding in eastern Washington: Geological Society of America Special Paper 144, 73 p.

Basham, P., and Adams, J., 1983, Earthquakes on the continental margin of eastern Canada; Need future large events be confined to the locations of large historical events?, *in* The 1886 Charleston earthquake and its implication for today: U.S. Geological Survey Open-File Report 83–842, p. 456–467.

Bentley, S. P., and Smalley, I. J., 1984, Landslips in sensitive clays, *in* Brunsden, D., and Prior, D. B., eds., Slope instability: New York, J. Wiley, p. 457–490.

Bowen, A. J., Normark, W. R., and Piper, D.J.W., 1984, Modelling of turbidity currents on Navy submarine fan, California continental borderland: Sedimentology, v. 31, p. 169–185.

Bretz, J. H., Smith, H.T.U., and Neff, G. E., 1956, Channeled Scablands of Washington; New data and interpretations: Geological Society of America Bulletin, v. 67, p. 957–1049.

Bridge, J. S., 1981, Hydraulic interpretation of grain size distributions using a physical model for bedload transport: Journal of Sedimentary Petrology, v. 51, p. 1109–1124.

Carling, P. A., 1983, Threshold of coarse sediment transport in broad and narrow natural streams: Earth Surface Processes and Landforms, v. 8, p. 1–18.

de Smitt, V. P., 1932, Earthquakes in the North Atlantic as related to submarine cables: EOS Transactions of the EOS American Geophysical Union, p. 103–109.

Dewey, J. W., and Gordon, D. W., 1984, Map showing recomputed hypocenters of earthquakes in the eastern and central United States and adjacent Canada: U.S. Geological Survey Miscellaneous Field Studies Map MF–1699, scale 1:2,500,000.

Doxsee, W. W., 1948, The Grand Banks earthquake of November 18, 1929: Ottawa, Publications of the Dominion Observatory, v. 7, no. 7, p. 323–335.

Emery, K. O., Uchupi, E., Phillips, J. D., Bowin, C. O., Bunce, E. T., and Knott, S. T., 1970, Continental rise off eastern North America: American Association of Petroleum Geologists Bulletin, v. 54, p. 44–108.

Emmel, F. J., and Curray, J. R., 1984, The Bengal submarine fan, northeastern Indian Ocean: Geo-Marine Letters, v. 3, p. 119–124.

Ericson, D. B., Ewing, M., Wollin, G., and Heezen, B. C., 1961, Atlantic deep-sea sediment cores: Geological Society of America Bulletin, v. 72, p. 193–286.

Fruth, L. S., 1965, The 1929 Grand Banks turbidite and the sediments of the Sohm Abyssal Plain [M.Sc. thesis]: New York, Columbia University, 258 p.

Hammond, F.D.C., Heathershaw, A. D., and Langhorne, D. N., 1984, A comparison between Shields threshold criterion and the movement of loosely packed gravel in a tidal channel: Sedimentology, v. 31, p. 51–62.

Hand, B. M., 1974, Supercritical flow in density currents: Journal of Sedimentary Petrology, v. 44, p. 637–648.

Haworth, R. T., and Keen, C. E., 1979, The Canadian Atlantic margin; A passive margin encompassing an active past: Tectonophysics, v. 59, p. 83.

Heezen, B. C., and Ewing, W. M., 1952, Turbidity currents and submarine slumps, and the 1929 Grand Banks earthquake: American Journal of Science, v. 250, p. 849–873.

Heezen, B. C., and Hollister, C. D., 1971, The face of the deep: New York, Oxford University Press, 659 p.

Heezen, B. C., Ericson, D. B., and Ewing, W. M., 1954, Further evidence for a turbidity current following the 1929 Grand Banks earthquake: Deep-Sea Research, v. 1, p. 193–202.

Higgins, 1930, in discussion following paper by Hodgson and Doxsee: Eastern Section of the Seismological Society of America, Proceedings of the 1930 Meeting, Washington, D.C., p. 79–81.

Hughes Clarke, J. E., 1986, Surficial morphology of Eastern Valley, Laurentian Fan: Geological Survey of Canada Open File 1425, 22 p.

Keith, A., 1930, The Grand Banks earthquake: Eastern Section of the Seismological Society of America, Supplement to the Proceedings of the 1930 Meeting, Washington, D.C., p. 1–9.

King, L. H., and MacLean, B., 1970, Origin of the outer part of the Laurentian Channel: Canadian Journal of Earth Sciences, v. 7, p. 1470–1484.

Komar, P. D., 1969, The channelised flow of turbidity currents with application to Monterey Deep-sea Fan Channel: Journal of Geophysical Research, v. 78, no. 18, p. 4544–4558.

—— , 1975, Supercritical flow in density current; A discussion: Journal of Sedimentary Petrology, v. 45, p. 747–749.

Kuenen, P. H., 1952, Estimated size of the Grand Banks turbidity current: American Journal of Science, v. 250, p. 874–884.

Kullenberg, B., 1954, Remarks on the Grand Banks turbidity current: Deep-Sea Research, v. 1, p. 203–210.

LaRochelle, P., Changnon, J. Y., and Lefebvre, G., 1970, Regional geology and landslides in marine clay deposits of eastern Canada: Canadian Geotechnical Journal, v. 7, p. 145–156.

Lonsdale, P., and Hollister, C. D., 1979, Cut-offs at an abyssal meander south of Iceland: Geology, v. 7, p. 597–601.

Lonsdale, P., Hollister, C. D., and Mayer, L., 1981, Erosion and deposition in interplain channels of the Maury Channel system, northeast Atlantic: Oceanologica Acta, v. 4, no. 2, p. 185–201.

Masson, D. G., Gardner, J. V., Parson, L. M., and Field, M. E., 1985, Morphology of upper Laurentian Fan using GLORIA long-range sidescan sonar: American Association of Petroleum Geologists Bulletin, v. 69, p. 950–959.

McGregor, B. Stubblefield, W. L., Ryan, W.B.F., and Twitchell, D. C., 1982, Wilmington submarine canyon; A marine fluvial-like system: Geology, v. 10, p. 27–30.

Meagher, L., 1984, Interpretation of Quaternary and Upper Neogene seismic stratigraphy on the continental slope off St. Pierre Bank: Geological Survey of Canada Open File 1077, 10 p.

Menard, H. W., 1964, Marine geology of the Pacific: New York, McGraw Hill Book Co., 271 p.

Middleton, G. V., 1966, Experiments on density and turbidity currents; I. Motion of the head: Canadian Journal of Earth Sciences, v. 3, p. 523–546.

Mutti, E., Ricci Lucchi, F., Seguret, M., and Zanzucchi, G., 1984, Seismoturbidites; A new group of resedimented deposits: Marine Geology, v. 55, p. 103–116.

Normark, W. R., Piper, D.J.W., and Stow, D.A.V., 1983, Quaternary development of channels, levees, and lobes on middle Laurentian Fan: American Association of Petroleum Geologists Bulletin, v. 67, p. 1400–1409.

Piper, D.J.W., and Aksu, A. E., 1987, The source and origin of the 1929 Grand Banks turbidity current inferred from sediment budgets: Geo-Marine Letters, v. 7, p. 177–182.

Piper, D.J.W., and Normark, W. R., 1982a, Effect of the 1929 Grand Banks earthquake on the continental slope off eastern Canada, *in* Current research, Part B: Geological Survey of Canada Paper 82–1B, p. 147–151.

—— , 1982b, Acoustic interpretation of Quaternary sedimentation and erosion on the channeled upper Laurentian Fan, Atlantic margin of Canada: Canadian Journal of Earth Sciences, v. 19, p. 1974–1984.

—— , 1983, Turbidite depositional patterns and flow characteristics, Navy submarine fan, California borderland: Sedimentology, v. 30, p. 681–694.

Piper, D.J.W., Stow, D.A.V., and Normark, W. R., 1984a, The Laurentian Fan; Sohm Abyssal Plain: Geo-Marine Letters, v. 3, p. 141–146.

Piper, D.J.W., Sparkes, R., Mosher, D. C., Shor, A. N., and Farre, J. A., 1984b, Seabed instability near the epicentre of the 1929 Grand Banks earthquake: Geological Survey of Canada Open File 1131, 29 p.

Piper, D.J.W., Shor, A. N., Farre, J. A., O'Connell, S., and Jacobi, R., 1985a, Sediment slides and turbidity currents on the Laurentian Fan; Side-scan sonar investigations near the epicentre of the 1929 Grand Banks earthquake: Geology, v. 13, p. 538–541.

Piper, D.J.W., Farre, J. A., and Shor, A. N., 1985b, Scotian Slope: Geological Society of America Bulletin, v. 96, p. 1508–1517.

Ryan, W.B.F., 1982, Interrelationship between oceanographic events and mass wasting of the seafloor, *in* Saxov, S., and Nieuwenhuis, J. K., eds., Marine slides and other mass movements: NATO Conference Series IV, Marine Science, v. 6, p. 263.

Stow, D.A.V., 1984, Laurentian Fan; Morphology, sediments processes, and growth pattern: American Association of Petroleum Geologists Bulletin, v. 65, p. 375–393.

Stow, D.A.V., and Bowen, A. J., 1980, A physical model for the transport and sorting of fine-grained sediment by turbidity currents: Sedimentology, v. 27, p. 31–46.

Uchupi, E., and Austin, J. A., Jr., 1979, The stratigraphy and structure of the Laurentian Cone region: Canadian Journal of Earth Sciences, v. 16, p. 1726–1752.

Walker, R. G., 1975, Generalized facies models for resedimented conglomerates of turbidite association: Geological Society of America Bulletin, v. 86, p. 737–748.

Winn, R. D., Jr., and Dott, R. H., Jr., 1977, Large scale traction produced structures in deep-water fan-channel conglomerates in southern Chile: Geology, v. 5, p. 41–44.

—— , 1979, Deep-water fan-channel conglomerates of Late Cretaceous age, southern Chile: Sedimentology, v. 26, p. 203–208.

Revised Manuscript Received November 1, 1986
Manuscript Accepted by the Society April 4, 1988

Printed in U.S.A.

Geological Society of America
Special Paper 229
1988

Basin plains; Giant sedimentation events

Orrin H. Pilkey, *Department of Geology, Duke University, Durham, North Carolina 27708*

ABSTRACT

The study of 13 modern basin plains ranging in size from the 200-km^2 Navidad Basin up to the giant (>100,000-km^2) Hatteras and Sohm Abyssal Plains has revealed that these features owe their existence to large-volume turbidity currents capable of covering the entire basin floor. Such convulsive events flatten out the topographic irregularities formed by the deposition of small flows between the big events. Giant events maintain the flat plain floor.

The largest event measured to date in modern basin plains is the Black Shell Turbidite of the Hatteras Abyssal Plain, which is at least 100 km^3 in volume and perhaps double that. In some instances, a single giant flow may arrive on the basin simultaneously from geographically widespread basin entry points, indicating that the initiating mechanism, probably an earthquake, was regional in scope.

INTRODUCTION

Basin plains are huge ponds of sediment at the end of the continental margin sediment transportation systems. In a very real sense, such plains may be viewed as lakes of sediment. Like water in a lake, the sediment column of any given point in a basin plain may be derived from any number of sources and directions.

During the past 15 years we have studied 13 basin plains in the western North Atlantic, Caribbean, and Mediterranean Basins (e.g., Bornhold and Pilkey, 1971; Bennetts and Pilkey, 1976; Elmore and others, 1979; Sieglie and others, 1976; Pilkey and Cleary, 1986; Pilkey, 1987). The studies have been based on closely spaced piston cores and have emphasized correlation of individual turbiditic units (e.g., Elmore and others, 1979), sand layer geometry (Pilkey and others, 1980), and elucidation of the processes of sedimentation responsible for basin-floor construction (Pilkey, 1988). The basin plains studied are shown in Figure 1.

Our understanding of sedimentation in modern basin plains began with the study of Ericson and others, 1961. Summaries of the status of knowledge of basin plains, mostly abyssal plains, include Heezen and Laughton (1963), Horn and others (1971), and Horn and others (1972), and Pilkey and Cleary (1986).

Basin plains are filled by turbidity currents as well as by hemipelagic sedimentation. Debris flows are a distant third in overall importance. In the basin plains studied, hemipelagic sediments usually make up less than 20 percent of the sediment volume. Turbidity currents are responsible for the flat basin floors, which usually exhibit slopes of less than 1:1,500. Such currents tend to debauch on the basin plain floor from canyon mouths (point sources) as opposed to line sources. Once a turbidity current arrives on the flat basin floor, it no longer erodes the sea floor to any significant degree. Contrary to widely held views based on studies of ancient rocks, basin plains are not necessarily distal portions of fans dominated by thin and relatively fine sediments (Pilkey, 1987). On the contrary, basin plains such as the Hatteras and Sohm Abyssal Plains are distinguished by the thickest and coarsest sands of any off-shelf sedimentary environments.

The purpose of this paper is to briefly review the evidence that basin plains contain the records of, and are a product of, convulsive sedimentary events. The lines of evidence are: (1) large sedimentary deposits, (2) widespread basin entry points for single events, and (3) the mechanics of flattening a plain.

LARGE EVENT DEPOSIT: THE HATTERAS ABYSSAL PLAIN

The Black Shell turbidite (Elmore and others, 1979) on the Hatteras Abyssal Plain was deposited about 16,000 yr ago during a time of lowered sea level. (The presence of blackened, beach-rounded, mollusk shell fragments led to the name "Black Shell turbidite.") It is at least 100 km^3 in volume and has been traced over a 44,000-km^2 area of the 180,000-km^2 Hatteras Abyssal Plain (Fig. 2). This, to our knowledge, is the largest volume turbidite correlated (maximum thickness is 400 cm) in a modern sedimentary environment. It is also the longest distance (500 km) over which a single catastrophic event has been traced, ancient or recent.

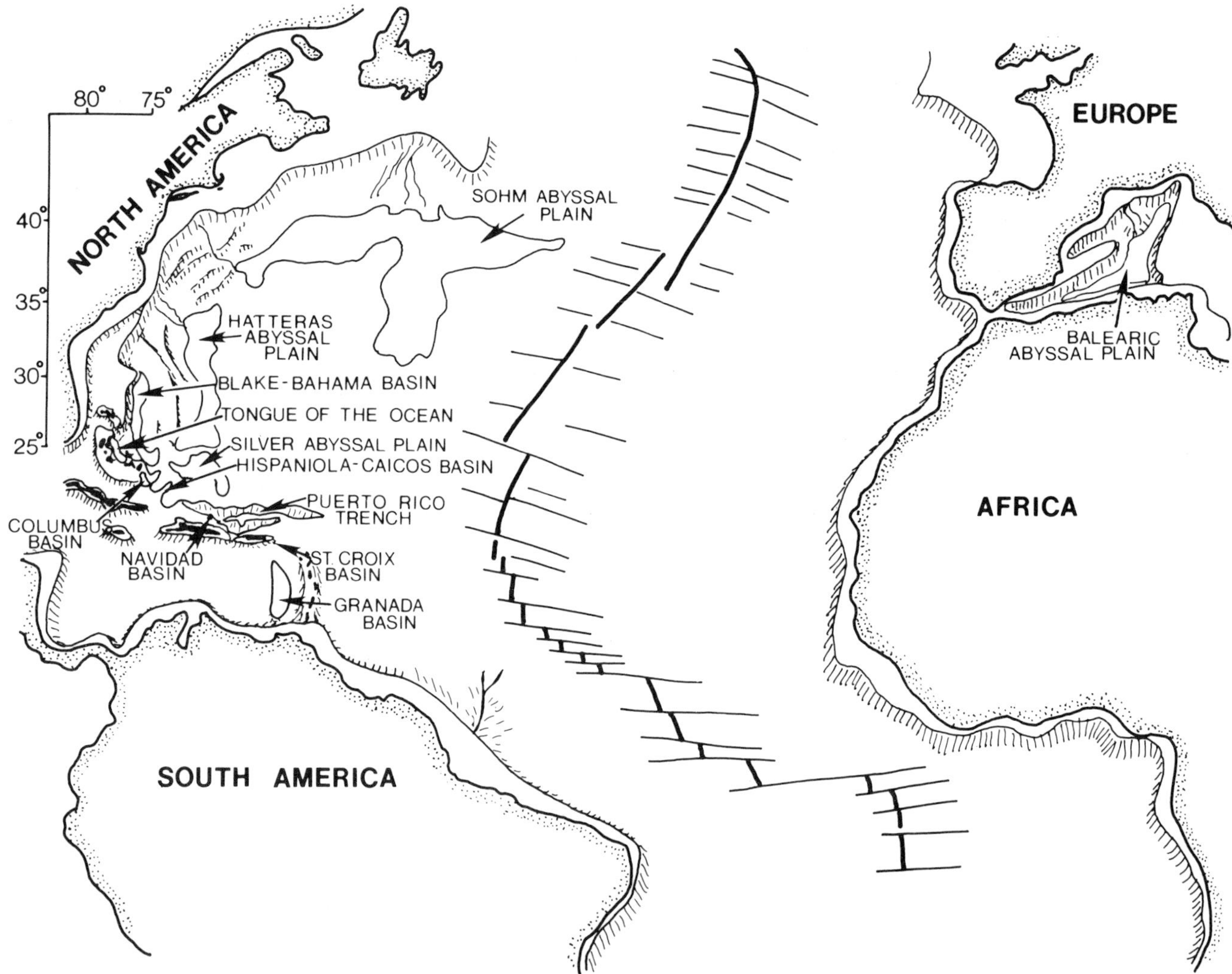

Figure 1. Names and locations of the basin plains studied.

The unit was correlated on the basis of: (1) core color stratigraphy, (2) the relative and absolute position of sand layers in the cores, (3) sand petrology and texture, and (4) macrofaunal assemblages. All of these core and turbidite correlation tools "broke down" in the distal portions of the abyssal plain where sand layers are absent. Hence, correlation of the event was not possible in the distal portions of the basin due to lack of a coarse fraction. Elmore and others (1979) speculate that the total volume of the deposit may be 200 km^3 or more.

Most of the sandstone that would result from the lithification of the Black Shell turbidite would comprise a wide range of graywackes. The basal sands contain between 8 and 30 percent mud. Median grain size of the basal sands decreases more or less regularly from a maximum of 0.4 mm near the source to 0.147 mm at the farthest extent of the sand-layer portion of the turbidite. Maximum-mean basal-sand grain size is near the center of the abyssal plain, but the maximum turbidite thickness tends to occur near the eastern margin of the abyssal plain.

Mineralogically the Black Shell Turbidite sands constitute a suite ranging from lithic arkose to quartz arenite. The variation in mineralogy or potential rock type is a function of grain size. As grain size decreases, distally and vertically, quartz grains and shell fragments decrease while feldspar increases. The differences are interpreted as solely due to sorting effects.

Prince and Ehrlich (1988) recognize three silt-sized, quartz-grain shape types. Each shape type can be traced to a separate major source of sediment contributing to the Black Shell Turbidity Current: (1) unconsolidated shelf-slope drape; (2) partially lithified pre-Pleistocene muds from the upper slope; and (3) upper continental rise silty muds derived at least in part from deposits laid down by the south-flowing Western Boundary Undercurrent.

The following scenario is suggested for the turbidity current event responsible for deposition of the Black Shell Turbidite. About 16,000 yr ago, slope oversteepening, a large spring flood, or most likely a large earthquake triggered a massive failure of

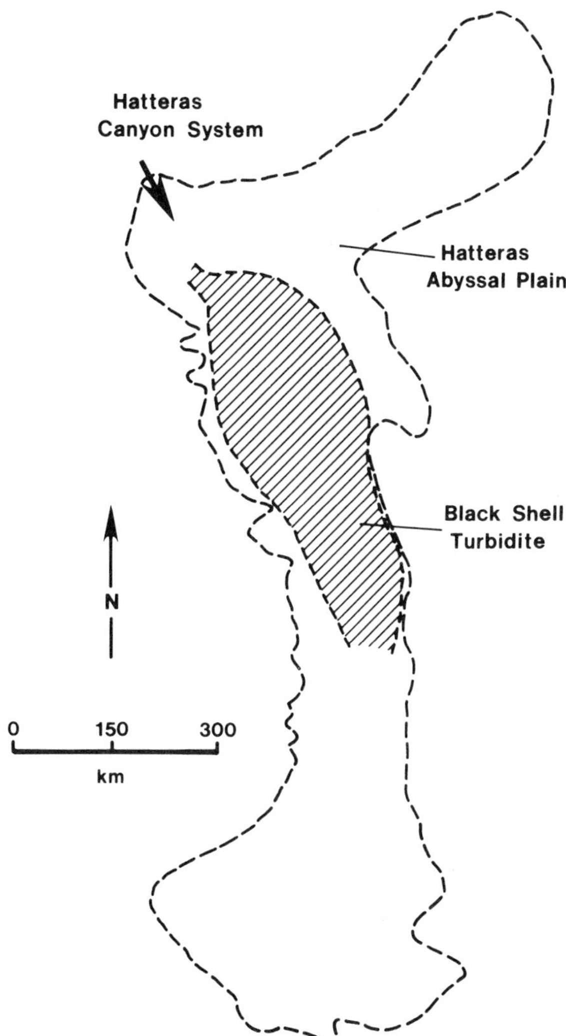

Figure 2. The known outline of the Black Shell Turbidite on the Hatteras Abyssal Plain. The turbidite has been traced for more than 500 km. It has a minimum volume of 100 km^3. It is an example of a large deposit resulting from a single event.

sediment on the upper continental slope and outer shelf off Cape Hatteras. The sediment involved was probably part of a small coalesced delta of the Susquehanna, Potomac, and Neuse Rivers.

Presumably the large slope failure evolved from slump to debris flow to turbidity current. Upon reaching the latter stage, the flow probably found its way simultaneously into several branches of the complex dendritic Hatteras Canyon system.

Just before arriving on the Hatteras Abyssal Plain, the flow was further constricted by the effect of the Hatteras Outer Ridge, which causes all the tributaries of the Hatteras Submarine Canyon system to coalesce into a single restricted channel. That is, the Black Shell event arrived on the plain via a single restricted channel. Thus, the Black Shell event arrived on the plain via a single point source.

The volume of the flow as recorded on the plain is sufficiently large to suggest that a notch must remain etched into the shelf break. Perhaps the notch is one of the small shelfbreak indentations made by the heads of two of the Hatteras Canyon tributaries (Newton and Pilkey, 1969). The excellent physical condition of some of the mollusk shell fragments in proximal sequences, including the lack of biodegradation of the beach-rounded shell material (Fig. 3), suggests the intriguing possibility that the slump actually incorporated a "bite" out of the shoreline that existed 16,000 yr ago (Pilkey and Curran, 1986). The fresh condition of some of the components suggests that the material was not derived from a continental shelf sediment cover, such as that on the present shelf off Cape Hatteras, with a highly biodegraded carbonate fraction (Pilkey and others, 1969). The most abundant bivalve species, *Mulina lateralis, Nucola proxima,* and *Spisula solidissima,* are today diagnostic of the shallow-ocean to estuarine zones of the Beaufort, North Carolina, area (Bird, 1970).

Large turbidity currents typically leave little record of their passage on the Mid-Atlantic U.S. Continental Rise. Rather, the rise is a zone of transportation of such events (Cleary and others, 1985), which leave their major imprint through deposition on the adjacent abyssal plain. The Black Shell Turbidite is no exception to this rule, since the few available piston cores from the Hatteras Canyon's system reveal no major turbidity current deposit of 16,000 yr B.P. Probably the Black Shell Turbidity Current did leave behind thin levee sands and perhaps contributed significantly to the volume of material being transported southward by the Western Boundary Undercurrent.

The Black Shell Turbidity Current entered the basin plain via the Hatteras Canyon system about one-third of the basin plain length "downstream" from its northern extremity. When the flow reached the mouth of the Hatteras Canyon, deposition did not take place immediately. In fact, initial deposition of coarse material took place a full 50 km southeast of the canyon mouth at the rim of the small Hatteras Fan (Cleary and Connolly, 1974). Behavior of the flow once it reached the abyssal plain is discussed by both Elmore and others (1979) and Prince and Ehrlich (1988). The coarse fraction of the flow was deposited on the western and central sections of the plain while the finer fractions are most abundant on the eastern margin. The Hatteras Abyssal Plain gently slopes to the south and the east, and it is along the eastern margin of the abyssal plain that the turbidite is thickest (Fig. 4). The ponding of the Black Shell turbidite along the eastern margin of the abyssal plain is an example of the basin-plain flattening effect of giant gravity flow events (Fig. 4).

As previously mentioned, the Black Shell turbidite is probably much larger in volume than the 100 km^3 actually measured over the correlatable portions of the flow. This flow was correlated over 500 km, but the largest flows in the Hatteras Abyssal Plain probably travel the entire 1,000-km length of the abyssal plain and then spill over into the adjacent Silver or Nares Abyssal Plains, adding another 200- or 500-km travel distance, respectively.

Major events such as the Black Shell Turbidity Current should not be assumed to be restricted to times of lowered sea

Figure 3. Shell material obtained from the proximal-zone Black Shell Turbidite near the mouth of the Hatteras Canyon System (Fig. 2). The roundness and lack of extensive biodegradation is a diagnostic characteristic of the beach environment.

level. The 1929 Grand Banks slump (Heezen and Ewing, 1952) may have involved as much as 400 km^3 of material. This slump resulted in a turbidity current that traveled 500 km to the Sohm Abyssal Plain, but the areal extent of the resulting turbidite is unknown.

The Black Shell event is probably much smaller in volume than the Contessa Bed (Ricci Lucchi, 1978) and some of the "megaturbidites" of the southern Pyrenees (e.g., Johns and others, 1981), which may also have been deposited on basin plains and represent "giant" sedimentary events.

REGIONAL INITIATION OF CONVULSIVE EVENTS: THE PUERTO RICO TRENCH AND THE BALEARIC ABYSSAL PLAIN

Careful examination of individual beds or layers that have been correlated on basin plain floors can reveal important evidence concerning the areal extent of the mechanisms initiating gravity flows. A case in point is a turbidite, less than 10 cm thick, found in the upper 50 cm of the southern Balearic Abyssal Plain sediment column (Schorsch, 1980). Texture and coarse fraction component analysis reveal that, although the event is a single correlatable layer, it was derived from three widely spaced sources (Fig. 5). Two of the sources were on the Algerian continental margin and a third was the Spanish continental margin. Although the turbidite does not represent a giant event volumetrically, whatever produced or initiated the gravity flow must have occurred over a wide geographic area, affecting both sides of the Mediterranean Basin simultaneously.

The basin plain in the floor of the Puerto Rico Trench provides another example of a gravity flow event of regional derivation (Doull, 1983). The largest correlatable coarse layer within piston coring range on this basin plain extends for at least 300 km with maximum thicknesses of close to 200 cm. Although small by Hatteras or Sohm Abyssal Plain standards, this turbidite represents a sizeable volume of material to be derived from relatively small source areas. The volume of this flow, called the Giant Turbidite, was first estimated by Connolly and Ewing (1967) to be 30 km^3, but much more detailed coring reveals a more likely volume of 2 km^3 (Doull, 1983).

Figure 6 is an isopach map of the Giant Turbidite on the Puerto Rico Trench basin plain. At least five basin entry points are suggested (see arrows on Fig. 6) on the basis of grain-size trends of basal sands and variations in event layer thickness. Apparently a turbidity current was produced by a very large seismic event, perhaps affecting the islands of Puerto Rico and Hispaniola and the Virgin Platform simultaneously. Material flowing into the western end of the trench most likely was derived from Hispaniola, and material at the eastern end of the trench came from the slope of the Virgin Platform. However, the bulk of the sediment of the Giant Turbidite almost certainly was derived from the insular margin of Puerto Rico.

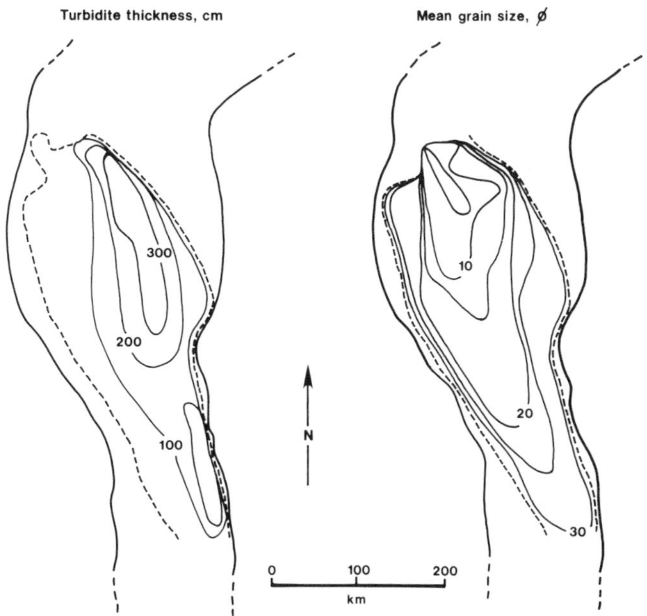

Figure 4. Thickness of the Black Shell Turbidite (L) and mean grain size of the basal sand of the turbidite in phi units (R).

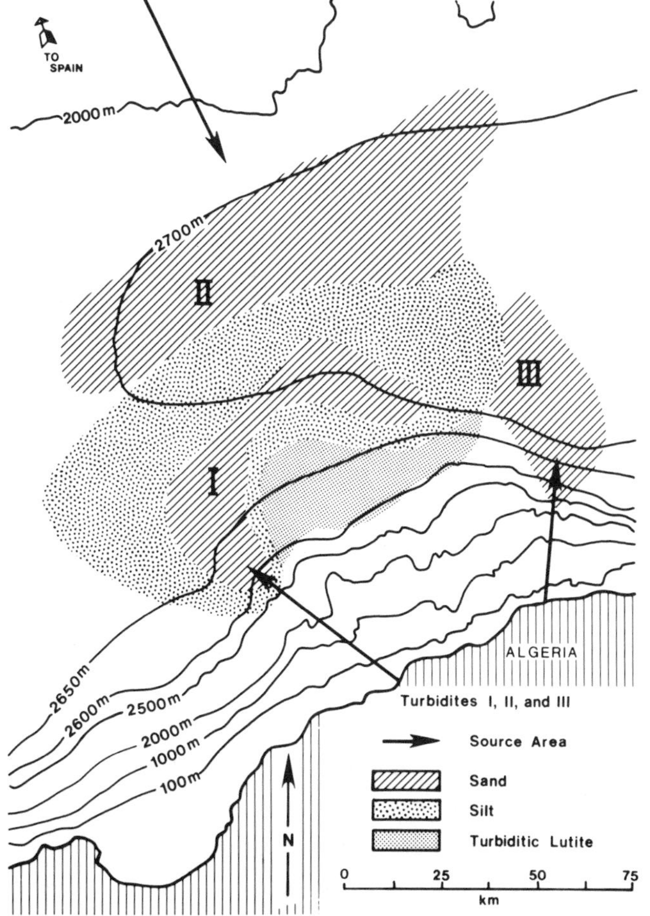

Figure 5. The shaded area represents a thin (<10 cm) coarse layer within the upper 50 cm of the sediment column of the Southern Balearic Abyssal Plain. The sand layer was derived from 3 separate and widely spaced sources presumably on both sides of the Mediterranean, as the result of a large regional earthquake.

BASIN PLAIN FLATTENING: GRENADA BASIN PLAIN

If a basin plain is fed only by small flows, each covering only a small portion of the plain, it follows that uneven deposition will result. The plain will no longer be a plain. In fact, most flows entering the basin plains we studied are far too small to cover the entire basin floor. It is the occasional large flow that "repairs" and flattens the depositional environment by preferential deposition in the basin lows. Such large flows in most basin plains are truly convulsive events.

Excellent examples of preferential deposition in topographic lows were noted in several basins. Individual large turbidites in the Hispaniola Caicos Abyssal Plain, the Puerto Rico Trench Basin Plain, and the Blake Bahama Abyssal Plain thicken by as much as 50 percent in depressions with as little relief as 10 m over a 20-km distance. The most striking example of such topographic control of turbidity current deposition is furnished by the Soft Turbidite in the Grenada Basin (Herrick, 1983). Figure 7 shows a comparison of the thickness of this turbidite and a map of the piston core depth to the base of the turbidite.

The core depth to the base of the unit can be considered to be a rough map of the shape of the sea floor prior to deposition of the Soft Turbidite. This assumes that the turbidity current did not erode the basin floor, which, on the basis of piston core stratigraphy, seems to be a valid assumption. Thus the 2-m regional depression shown by hachures in Figure 7 is presumed to be the result of preferential small turbidite deposition to the north and south of the depressed area. When the Soft Turbidite was deposited, the minor irregularities in the shape of the sea floor disappeared.

As a first attempt at estimating the volume of sedimentary deposits produced by the convulsive events required to maintain basin plain, assumptions can be made that: (1) the event must cover the entire basin floor and (2) the average thickness of the minimum-sized turbidite capable of covering the required areal extent is 1 m. Using these assumptions, minimum-flow volumes for each of the basin plains studied are given in Table 1. Required volumes range from 0.2 km^3 for Navidad Basin to 600 km^3 for the giant Sohm Abyssal Plain.

Basin-filling events will be required at widely varying frequencies depending on the widely varying rates of sedimentation in basin plains. The basic requirement is that giant events be of sufficient frequency to fill in the lows between regions of relatively rapid accumulation of small flows. In other words, if basin-flattening events are not sufficiently frequent, fan-like features will begin to build up at the basin entry points of turbidity currents. Considering known rates of sedimentation in abyssal plains,

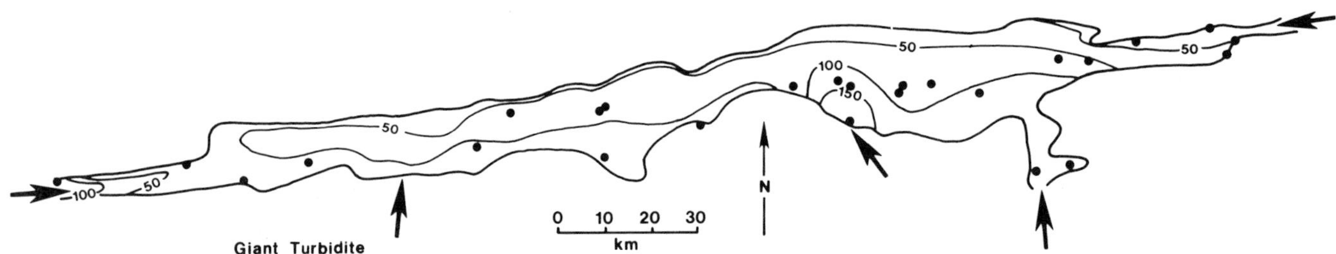

Figure 6. Thickness of The "Giant" Turbidite on the basin plain on the floor of the Puerto Rico Trench was derived simultaneously from 5 point sources, an indication that sediment was set in motion simultaneously over a large area. Arrows show basin entry points. Thicknesses are in cm.

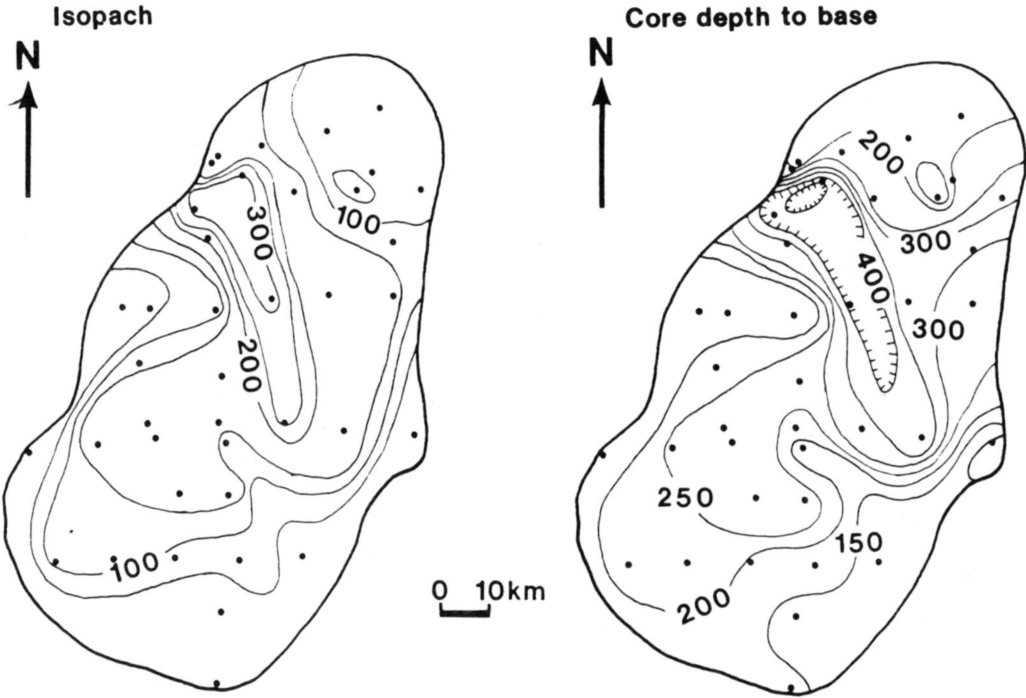

Figure 7. Shown (L) is an isopach map of the "Soft" Turbidite from the Grenada Basin (see Fig. 1) and a map of the core depth to the base of the turbidite (R). The similarity of the two may reflect control of the thickness of the turbidite deposits on a basin plain floor by preexisting sea floor topography. In this case, a large flow, capable of covering the entire basin floor, smoothed out the basin plain floor. Thicknesses of core depths are in cm.

TABLE 1. ESTIMATE OF THE MINIMUM SEDIMENT VOLUME OF THE GIANT EVENT REQUIRED TO MAINTAIN THE BASIN PLAINS NOTED

Basin Plain	*Minimum Volume Event* (km^3)
Sohm	600
Hatteras	180
Balearic	
1. Algero-Provencal	170
2. South-Balearic	30
Grenada	20
Silver	19
Blake Bahama	15
Columbus	9
Tongue of the Ocean	6
Puerto Rico Trench	4
St. Croix	0.7
Navidad	0.2

it is probable that giant events occur at least every 50,000 yr in the largest basins and more frequently in smaller basins.

SUMMARY OF CONCLUSIONS

Basin plains are products of giant sedimentary events. The evidence supporting this is as follows:

(1) There are very large bodies of sediments (of the order of 100 km^3 or more) on basin plains that were emplaced "instantaneously."

(2) In some cases the sediment bodies emplaced by single events arrived on the basin plains simultaneously from widely spaced basin entry points. Such instances may reflect derivation from very large flow-initiating earthquakes/tsunamis/storms of regional extent.

(3) The mechanics of basin filling are such that a flat plain can only be formed if occasional events are large enough to cover the entire basin plain and fill in topographic lows produced by uneven rates of sedimentation of small events.

REFERENCES CITED

Bennetts, K., and Pilkey, O. H., 1976, Characteristics of three turbidites, Hispaniola–Caicos Basin: Geological Society of America Bulletin, v. 87, p. 1291–1300.

Bird, S. O., 1970, Shallow-marine and estuarial benthic molluscan communities from the area of Beaufort, North Carolina: American Association of Petroleum Geologists Bulletin, v. 54, p. 1651–1676.

Bornhold, B. D., and Pilkey, O. H., 1971, Bioclastic turbidite sedimentation in Columbus Basin, Bahamas: Geological Society of America Bulletin, v. 82, p. 1341–1354.

Cleary, W. J., and Connolly, J. R., 1974, Hatteras deep-sea fan: Journal of Sedimentary Petrology, v. 44, p. 1140–1154.

Cleary, W. J., and others, 1985, Wilmington Fan, Atlantic Ocean, *in* Bouma, A. H., Barnes, N. E., and Normants, W. R., eds., Submarine fans and related turbidite sequences: New York, Springer-Verlag, p. 157–164.

Connolly, J. R., and Ewing, M., 1967, Sedimentation in the Puerto Rico Trench: Journal of Sedimentary Petrology, v. 37, p. 44–59.

Doull, M. E., 1983, Turbidite sedimentation in the Puerto Rico Trench Abyssal Plain [M.S. thesis]: Durham, North Carolina, Duke University, 124 p.

Elmore, R. D., Pilkey, O. H., Cleary, W. J., and Curran, H. A., 1979, Black Shell turbidite, Hatteras Abyssal Plain, western Atlantic Ocean: Geological Society of America Bulletin, part 1, v. 90, p. 1165–1176.

Ericson, D. B., and others, 1961, Atlantic deep sea sediment cores: Geological Society of America Bulletin, v. 72, p. 193–286.

Heezen, B. C., and Ewing, M., 1952, Turbidity currents and submarine slumps and the 1929 Grand Banks earthquake: American Journal of Science, v. 250, p. 849–878.

Heezen, B. C., and Laughton, A. S., 1963, Abyssal plains, *in* Hill, M. N., ed., The sea, v. 3: New York, John Wiley and Sons, p. 312–364.

Herrick, D., 1983, Turbidite sedimentation in the Granada Basin Plain [M.S. thesis]: Durham, North Carolina, Duke University, 135 p.

Horn, D. R., and others, 1971, Turbidites of the Hatteras and Sohm Abyssal Plains, western North Carolina: Marine Geology, v. 11, p. 287–323.

Horn, D. R., Ewing, J. I., and Ewing, M., 1972, Graded-bed sequences emplaced by turbidity currents north of 20°North in the Pacific, Atlantic, and Mediterranean: Sedimentology, v. 18, p. 247–275.

Johns, D. R., Mutti, E., Rosell, J., and Sequret, M., 1981, Origin of a thick, redeposited carbonate bed in Eocene turbidites of Hecho Group, south-central Pyrenees, Spain: Geology, v. 9, p. 161–164.

Newton, J. G., and Pilkey, O. H., 1969, Topography of the continental margin off the Carolinas: Southeastern Geology, v. 10, p. 87–92.

Pilkey, O. H., 1987, Sedimentology of abyssal plains, *in* Weaver, E., ed., Geology and geochemistry of abyssal plains: Geological Society of London Special Publication, no. 31, p. 1–12.

Pilkey, O. H., and Cleary, W. J., 1986, Turbidite sedimentation in the northwestern Atlantic Ocean Basin, *in* Vogt, P. R., and Tucholke, B. E., The western North Atlantic region: Boulder, Colorado, Geological Society of America, The Geology of North America, v. M, p. 437–450.

Pilkey, O. H., and Curran, H. A., 1986, Molluscan shell transport—you ain't seen nothin' yet: Palaios, v. 1, p. 197.

Pilkey, O. H., and others, 1969, Aspects of carbonate sedimentation on the Atlantic continental shelf off the southern United States: Journal of Sedimentary Petrology, v. 39, p. 744–768.

Pilkey, O. H., Locker, S. D., and Cleary, W. J., 1980, Comparison of sand-layer geometry on flat floors of ten modern depositional basins: American Association of Petroleum Geologists Bulletin, v. 64, no. 6, p. 841–856.

Prince, C. M., and Ehrlich, R., 1988, Vertical provenance changes within the Black Shell turbidite (in press).

Ricci Lucchi, F., 1978, Turbidite dispersal in a Miocene deep-sea plain; The Marnoso-arenacea of the northern Apennines, *in* van Loon, A. J., ed., Keynotes of the MEGS-II (Amsterdam, 1978): Geologie en Mijnbouw, v. 57, p. 559–576.

Schorsch, L., 1980, The Orleansville turbidite, southern Balearic Basin, western Mediterranean Sea [M.S. thesis]: Durham, North Carolina, Duke University, 160 p.

Sieglie, G. A., Froelich, P. N., and Pilkey, O. H., 1976, Deep sea sediments of Navidad Basin: Correlation of sand layers: Deep-Sea Research, v. 23, p. 89–101.

MANUSCRIPT ACCEPTED BY THE SOCIETY APRIL 4, 1988

Printed in U.S.A.

Geological Society of America
Special Paper 229
1988

Large-scale bedforms in boulder gravel produced by giant waves in Hawaii

George W. Moore, *Department of Geology, Oregon State University, Corvallis, Oregon 97331*
James G. Moore, *U.S. Geological Survey, 345 Middlefield Road, Menlo Park, California 94025*

ABSTRACT

Approximately 105,000 yr ago (based on uranium-series dating), waves in a giant wave train swept up to an elevation of about 375 m on the island of Lanai. The waves deposited the Hulopoe Gravel, which near the present shoreline consists of basalt boulders, coral fragments, and calcareous beachrock slabs, and near the upper limit of the deposit consists of sand and shell fragments. The maximum heights of similar but lower deposits on nearby islands, when adjusted for their estimated subsidence due to volcanic loading during the past 105,000 yr, indicate that the source of the wave was about 50 km southwest of Lanai. We hypothesize that failure and downward movement of the huge Lanai submarine landslide created an ocean disturbance, which produced waves that rushed across the Lanai reef and beach, picked up limestone and lava fragments, and deposited them high on the island as the Hulopoe Gravel. Backwash from the waves stripped soil and rock from the islands and carried much of it to the sea.

The Hulopoe Gravel is 8 m thick in a gulch 200 m inland from the Lanai shoreline, where it consists of three beds, successively, 2, 4, and 2 m thick. These beds are considered to have been laid down by successive waves in the wave train. Each bed consists of two units: a lower unit of basalt and limestone boulders, cobbles, and sand, and an upper bimodal unit of large basalt boulders with a pebbly sand matrix. These subunits are inferred to be deposited from the runup and backwash of each wave.

At the upper surface of the Hulopoe Gravel, basalt boulders averaging 0.5 m in diameter are arranged in dunelike ridges about 1 m high and 10 m apart. Nearby, where young streams have cut into and exposed the lower beds of the Hulopoe, clasts at the boundaries between the beds are locally imbricated and dip landward. We interpret these features as aspects of torrential flow and crossbedding created during the high-speed backwash of the great waves.

INTRODUCTION

In 1844, Couthouy reported clasts of coral at anomalously high positions on the Hawaiian Islands, and in 1940, Stearns published a geologic map and description of these deposits on the Island of Lanai. Stearns showed that coral and shell fragments extend up to a maximum elevation of 326 m and that the soil is stripped off for some distance above that. At lower levels on Lanai, he mapped locally continuous beds of boulder gravel, in places cemented by calcite, that consist of clasts that are about 95 percent basalt, both vesicular and dense, and 5 percent honeycomb coral, coralline algae, and calcareous beachrock. Stearns (1940) suggested that the limestone boulders were deposited along ancient shorelines that have since been uplifted.

Tide-gauge records and the radiocarbon dating of drowned coral reefs have shown that the southeastern Hawaiian Islands subside too fast for ancient shorelines resulting from worldwide high stands of the sea to be still emergent (Moore, 1987). This knowledge led us to reexamine the limestone boulders on Lanai, and we found that they are part of a continuous formation, the Hulopoe Gravel, that thins from near the present shoreline to the last vestiges discovered by Stearns high up on the island (Moore and Moore, 1984).

The Hulopoe Gravel is a well-exposed deposit from a convulsive geologic event. In a previous paper we showed that a great wave must have laid down the deposit (Moore and Moore,

1984). In this study we describe the sedimentary structures of the deposit, including dunelike boulder ridges on its upper surface, and evaluate a hypothesis of wave generation by sea disturbances at the head of the newly discovered Lanai submarine landslide, about 50 km southwest of Lanai (Fig. 1).

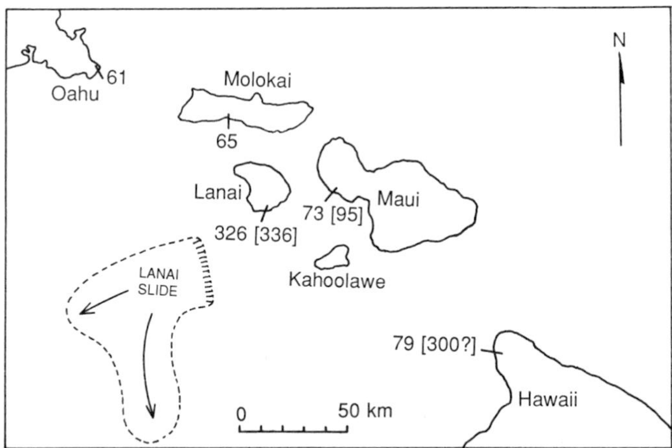

Figure 1. Index map of the southeastern Hawaiian Islands showing the maximum elevation in meters of presumed wave-deposited limestone clasts and the position of the Lanai slide, the inferred source of the wave train. Clast elevations in brackets have been adjusted to their original depositional level, using a subsidence curve inferred from tide-gauge records.

EROSION

The waves that left stratified deposits in the gulches and formed boulder dunes on the interfluve surfaces were vigorously erosive. On the south side of Lanai, about 2 m of soil and weathered basalt were removed in a belt more than 2 km wide. In contrast, the remaining stratified deposits and the boulder dunes make up a volume perhaps one tenth of that removed.

In places, the waves left no deposit at all; elsewhere, an incomplete section indicates that beds left by some waves were eroded and removed by others. The later waves, particularly those washing downhill on the upland surfaces between the gulches, winnowed preexisting deposits and reshaped the boulder dunes.

SEDIMENTOLOGIC FEATURES

The Hulopoe Gravel occurs in patches on the flanks of the Island of Lanai, and the thickest, highest, and most continuous patch covers about 2 km^2 along the south coast of the island around Hulopoe Bay (Fig. 2). The gravel is best preserved near the present shore and in gulches, where it reaches a maximum thickness of about 10 m. On upland areas between the gulches, the relation of the gravel to scattered basalt bedrock outcrops indicates that the gravel is only a few clast diameters thick. Farther upslope, the deposit thins to one clast thick, and beyond that the particles become more and more widely spaced and smaller until ultimately the last recognizable remnant of the formation is a calcareous sand that fills bedrock cracks.

The base of the Hulopoe Gravel usually rests on relatively unweathered basalt, despite the fact that at higher elevations where pineapples are grown, the bedrock is weathered to a reddish brown sequence of soil and saprolite 2 to 3 m deep. In a few places, a reddish brown fossil soil is preserved in pockets below the gravel, but apparently the process that deposited the gravel removed the soil and weathered rock in most places.

The limestone clasts in the Hulopoe Gravel are conspicuous because they are white, but they constitute only about 5 percent of the formation, which is mainly composed of basalt boulders. Limestone clasts are usually absent from the surface of the gravel between the gulches, but rarely—and usually along the highest parts of the interfluves—slabby limestone boulders and cobbles that are deeply grooved by surficial dissolution lie scattered about.

The most extensive deposits of the gravel occur in the gulches, which existed in approximately their present state before the gravel was deposited. Subsequent partial erosional excavation of these gulches has created excellent exposures in which the gravel stratigraphy is well preserved. The gulches served as sediment traps for the gravel mobilized by wave action.

At the type section of the Hulopoe Gravel in the gulch at the head of Kapihaa Bay, about 200 m from the shore and 22 m above sea level, a complete section, including fresh underlying basalt, is exposed (Fig. 3). At this place, the formation consists of three beds. The lower bed is about 2 m thick, the middle 4 m, and the upper 2 m. The deposits imply that no other waves reached an elevation of 22 m, and we assume that the wave that laid down the thickest (middle) bed is the one that swept farthest up onto the island to lay down the highest deposit there and to strip off the soil.

The limestone clasts in the Hulopoe Gravel are mainly confined to the lower third of the beds. The larger particles of all rock types are clast supported rather than matrix supported, indicating deposition from a water suspension or bedload rather than from a debris flow. These lines of evidence support the supposition that the limestone-bearing lower parts of the beds were deposited during the runup of the waves, and the limestone-free upper parts of the beds during the subsequent rundown.

The beds show an overall inverse grading, in which the largest basalt clasts are in the upper two-thirds of each bed. On modern beaches, which are also characterized by flow reversal, similar grading occurs (Clifton, 1969). In the limestone-free upper part of a bed, which is distinctly bimodal in size distribution, basalt boulders are enclosed in a silty pebbly matrix that in places contains abundant fine-grained marine debris, including forams and fragments of sea-urchin spines.

Many of the subrounded boulders as large as 1.5 m in diameter at the tops of the beds were formed as residual boulders in the weathered zone of lava flows. Roadcuts through the deep soil above the level of the Hulopoe Gravel reveal many similar rather fresh basaltic masses that have been rounded by spheroidal

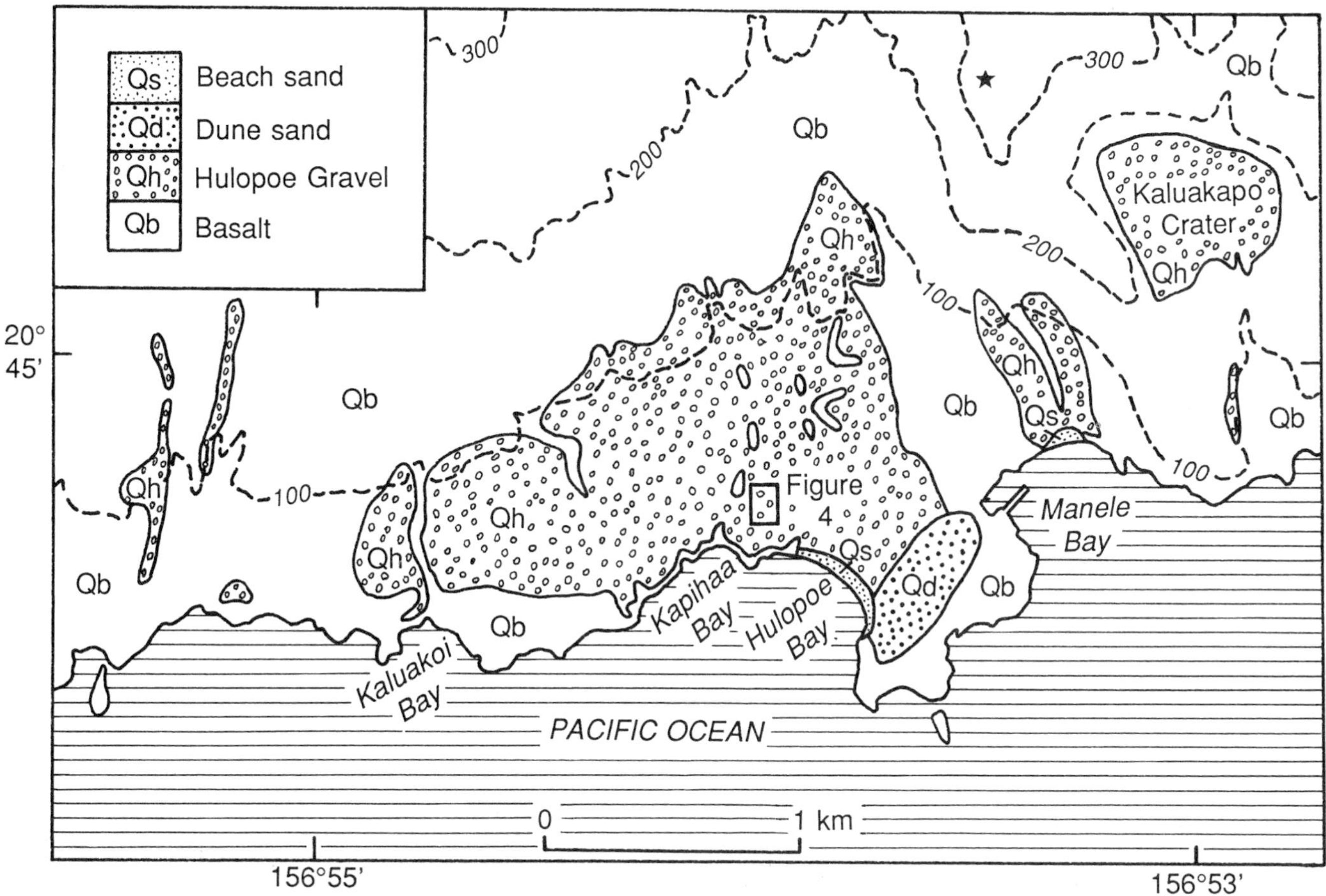

Figure 2. Geologic map of the south coast of the island of Lanai showing the distribution of the Hulopoe Gravel where it is more than about 1 m thick. Topographic contour interval 100 m. Star at upper right at 326 m marks the highest known calcareous sand preserved in a crack in basalt.

weathering and are enclosed in weathered bedrock. Apparently the waves freed such boulders from the soil, moved them downslope during the backwash, and deposited them in the upper parts of the beds.

In general, the strata are obscurely bedded, but some clasts are imbricated where downwash prevailed at the tops of the beds (see especially the interval beneath the limestone clasts at the base of the middle bed in Fig. 3). In the imbrication, most clearly shown by small boulders and large cobbles, the clasts dip upslope (upcurrent), presumably because clasts with that orientation are more stable than those that tilt up into the current and thus present a less streamlined face to the flowing water (Rusnak, 1957).

An erosion surface lies at the base of the middle bed, directly above the imbricated clasts at the top of the lowest bed. Because of the imbrication, the lowest bed was rough at the time the middle bed began to be deposited. Limestone fragments and sand then filled in the interstices at the upper surface of the well-sorted lowest bed. After that, a planar erosion surface formed in the sand at approximately the level of the tops of the larger clasts of the lowest bed. This planar surface was overlain by the limestone-bearing gravel of the middle bed. The surface seems to represent a brief erosional interval when the high-speed onshore wave that deposited the lower part of the middle bed reached maximum velocity. Except for the sediment filling the voids at the top of the lowest bed, the wave at that time carried its entire load and smoothed off the substratum.

AGE OF THE GRAVEL

The age of deposition of the Hulopoe Gravel can be estimated by its degree of cementation, by the weathering of basalt boulders on its upper surface, by its molluscan fossils, and by uranium-series dating of coral clasts within it.

The gravel is locally well cemented, implying some antiquity. The calcite cement is confined to the parts of beds containing limestone fragments and was probably derived from them by dissolution. The cement contains silt grains and apparently was deposited from water that percolated through the gravel after deposition. All the limestone fragments are transported clasts, and we saw no evidence for cementation by the growth of calcareous algae.

The basalt boulders on the upper surface of the Hulopoe Gravel are weathered to reddish brown, whereas the basalt

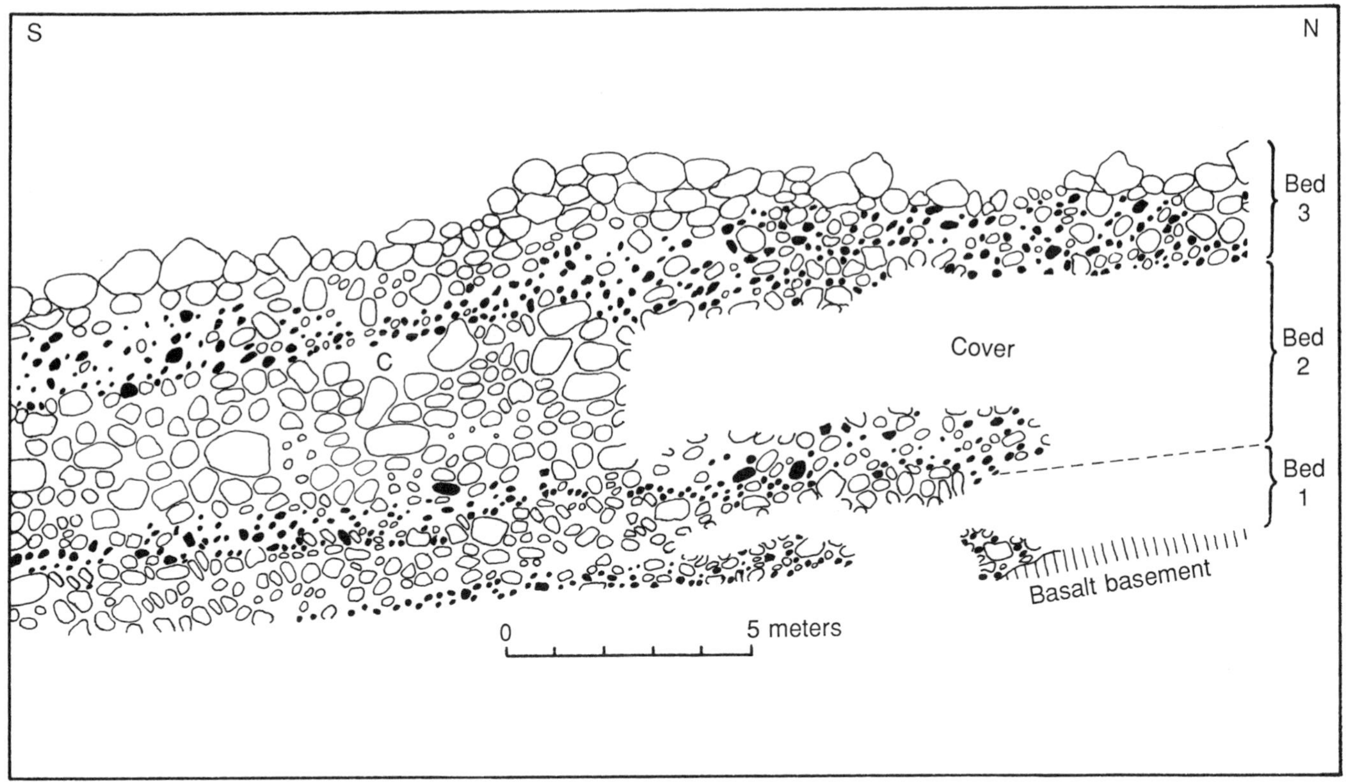

Figure 3. Section of the Hulopoe Gravel at its type locality, 200 m from the shore, on the west wall of the gulch at the head of Kapihaa Bay, Lanai. The elevation of the base of the section in the center of the profile is 22 m. Basalt clasts are white and limestone clasts are black. The size of the smaller limestone clasts is exaggerated for clarity.

boulders on the modern beach are medium gray. The weathering does not extend deeply into the Hulopoe boulders, however, and the rocks are still sound. The discoloration is about 0.5 mm deep on dense basalt and 1 cm deep on vesicular basalt.

Molluscan shells are scattered throughout the Hulopoe Gravel. Most are fresh and not beachworn, and some, such as *Cypraea, Conus, and Strombus,* faintly retain their original coloration. Table 1 cites the species in two collections identified and interpreted by Ellen J. Moore (written communication, 1985). Most species in the molluscan collections are ones that live today in Hawaii intertidally to a depth of 20 m, but *Strombus dentatus* lives from 70 to 80 m, indicating a mixed-depth assemblage for the Hulopoe fossils. *Strombus* (*Gibberulus*) *gibberulus gibbosus,* which lives today in the more equatorial Marshall and Line Islands, is now locally extinct at the Hawaiian Islands, implying a somewhat warmer water temperature and a Pleistocene age for the assemblage.

H. H. Veeh dated a coral fragment from an elevation of 6 m on the island of Hawaii by the uranium-series method and found it to be 110,000 ± 10,000 yr old (Stearns, 1973). Three new samples of aragonitic coral from elevations above 100 m on Lanai have yielded similar uranium-series ages by the ^{230}Th/^{234}U method (Barney J. Szabo, written communication, 1984). The ages of two samples from Kawaiu Gulch, which are little altered as indicated by only a small calcite contamination (Table 2), indicate that the great wave train struck the southeastern Hawaiian Islands about 105,000 yr ago.

BEDFORMS ON UPLAND SURFACES

Over broad areas near the coast of Lanai, large-scale boulder-gravel bedforms cover gently sloping areas between the gulches. They are best developed at elevations of about 20 to 40 m, where the slope is typically 7°. The bedforms consist of branching dunelike gravel ridges, generally 1 m high with a spacing of 10 m, that are roughly transverse to the slope of the land (Fig. 4).

The basalt clasts that make up the bedforms average 0.5 m in diameter and rarely exceed 1.5 m. The boulder ridges are asymmetric in cross section, with the steeper slopes facing the coast: the downhill (lee) surfaces slope about 35°, and the uphill (stoss) surfaces are nearly flat. A cross section of such a wavelike gravel bedform is illustrated at the top of Figure 3, where canyon fill approaches the level of the adjacent surface.

Some of the gravel waves trend nearly parallel to the slope direction and presumably are longitudinal to the flow. In places the high points of the transverse gravel waves are aligned so as to produce a downslope bar (Fig. 4). These irregularities seem to be related to the positions of bedrock outcrops. The outcrops are about the same height as the gravel waves and invariably incor-

TABLE 1. MOLLUSCAN FOSSILS FROM THE HULOPOE GRAVEL, LANAI

U.S.G.S. Cenozoic locality M8841, gulch at the head of Kapihaa Bay, 200 m inland from shore, 22 m elevation, 20°44'7"N, 156°54'1"W, Lanai, Hawaii

Gastropods:
- *Cerithium* sp. cf. *C. mutatum* Sowerby
- *Conus pulicarius* Hwaas
- *Cypraea* sp. cf. *C. cicercula* Linné
- *Drupa (Ricinella) rubusidaeus* Roding
- *Mitra* sp.
- *Modulus?* sp.
- *Strombus dentatus* Linné
- *Strombus (Gibberulus) gibberulus gibbosus* (Roding)
- *Strombus helli* Keiner
- *Terebra gouldi* Deshayes
- *Terebra* sp. a
- *Terebra* sp. b
- *Terebra* sp. c
- *Turbo* sp.
- Turbinid operculum

Pelecypods:
- *Codakia punctata* (Linné)
- *Spondylus?* sp.

U.S.G.S. Cenozoic locality M8475, gulch at the head of Kaluakoi Bay, top of seacliff, 20°44'4"N, 156°54'8"W, Lanai, Hawaii

Gastropods:
- *Cellana* sp. cf. *C. sandwicensis* (Pease)
- *Cerithium mutatum* Sowerby
- *Conus* sp. cf. *C. imperialis* Linné
- *Conus* sp. a
- *Conus* sp. b
- *Cymatium (Septa)* sp. cf. *C.(S.) gemmatum* (Reeve)
- *Cyprae* sp.
- *Mitra (Nebularia?)* sp.
- *Sabia conica* (Schumacher)
- *Strombus (Gibberulus) gibberulus gibbosus* (Röding)
- Turbinid operculum

Pelecypods:
- Unidentified ostreid
- Unidentified spondylid?

porate either a longitudinal or transverse bedform, most commonly on the upslope side.

The general distribution of the bedrock outcrops indicates that the gravel layer of the upland areas attains a maximum thickness of about 2 m, in contrast to the maximum thickness of 10 m in the adjacent gulches. In the general vicinity of the type locality of the Hulopoe Gravel, gravel also accumulated in longitudinal bars on the east shoulders of downslope ridges (the west sides of valleys), possibly because the angle of incidence of the waves was oblique to the coast from the southwest.

The gravel is absent at the tops of coastal headlands (Fig. 2). At such places the basalt bedrock, reddish brown but otherwise sound like the boulders, is commonly stripped down to scoriaceous flow tops. Possibly the deposits are absent in these zones because the final depositional stages of the backwash were confined to lower ground between the headland ridges.

The aboriginal Hawaiians used the gravel dunes in house construction (Fig. 4). In building their small houses, they commonly adapted the steep lee slope of a dune to be the back wall

TABLE 2. URANIUM-SERIES AGES OF ARAGONITIC CORAL CLASTS IN THE HULOPOE GRAVEL, LANAI

Sample locality	Coordinates (N lat, W long)	Elevation (meters)	Calcite (percent)	Age (years)
Kawaiu Gulch	20°46'0",156°51'6"	115	<3	108,000±5,000
Kawaiu Gulch	20°45'9",156°51'4"	120	<3	101,000±4,000
Kaluakapo Crater	20°45'2",156°52'9"	155	5	134,000±7,000

and framed the side walls by stacking boulders in lines about 1 m high. The superstructure of the houses, now gone, is believed to have been assembled from vine-tied sticks thatched with leaves or tied bundles of grass (Emory, 1924). Nestled below the dunes, the occupants had both wind protection and a clear outlook toward the sea.

Waterlaid dunes are common elsewhere in the world on the bottoms of rivers and in tidal inlets (Rubin and McCulloch, 1980). In several places, floods have formed gravel dunes that resemble those on Lanai, and one of the most carefully studied gravel dunefields is that created in the channeled scabland of Washington by the breakout of glacial Lake Missoula (Baker, 1973). The boulder size in the scabland is similar to that on Lanai, but the spacing of the Washington gravel waves averages about 70 m (versus 10 m for Lanai) chiefly because of a greater water depth during the formation of the final dunes that are preserved.

The fact that the lee slopes of the gravel dunes are on their downslope sides indicates that the flow that controlled their final shape was the rundown of the last wave to cross them. Evidence from the gravel stratigraphy in the gulch at the head of Kapihaa Bay, adjacent to the dunefield, suggests that the third wave was the last to reach the level of the dunes (Fig. 3). The third wave deposited a layer about half as thick as the most energetic second wave, so we assume that it went up less high on Lanai. In the calculations that follow, we assume that it went up roughly 190 m, or half as high as the second wave, an intermediate estimate that takes into account both a four-fold reduction of water flow and a relatively great sediment availability. Maximum runup of tsunamis exceeds incident half-wave height by about a factor of 2.6 (Goring, 1978). Using this factor with the 190-m runup, and taking into account the 30-m elevation of the center of the dunefield, we suggest that the water depth over the dunefield during the third wave reached about 40 m.

Whether waterlaid dunes will form depends on the water velocity, on the availability of sediment, and on an adequate duration of flow. When these factors are sufficient, the equilibrium spacing of the dunes is mainly controlled by the depth of the water. Richards (1980) has proposed that the average spacing of such dunes equals the water depth times 2π. If this relationship applies, the dunes on Lanai, which have a spacing of 10 m (although it was probably different during previous waves and during earlier stages of the third wave), took their final form when the rundown water from the third wave was only about 1.6 m deep. This was a time when the rundown had a high

Figure 4. Plan view of gravel bedforms north of the point between Kapihaa and Hulopoe Bays, Lanai. Relief is about 0.5 m between the gravel-wave crests (lines of black dots) and their flanks (diagrammatic gravel pattern) and about another 0.5 m between the flanks and the troughs (unpatterned areas). Crosshatch pattern indicates basalt bedrock; U-shaped symbols are aboriginal house walls; the elevation at the bottom of the map is about 20 m and at the top about 40 m.

velocity, immediately before the height of the bedforms themselves began to interfere with the flow.

POSTDEPOSITIONAL SUBSIDENCE

The maximum height of soil stripping on Lanai is 365 m (Stearns, 1940). Tide-gauge records indicate that, because of volcanic loading, Hilo, Hawaii, is subsiding at about 2.4 mm/yr, and Kahului, Maui, at 0.3 mm/yr (Moore, 1987). Extrapolation from these values indicates that the type locality of the Hulopoe Gravel on Lanai is subsiding at 0.1 mm/yr. This rate equals a subsidence of about 10 m during the 105,000 yr since the time of the Hulopoe waves, a conclusion supported by the fact that the Waimanalo (Sangamon) terrace and fossil reefs of approximately that age, which stand at an elevation of 7 m on the more stable island of Oahu, are not found above sea level on Lanai. The level of soil stripping and the amount of subsidence together indicate that the highest runup at Lanai extended to an elevation of 375 m relative to sea level at the time of the waves. In addition, if the sea level was not at the highest stand of the Sangamon interglaciation when the waves occurred, the runup could have been as much as 50 m higher.

SOURCE OF THE WAVE TRAIN

The elevations of the highest known limestone clasts on the various Hawaiian Islands, and hence approximately the heights of wave runup there, give clues as to the position of the source of the Hulopoe waves (Moore and Moore, 1984). High-level clasts are not known on Kauai, or on the other more distant islands to the northwest beyond Oahu, indicating that the source of the waves was southeast of Oahu. The rate of subsidence of the southeastern islands and the age of the Hulopoe Gravel provide an estimate of the original height of the highest limestone clasts on the islands (Fig. 1). The distribution of this array of wave heights—highest on Lanai and successively lower on more distant islands—implies a source southwest of Lanai and not very distant.

The pattern of soil stripping on the islands, which reaches somewhat higher elevations than the highest limestone clasts, is a more continuous data set and hence can add additional refinement to the inferred source of the wave train. Stearns (1940) reported that the soil is stripped to 365 m on Lanai and about 240 m on the island of Kahoolawe, which lies 40 km southeast of Lanai. Erosion younger than the Hulopoe Gravel, especially during the heavier rainfall of the last (Wisconsin) glaciation, has cut gullies in the high-level soil and weathered bedrock. We mapped lines bounding the lowest preserved soil on Lanai and Kahoolawe, disregarding the reentrants caused by the gullying. In detail, the limit of soil stripping is not a horizontal plane but varies in elevation around the islands. The highest levels of soil stripping by wave action are on the south side of Lanai and on the west side of Kahoolawe (Fig. 5). On the back sides of both islands, opposite to these areas of high soil stripping, the level of stripping declines to below an elevation of 100 m. Within the limits of variations caused by irregularities in bathymetry and topography around the islands, the generally high and low areas of stripping on each island define a line that would be expected to point toward the source of the wave train. The intersection of such lines from these two islands, modulated also by the maximum height on the islands, indicates a source of 50 km southwest of Lanai (Fig. 5).

The decline in the levels of soil stripping on the front and back sides of Lanai and Kahoolawe can be used to estimate the wavelength of the waves. Very small waves would be almost entirely eliminated in the lee of an island, whereas waves with a wavelength much longer than the dimensions of the island would impinge uniformly around its perimeter (Smith and Sprinks, 1975). The fact that the stripping occurs on both the front and back sides of the islands, but the height of stripping is dissimilar, indicates that the wavelength was about the size of the islands, or 10 to 30 km.

ORIGIN OF THE WAVES

The relatively local source for the wave train, as indicated by the distribution of the onshore deposits and by the soil stripping, eliminates a tsunami generated by a distant tectonic event as a possible cause for the waves. Also, such tsunamis have a runup height of no more than 10 to 20 m (McCulloch, 1985), an order of magnitude less than the waves that deposited the Hulopoe Gravel and stripped the soil on Lanai. Several other possible origins can be considered for the waves: (1) a submarine volcanic explosion; (2) the impact of a meteorite in the ocean; and (3) a local submarine landslide.

The most likely cause of a major nearby volcanic explosion sufficient to generate such giant waves would be a volcano-tectonic event, such as caldera collapse of a volcano near sea level, which could conceivably permit sea-water access to a shallow magma chamber and trigger an explosion. But such an event (more likely with volcanoes of the stage of development seen at the island of Hawaii than at Lanai) would have left extensive onshore pyroclastic deposits approximately contemporaneous with the wave deposits. No such deposits have been found, and no evidence has been seen on the sea floor near Lanai for an active shallow volcano.

A meteorite impact in water deep enough that it would not shower identifiable ejecta from the sea floor onto the islands could have created a wave of the 375-m height inferred for Lanai. Available sea-floor imaging has revealed no features in this area suggestive of meteorite disturbance, but a meteorite that totally dissipated its energy in the water column would be difficult to discount. We rely on newly discovered definite evidence for an alternative cause—a giant submarine landslide—to provide an explanation for the great waves.

The Hawaiian Ridge is flanked by some of the steepest major slopes in the world, and sea-floor imaging of the slope southwest of Lanai with the Gloria sidescan sonar system has identified a huge submarine landslide, the Lanai slide (Normark and others, 1987). The Lanai slide begins about 40 km southwest of the island of Lanai, and its deposits are found more than 70 km downslope from its head. The slide, about 40 km wide at the top, begins in water less than 2,000 m deep, and the area of its toe extends to a depth of 4,500 m (Fig. 5). A large part of the slide extends toward the south, and the bathymetry suggests that another part, not yet fully imaged, extends westward toward Lanai Deep. The lower part of the slide is pervasively hummocky and covered by many blocks 200 m across, with the largest exceeding 1 km. The fact that the slide shows these numerous, rather evenly spaced hummocks, similar to those on the fast-moving 1980 Mount St. Helens landslide, implies to us that the Lanai landslide was fast moving and hence capable of producing a catastrophic water disturbance.

We infer that the waves that impinged against the islands were of the type that have been called backfill waves by McCulloch (1966). Backfill waves form when an underwater slide moves downward and forms a depression in the water surface above the head of the slide; the water that rushes in to fill the depression overshoots and propagates as a packet of waves that radiate outward.

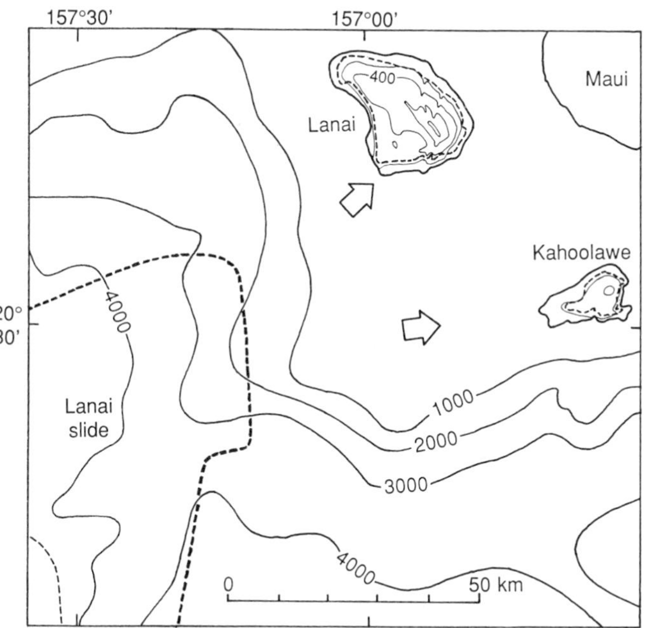

Figure 5. Relationship of the Lanai submarine slide (heavy dashed line) to areas of soil stripping on the islands of Lanai and Kahoolawe. Light dashed lines, interpreted from aerial photographs, enclose the lowest remaining patches of unstripped soil on those islands. Arrows indicate apparent directions of the soil-stripping waves; offshore contour interval is 1,000 m, onshore interval 200 m.

The horizontal extent of a packet of waves created at the head of a submarine landslide would be comparable to the width of the sea-floor depression that caused it (Wiegel, 1970). The depressed head of the slide off Lanai is about 40 km across, implying a similar size for the wave disturbance. We suggest that the three major waves that left deposits on Lanai constituted the packet and had a wavelength of the order of one-third of 40 km, or 13 km.

The velocity of a wave is equal to or greater than the square root of the product of the water depth and the acceleration due to gravity (9.8 m/s^2). For the depth of approximately 2,000 m at the head of the Lanai slide, this relationship gives a minimum initial velocity of 140 m/s. When combined with the wavelength of the individual waves that deposited the three beds of the Hulopoe Gravel, this velocity yields a wave period of 90 s. Hence, considering the uncertainties inherent in analyses of this sort, the three waves in the wave train that crashed against the southeastern Hawaiian Islands 105,000 yr ago arrived, to a first approximation, at intervals of roughly one and a half minutes.

The evidence that three waves made up the wave train (as recorded by the three gravel beds trapped in narrow gulches), that the wavelength was of the order of 10 to 30 km (as suggested by soil-stripping patterns on the islands of Lanai and Kahoolawe), and that the source was about 50 km southwest of Lanai (as

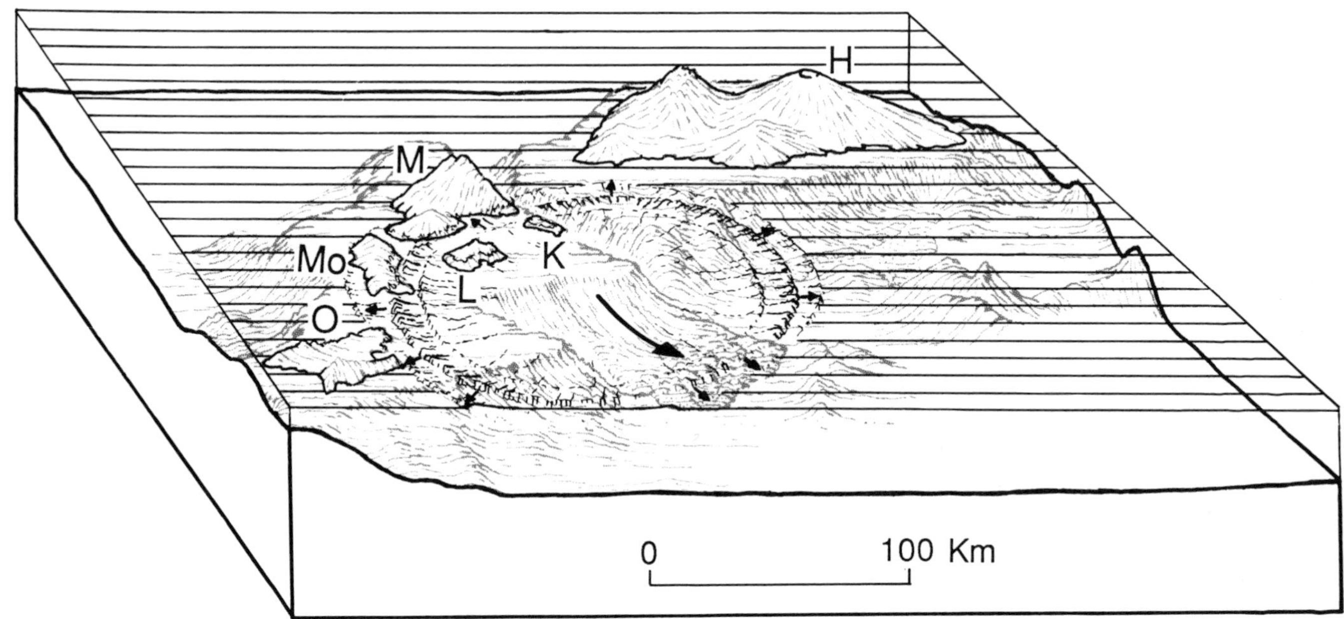

Figure 6. View toward the east of waves on the sea surface (small arrows) radiating from an area above a sea-floor disturbance caused by the Lanai submarine landslide (large arrow). The wave front has passed the islands of Lanai (L) and Kahoolawe (K), is impinging on Oahu (O), Molokai (Mo), and Maui (M), and has not yet reached Hawaii (H). Adapted from a diagram by Tau Rho Alpha.

indicated by the highest wave effects on the various islands) all help to paint a remarkably clear picture of the violent events that occurred on that day (Fig. 6).

The giant waves rushed across the shelf toward the islands at a speed of about 140 m/s. On Lanai, the first wave ran up to a height of about 190 m, briefly hesitated, then accelerated back toward the coast, carrying with it soil that had taken more than a million years to form and churning the residual boulders into a pattern of gravel dunes.

About a minute and a half later, the second and highest wave hit. It swept up to an elevation of 375 m and, except in the deep gulches, picked the gravel back up into suspension. As the wave front coursed its way around the sides of the island, it stripped the terrain down to a scabland surface, lower at the headlands but almost to its initial runup height in embayments (Fig. 5).

When the third and final large wave hit, the first wave was just leaving the northeast coast of the island, 26 km away, and speeding northward at 500 km/hr. The third wave ascended to about 190 m and again took the boulders into suspension. Then, as it accelerated back downward in what would be the final convulsive event of this magnitude to affect Lanai for at least 105,000 yr, it carried away most silt that still remained from the former soil and began to mold the boulders into their final bedforms. When the water reached its final effective velocity at a depth of only 1.6 m, just before its last vestiges sank into the permeable gravel, it shaped the gravel dunes into the 10-m spacing that has remained until now as a fossilized record of this great event.

The 375-m wave runup in Hawaii is the highest attributed to a submarine landslide, although an even higher runup occurred in 1957 at Lituya Bay, Alaska, when a rockfall from land plunged into a fiord and produced a wave that stripped the forest to an elevation of 525 m (Miller, 1960). Major tsunamis from submarine slides, commonly triggered by earthquakes, have also occurred in 1692 at Port Royal, Jamaica (Link, 1960); in 1927 at Sagami Bay, Japan (Shepard, 1933); in 1964 at Seward, Valdez, and Whittier, Alaska (Coulter and Migliaccio, 1966; Lemke, 1967; Kachadoorian, 1965); in 1975 at Moon Bay, British Columbia (Murty, 1979); and in 1979 at Nice, France (Gennesseaux and others, 1980). A tsunami in 1771 at Ishigaki Island, Japan, carried coral clasts to an elevation of 85 m (Miyoshi, 1977).

CONCLUSIONS

High-level marine deposits that have been previously mapped and described on some of the Hawaiian Islands can best be explained as the traces of giant waves and not as relics of preexisting shorelines carried up by tectonically rising volcanoes. The 375-m wave runup on Lanai was apparently caused by a massive submarine landslide off the southwest flank of Lanai Volcano. Coral fragments in the marine deposits yield a uranium-series age of 105,000 yr. That age for the occurrence of the wave is compatible with the thickness of weathering rinds on the basalt clasts, the degree of cementation of the gravel deposits, the extensive dissolution of surface-exposed limestone clasts, and the age of molluscan fossils contained in the wave deposits. The presence of some fossils of species that live today at depths of 70 to 80 m may indicate the drawdown associated with the high runup of the waves.

The waves that washed up the south slope of Lanai were highly erosive and moved a much greater volume of soil and weathered basalt back into the sea than they reworked and left on land. This removed material should be a conspicuous element in the deep-sea sedimentary section off Lanai. Despite the sheetlike character of the waves, they deposited poorly sorted material in favorable depressions as they moved both uphill and down. Such sites include preexisting gulches and at least one volcanic crater.

The features of these fortuitous deposits are important in understanding this convulsive geologic event, as well as in identifying other similar deposits that must surely exist elsewhere.

The deposits in gulches within 300 m of the coast attain thicknesses of 10 m and commonly consist of 1 to 3 beds, each bed a doublet made up of two parts. The lower part was deposited by the wave washing uphill, and it is enriched in organic debris from the sea floor. The upper part was deposited by the wave washing downhill and is enriched in upland-derived material.

The lower part of each bed is a boulder gravel, commonly with a sandy matrix, especially rich in marine calcareous debris near the bottom. It contains clasts of coral, corralline algae, and calcareous beachrock, as well as water-worn and angular basaltic fragments. The white calcareous fragments are highly visible in outcrop, yet they account for only about 5 percent of this interval.

The upper, downwash-produced part of each bed is made up chiefly of large, clast-supported, subrounded to subangular basalt boulders. The large clasts primarily formed as residual basalt boulders that survived weathering, but they also include fragments from the scoriaceous and clinkery flow tops of lava flows, as well as some basalt boulders carried up from the beach. The unit shows a stronger bimodal size distribution than the lower part of the bed. The matrix is silty to pebbly sand derived primiarily from weathered basalt and the overlying soil layers. This matrix material, where it has not been leached, contains thoroughly admixed small marine fossils, including foraminifers, crustacean appendages, echinoid spines, worm tubes, and tiny snails. These microfossils were carried up from the offshore reef and platform and remained suspended in the wave as it washed back downslope. They are a usefully diagnostic feature of wave deposits that carry few marine materials of larger size.

Although the layered deposits in the gulches are discontinuous and incomplete, they record a more complete history of the passage of the waves than do the deposits on the interfluve surfaces, where the primary traces of wave passage are irregular, branching, dunelike boulder ridges that are usually transverse to the downslope-moving water. The dunes are made up of boulders that average 0.5 m in size, and the ridges are about 1 m high and 10 m apart.

During the half century before this study began, these high-level marine deposits were assumed to represent marine material at preexisting shorelines that had been tectonically uplifted to their present position. The preponderance of the evidence now, however, favors the concept that a convulsive geologic event—a series of waves of extraordinary magnitude—was responsible for these deposits. The recurrence interval of such catastrophies is very low, but the study of unusual phenomena is a critical element of the scientific process (Kuhn, 1962).

ACKNOWLEDGMENTS

We are indebted to Tau Rho Alpha for the isometric diagram, Ellen J. Moore for analyses of molluscan faunas, Alan L. Schweitzer for suggestions above wave properties, and Barney J. Szabo for uranium-series dates. The manuscript was improved as a result of critical reviews by David A. Cacchione, H. Edward Clifton, David M. Rubin, and Peter H. Stauffer.

REFERENCES CITED

Baker, V. R., 1973, Paleohydrology and sedimentology of Lake Missoula flooding in eastern Washington: Geological Society of America Special Paper 144, 79 p.

Clifton, H. E., 1969, Beach lamination; Nature and origin: Marine Geology, v. 7, p. 553–559.

Coulter, H. W., and Migliaccio, R. R., 1966, Effects of the earthquake of March 27, 1964, at Valdez, Alaska: U.S. Geological Survey Professional Paper 542–C, 36 p.

Couthouy, J. P., 1844, Remarks on coral formations in the Pacific: Boston Journal of Natural History, v. 4, p. 137–162.

Emory, K. P., 1924, The island of Lanai; A survey of native culture: Bishop Museum Bulletin 12, 139 p.

Gennesseaux, A. M., Mauffret, A., and Pautot, G., 1980, Les glissements sous-marins de la pente continentale nicoise et la rupture de cables en mer Ligure (Mediterranee occidentale): Paris Academie des Sciences Comptes Rendus, v. 290, ser. D, p. 959–962.

Goring, D. G., 1978, Tsunamis; The propagation of long waves onto a shelf: California Institute of Technology Keck Laboratory Report KH–R–38, 337 p.

Kachadoorian, R., 1965, Effects of the earthquake of March 27, 1964, at Whittier, Alaska: U.S. Geological Survey Professional Paper 542–B, 21 p.

Kuhn, T. S., 1962, The structure of scientific revolutions: Chicago, Illinois, University of Chicago Press, 210 p.

Lemke, R. W., 1967, Effects of the earthquake of March 27, 1964, at Seward, Alaska: U.S. Geological Survey Professional Paper 542–E, 43 p.

Link, M. C., 1960, Exploring the drowned city of Port Royal: National Geographic Magazine, v. 117, no. 2, p. 151–183.

McCulloch, D. S., 1966, Slide-induced waves, seiching, and ground fracturing caused by the earthquake of March 27, 1964, at Kenai Lake, Alaska: U.S. Geological Survey Professional paper 543–A, 41 p.

——, 1985, Evaluating tsunami potential: U.S. Geological Survey Professional Paper 1360, p. 375–500.

Miller, D. J., 1960, Giant waves at Lituya Bay, Alaska: U.S. Geological Survey Professional Paper 354–C, p. 51–83.

Miyoshi, H., 1977, The most amazing tsunami in history: International Tsunami Information Center Newsletter, v. 10, no. 1, p. 1–3.

Moore, J. G., 1987, Subsidence of the Hawaiian Ridge: U.S. Geological Survey Professional Paper 1350, p. 85–100.

Moore, J. G., and Moore, G. W., 1984, Deposit from a giant wave on the island of Lanai, Hawaii: Science, v. 226, p. 1312–1315.

Murty, T. S., 1979, Submarine slide-generated water waves in Kitimat Inlet, British Columbia: Journal of Geophysical Research, v. 84, p. 7777–7779.

Normark, W. R., Lipman, P. W., Wilson, J. B., Jacobs, C. L., Johnson, D. P., and Gutmacher, C. E., 1987, Preliminary cruise report, Hawaiian Gloria Leg 2, F6-86-HW, November 1986: U.S. Geological Survey Open-File Report 87-298, 42 p.

Richards, K. J., 1980, The formation of ripples and dunes on an erodible bed: Journal of Fluid Mechanics, v. 99, p. 597–618.

Rubin, D. M., and McCulloch, D. S., 1980, Single and superimposed bedforms; A synthesis of San Francisco Bay and flume observations: Sedimentary Geology, v. 26, p. 207–231.

Rusnak, G. A., 1957, The orientation of sand grains under conditions of "unidirectional" fluid flow: Journal of Geology, v. 65, p. 384–409.

Shepard, F. P., 1933, Depth changes in Sagami Bay during the great Japanese earthquake: Journal of Geology, v. 41, p. 527–536.

Smith, R., and Sprinks, T., 1975, Scatterings of surface waves by a conical island: Journal of Fluid Mechanics, v. 72, p. 373–384.

Stearns, H. T., 1940, Geology and ground-water resources of the islands of Lanai and Kahoolawe, Hawaii: Hawaii Division of Hydrography Bulletin 6, 177 p.

——, 1973, Potassium-argon ages of lavas from the Hawi and Pololu Volcanic Series, Kohala Volcano, Hawaii; Discussion: Geological Society of America Bulletin, v. 84, p. 3483–3484.

Wiegel, R. L., 1970, Tsunamis, *in* Wiegel, R. L., ed., Earthquake engineering: Englewood Cliffs, New Jersey, Prentice-Hall, p. 253–306.

Manuscript Accepted by the Society April 4, 1988

Printed in U.S.A.

Geological Society of America
Special Paper 229
1988

Sedimentological consequences of two floods of extreme magnitude in the late Wisconsinan Wabash Valley

Gordon S. Fraser and Ned K. Bleuer, *Indiana Geological Survey, 611 N. Walnut Grove, Bloomington, Indiana 47405*

ABSTRACT

Two floods of extreme magnitude occurred in the Wabash Valley during late Wisconsinan time. The first flood occurred when meltwater that was stored in disintegrating ice to the north was rapidly released into the valley at a point near Delphi Indiana. During the period of gradually increasing flow, a coarsening- and thickening-upward sequence was deposited, and at peak flow, crossbedded cobble gravels accumulated in sets as much as 5 m thick. Trough crossbeds were deposited along narrow reaches of the valley where flow depths and velocities were greatest. Along wider reaches, however, where velocities and depths were less, tabular and convex upward crossbeds were deposited, and in the widest parts of the valley, where the flow was too shallow to maintain equilibrium bedforms, planar-bedded gravels accumulated.

The second flood occurred when a morainal dam was breached and glacial Lake Maumee drained out of the Lake Erie basin and into the valley at Fort Wayne. During initial stages the flood eroded a deep trench in older outwash in the valley, sculpted bedrock outcrops in the valley floor, and carved flutes in uplands where the flow topped valley margins. Deposition occurred during the waning stage, when disorganized cobble gravels in lee-eddy, pendant, and expansion bars were deposited, a train of megaripples was established, and sediments accumulated in the lee of large-scale obstructions on the channel floor.

During both floods, flow strength declined quickly. Waning stage of the first flood is marked only in the uppermost parts of the sequence, where increased sand content, channel-fills of sand and fine gravel, and diminished bed-set thickness occur. In the upper parts of the bars deposited during the second flood, sediments fine upward markedly and are organized into well-defined planar beds.

INTRODUCTION

A convulsive event is one that produces a brief but violent disturbance in the normal course of events. In geology, the term is normally restricted to events with regional or even global impact (Clifton, this volume). An important criterion for establishing the occurrence of a convulsive event, therefore, is recognition of its products over a significantly large area.

The event must also stand out from background processes by the magnitude of its effects and the rarity of its occurrence or by the singular nature of its mechanism. Because of their ability to substantially alter normal sedimentary processes, products of convulsive events, therefore, stand out from those of the normal regime because of their composition or the scale of their components.

The nature of glaciers and their deposits, however, makes application of these criteria difficult in glaciated terrains. Glacial deposits are normally heterolithic, and the appearance of anomalous mineralogical components can easily be lost in the array. Likewise, the size of the components of glacial deposits may be grossly outsized in comparison to the size of products of nonglacial depositional processes. Cobble-boulder gravels, for example, are common constituents of Wabash Valley outwash, especially at points of lateral entry into the river, and cross sets as much as 2 m thick were deposited at the base of the Wisconsinan sequence when the addition of glacial meltwater greatly increased discharge in the system (Fraser and others, 1983; Fraser and Fishbaugh, 1986).

But convulsive events are unique occurrences that normally are of only brief duration. Not only are their products unique, but they tend to make their appearance abruptly in depositional sequences, and it was on this basis that the occurrence of two floods of extreme magnitude were recognized in the Wisconsinan deposits of the Wabash Valley. During both events the morphology of the river and adjacent uplands was substantially altered, and deposits radically different from those that enclose them were left in the same stratigraphic position in the depositional sequence everywhere in the valley. The purpose of this paper is to describe the products of these two floods, to propose a source for their water, and to point out those features of their deposits that are indicative of the structure of their flows.

GEOGRAPHIC SETTING

Upstream from the Great Bend of the Wabash at Williamsport/Attica, the Wabash Valley is a narrow, northeast-to-southwest-trending trough that was cut primarily in till during late Wisconsinan time. At Williamsport, however, the valley turns to the south (Fig. 1) and occupies a Tertiary bedrock valley. South of Terre Haute this bedrock valley widens abruptly to several times the width of the upstream reaches.

The modern Wabash River diverges from the late Woodfordian channel at Huntington where it swings to the southeast and drains an intermorainic lowland between the Wabash and Salamonie Moraines (Fig. 1). During late Wisconsinan time, however, a major drainageway was established in a channel that extends northeastward from the present valley between Huntington and Fort Wayne, where it cuts through the Fort Wayne Moraine. This channel is presently occupied by the Little River, which drains to the west into the Wabash River basin.

GEOLOGIC FRAMEWORK

During much of late Wisconsinan time, the stretch of river upstream from Terre Haute flowed within a zone of interaction between northern-source ice of the Lake Michigan Lobe and eastern-source ice of the Huron-Erie Lobe. Because the Wabash River flowed in a zone of interaction between these two ice sheets, it received meltwater and sediments from both during this period (Bleuer and others, 1982).

During the latter stages of the late Wisconsinan, however, northeastern Indiana was overridden by a final advance of eastern-source ice. This ice lobe advanced to a terminal position at the Union City Moraine and then began a retreat that deranged the preexisting drainage and left the well-defined recessional moraines of the east-central ridge moraine province. Elements of the preexisting drainage can still be recognized as discontinuous linear gravel deposits, muck-filled troughs, and short stream segments that have been incorporated into the subsequent drainage net (Ballard, 1985). Apparently because of its large size, the Wabash River was able to maintain its identity through this last ice advance, and except for that part of the modern stream that was diverted at Huntington into an intermorainic trough, it emerged from the Wisconsinan as a coherent channel.

ALLUVIAL SEDIMENTATION

Pre-flood glacio-fluvial sedimentation

During much of the early part of latest Wisconsinan time, the Wabash River acted as a conduit for meltwater and debris from both eastern- and northern-source ice. Waste products of these glaciers entered the valley at points through tributaries, across outwash fans that prograded into the valley, and along broad fronts where ice margins were adjacent to the valley (Fraser, 1982; Fraser and others, 1983; Fraser and Fishbaugh, 1986). On entry into the valley from these lateral sources, these sediments were redirected downvalley and accumulated as a broad, braided, alluvial plain.

Proximal-to-distal changes in grain size and bed configuration are typical for most rivers, but because of the numerous lateral sources spaced along the course of the Wabash River, no such relationship can be confidently defined. Instead, longitudinal bar deposits are juxtaposed vertically with transverse bar deposits, and both show evidence of extensive modification during low flow by channels of various scale (Fraser and Fishbaugh, 1980). Apparently, the streambed consisted of widely varying sediment types that responded differently to changing flow conditions. The resultant deposits consist of an array of cobble gravels in crude horizontal beds, sand and fine gravel in tabular crossbeds, and trough-crossbedded sand (Fraser and others, 1983). The valley was aggrading rapidly during this period, and the channel configuration was in constant flux in response to rapid and extreme variations in discharge imposed by the hydrology of the adjacent glaciers. The resultant alluvial fill, therefore, for the most part consists of only the erosional remnants of these bar and channel deposits.

Jökulhlaup

Products. Sediments of radically different character appear abruptly in the upper part of the alluvial fill of the Wabash Valley, signaling a profound change in sedimentary style. These sediments consist most commonly of cobble gravels in cross sets as much as 5 m thick (Fig. 2a). They overlie the braided-stream alluvium with an erosional contact normally marked by a boulder lag in most places (Fig. 3), and they occur in the same stratigraphic position in outcrops from Terre Haute upstream to Delphi, Indiana.

The deposits, however, do not occur upstream from the Tippecanoe Fan at Delphi. The sand and gravel deposits of the Tippecanoe Fan form a broad outwash plain that slopes gently to the south and intersects the Wabash Valley along a wide front (Fraser, 1982) (Fig. 4). Outwash sediments of the fan rest on a highly irregular surface eroded into northern-source tills (Fig. 4). Sand- and gravel-filled troughs as much as 30 m deep occur under the fan, and till ridges project above the fan surface as low hills. The troughs and ridges together produce a fluted topography on the fan surface that is readily visible on satellite imagery

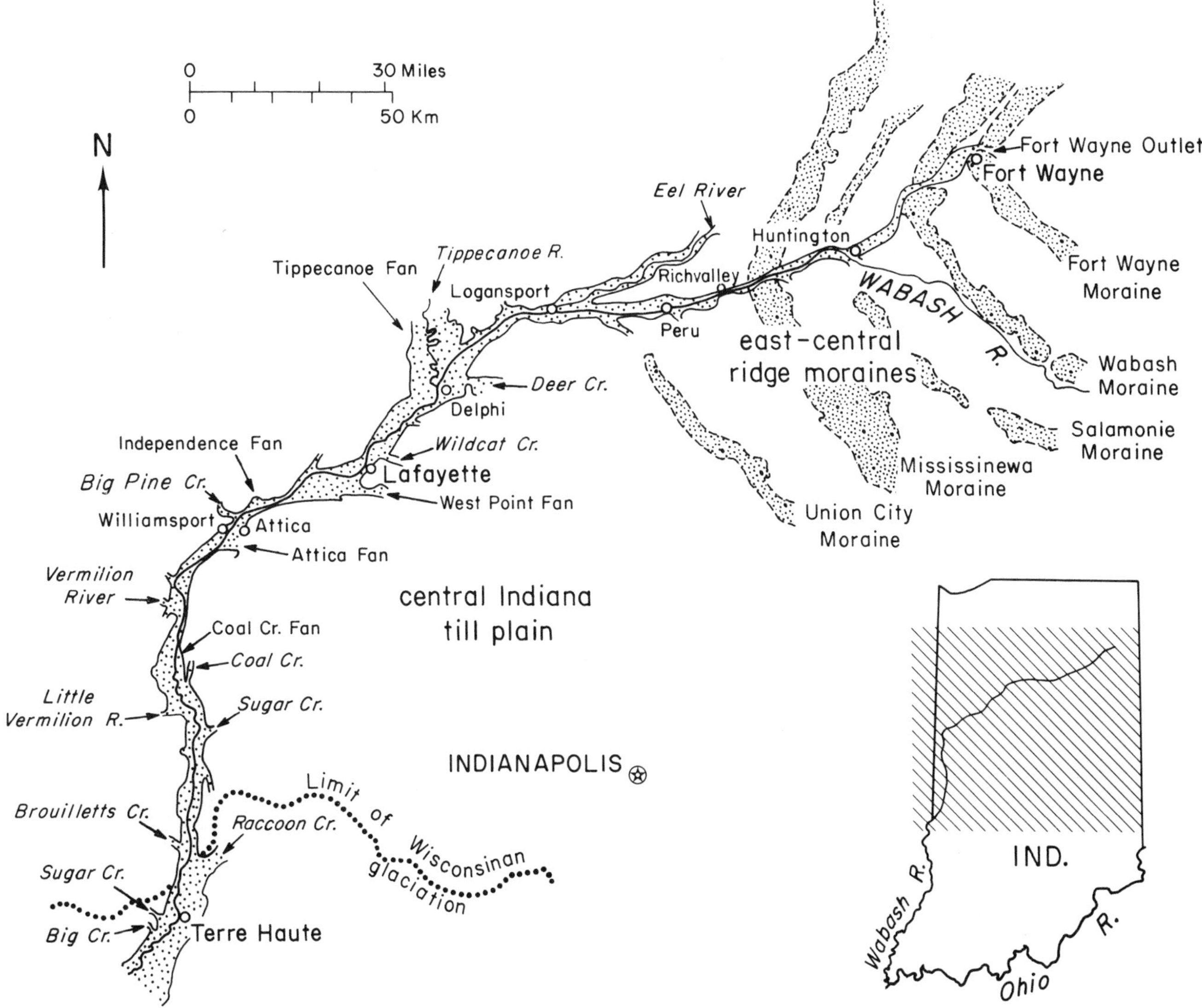

Figure 1. Map of the study area showing sources of lateral input to the Wabash Valley during the late Wisconsinan (arrows), major physiographic features, and location of the channel of the modern Wabash River and the Pleistocene drainageway between Huntington and Fort Wayne Indiana.

(Fig. 5). The fluted topography extends northward until it merges into a terrain composed primarily of till interspersed with sediments deposited in intraglacial channelways and lakes formed in the disintegrating ice.

Basal deposits of the fan consist of cobble gravels in megasets typical of those found in the valley, but within the fan the larger clasts consist predominantly of northern-source till that probably was eroded from the till underlying the fan (Fig. 2b). The upper part of the deposit consists of granular sands in trough sets as much as 10 m thick and 100 m wide. Those larger clasts that occur in the upper part consist almost entirely of till balls; these are normally found only in gravel lenses that are restricted to trough axes (Fig. 2c).

Interpretation. We contend that the Tippecanoe Fan and the megasets of crossbedded cobble gravels in the Wabash Valley resulted from the catastrophic release into the valley of meltwater stored intraglacially in stagnating ice to the north. Such flows, termed "jökulhlaups" in Iceland, occur occasionally in association with subglacial volcanic eruptions. Most, however, are the result of rapid drainage of intraglacial lakes or ice-dammed lakes (Thorarinsson, 1953; Stone, 1963; Fahnestock and Bradley, 1973; Theakstone, 1978).

Perhaps the most compelling evidence for invoking an extreme flow event to explain the megasets in the Wabash Valley is the size and geometry of the sets coupled with the coarseness of their sediments. Coarse sediments can be deposited in large-scale

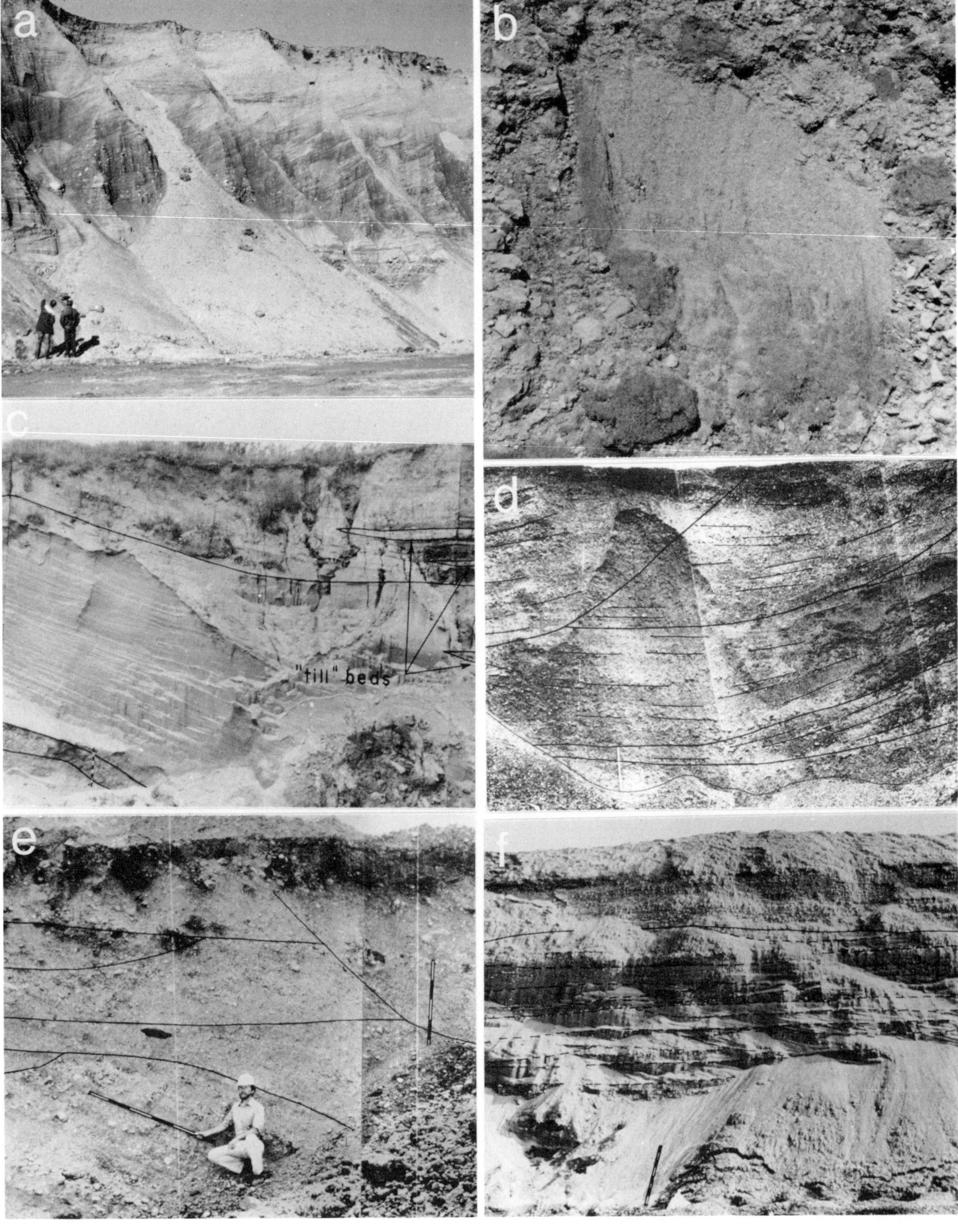
a
b
c
d
"till" beds
e
f

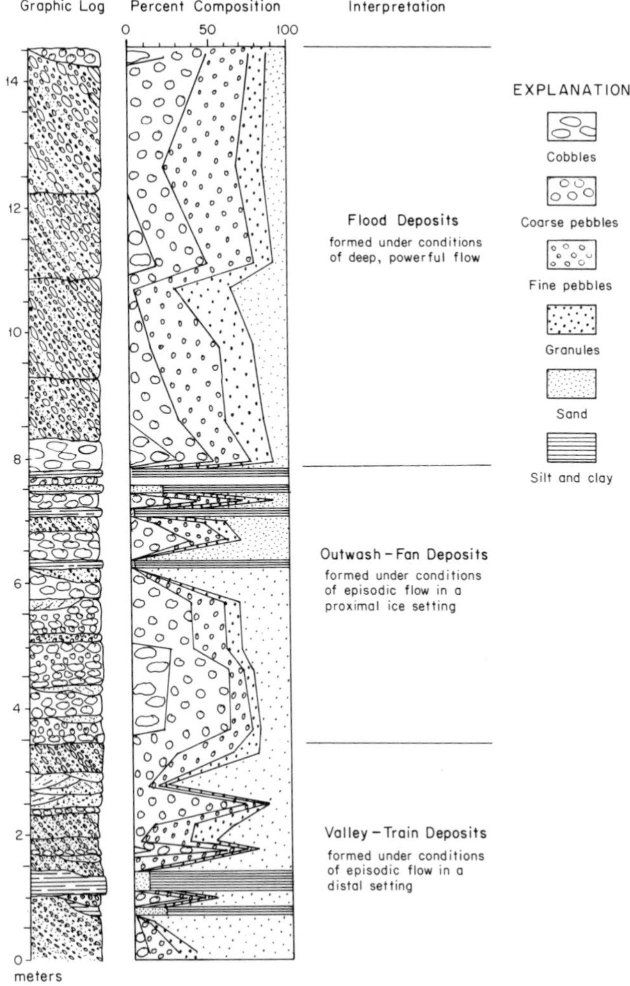

Figure 3. Stratigraphic sequence at Attica, Indiana, showing internal structures and grain-size distribution of jökulhlaup (flood) deposits and underlying valley-train and outwash-fan sediments (after Fraser and others, 1983).

Figure 2. Outcrop photographs and photomosaics of jökulhlaup deposits in the Wabash Valley: (a) large-scale cross sets overlying outwash sediments of the Attica Fan, Neal Gravel Co. pit south of Attica; (b) till clasts in basal deposits of the Tippecanoe Fan; till boulder is about 1 m long, Ryder Sand and Gravel Co. pit west of Delphi; (c) large-scale trough set in the upper part of the deposits of the Tippecanoe Fan, meter stick for scale in lower left, Ryder Sand and Gravel Co. pit west of Delphi; (d) large-scale trough sets in cobble gravels, these are the most typical of jökulhlaup deposits, meter stick for scale in lower left, Parke County gravel pit east of Montezuma; (e) planar and convex-upward foresets typical of those found at wide valley reaches, Triple-G Gravel Co. pit northwest of Waterman; (f) horizontally bedded cobble gravels overlying braided-plain alluvium at Terre Haute, meter stick for scale in bottom center, Martin Marietta gravel pit at Terre Haute.

crossbeds by several mechanisms, including deposition on avalanche faces of deltaic foresets or on nonregime, large-scale bedforms (macroforms). The occurrence of trough crossbeds with tangential basal contacts in the deposits (Fig. 2d), however, suggests that the sets were formed by equilibrium bedforms produced in flows fast and deep enough to initiate and maintain three-dimensional dunes by transporting gravel in at least intermittent suspension past the point of flow separation and by eroding cobble-sized material at the point of flow reattachment.

Not all megasets are trough shaped, however. At wider parts of the valley, the sets are tabular or convex upward and 1 to 2 m high, probably in response to decreasing flow depth and velocity (Fig. 2e). At Terre Haute, where the valley widens threefold (Fig. 1), the sediments stratigraphically equivalent to the megasets consist of cobble gravels in crudely developed horizontal beds (Fig. 2f). Flow velocity at this point must have been sufficient to transport a bedload of coarse gravel, but flow depths were insufficient to permit organization of the gravels into dunes.

Not only was the flow of extreme magnitude, but it also was neither a recurring event, nor a series of local events that produced similar products that were able to affect the whole valley cumulatively over time. The profound difference in sediment character that exists between the underlying braid-plain alluvium and the overlying megasets indicates that normal sedimentary processes were interrupted by a flow of greatly different magnitude and style. The flow, therefore, must have been an extraordinary one even for glacial settings in which the annual nival floods can release large quantities of meltwater in a short time (Ashley and others, 1985). That the flow was not a recurring event is indicated by the restriction of these sediments to the same stratigraphic position in the valley-fill sequence.

The farthest upstream occurrence of the megasets is at the toe of the Tippecanoe Fan where it abuts the Wabash Valley at Delphi. It is probable, therefore, that the flow entered the valley at that point. The ultimate origin of the flow, however, cannot be as well defined.

North of Monticello the fluted topography of the fan is replaced by a centripetal drainage pattern formed by the Tippecanoe River and its first-order tributaries. The tributaries radiate from the trunk stream along relatively straight courses to the periphery of the drainage basin, where higher-order tributaries form a complex network of highly irregular stream courses (Fig. 6). The pattern formed by these minor streams is a consequence of the irregular topography at the periphery of the basin that was formed from stagnating ice. The landscape is dominated by circular or linear collapse features, including lakes, former lake basins filled with muck or peat, and ice-contact glaciofluvial deposits. These collapse features were formed in an ice sheet saturated with water that filled intraglacial voids. The genetic association of the fluted topography on the fan surface and the underlying erosional troughs, together with the flood sediments in the fan and in the Wabash Valley, suggests that a catastrophic release of this meltwater eroded the till plain north of the valley, remolded the coarse alluvium in the valley into large-scale

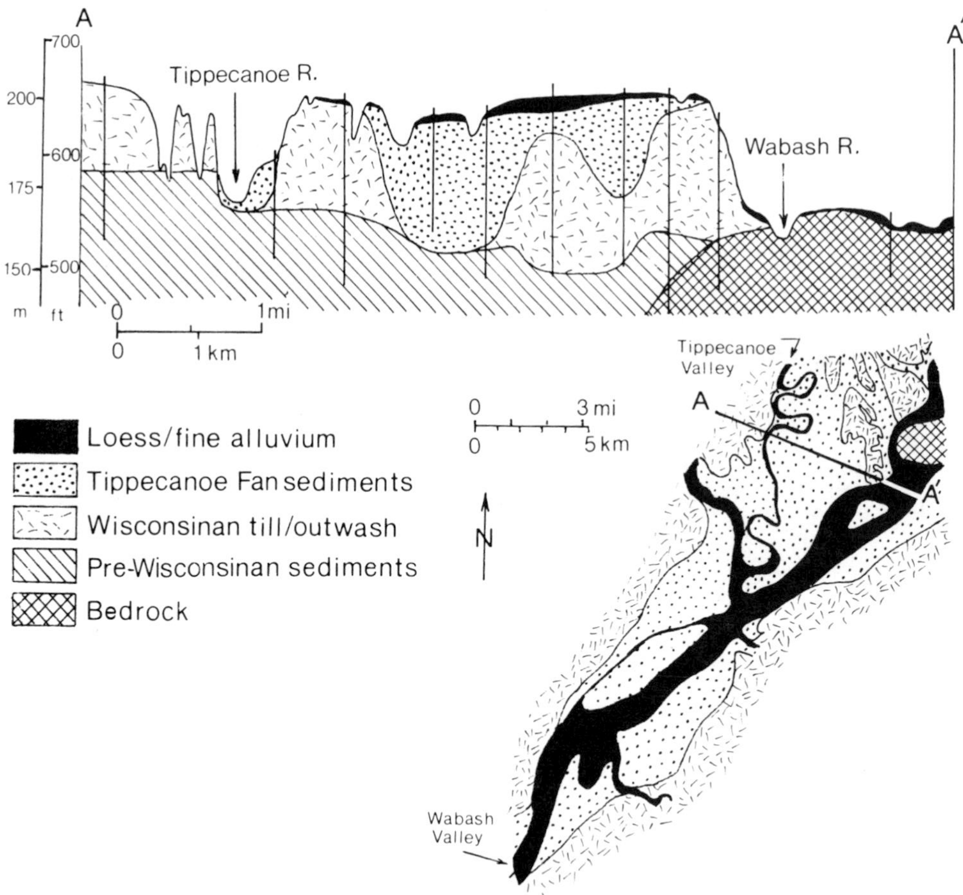

Figure 4. Cross section through the Tippecanoe Fan showing the internal arrangement of sediment facies and map of Tippecanoe Fan area showing surficial geology and location of cross section (after Fraser, 1983).

dunes, and deposited the fan sediments in a back-filling sequence during waning flow.

Except for the outwash sediments of the Tippecanoe Fan, the only evidence of major outlets for waste products of the ice sheet north of the Wabash Valley occurs more than 50 km southwest of Delphi, where large bodies of outwash occur in the Independence Fan (Fraser, 1982, 1983) and in Big Pine Creek. Drainage from the glacier may have been prevented by a bedrock high, paralleling the northwest side of the valley, that forced interior drainage on the ice sheet. Intraglacial water may have accumulated until a threshold was exceeded and meltwater was released through the only significant break in the ridge, which is a bedrock valley that underlies the lower reaches of the Tippecanoe River (Fig. 7). This bedrock valley may, therefore, have localized drainage from the ice and directed it into the Wabash Valley near Delphi.

As the flow from the glacier emerged from the ice, it may have been analogous to the jökulhlaups of Iceland that spread over outwash fans and undergo rapid flow expansion and deceleration. Once the flow entered the Wabash Valley, however, its more likely modern counterparts may be the annual flood from Lake George down the Knik River in Alaska (Bradley and others, 1972; Fahnestock and Bradley, 1973) or the periodic floods down the Alsek Valley from neoglacial Lake Alsek in northwestern Canada (Clague and Rampton, 1982). Flow depths and velocities are maintained during these floods because they are confined to valleys and do not undergo pronounced flow expansion.

Lake George is an ice-dammed lake impounded in a valley by the Knik Glacier. Until 1966, the dam was breached annually in the summer, and discharges as much as 20 times the summer flow resulted. Echograms of the channel profile during flood flow reveal dunes on the channel floor that have amplitudes of about 3 m and wavelengths of about 30 m. The scale of the bedforms was largest along narrow reaches where water depth was greatest and flow velocity fastest. Water depth and flow velocity declined along wider reaches, and bedforms were significantly smaller (Fahnestock and Bradley, 1973).

Large-scale bedforms also occur on the floor of the Alsek Valley, which was periodically flooded by the draining of neogla-

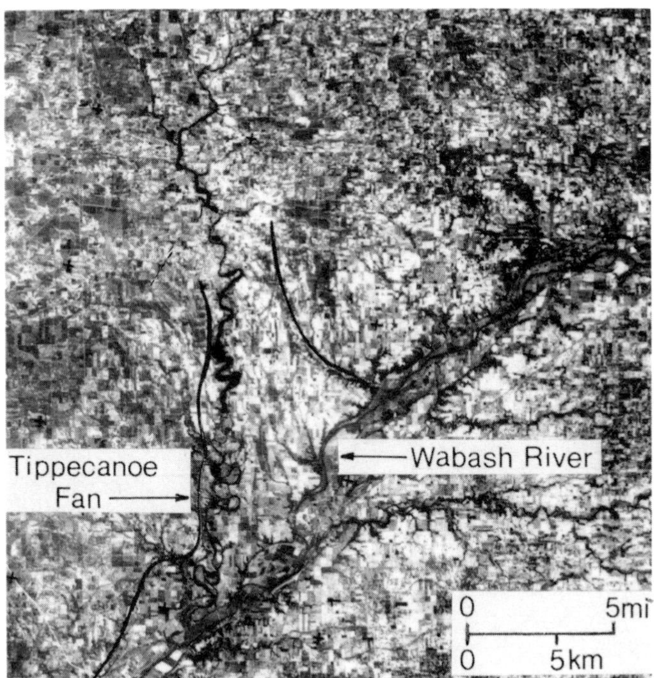

Figure 5. Landsat RBV imagery of the Tippecanoe Fan showing the fluted topography of the fan surface.

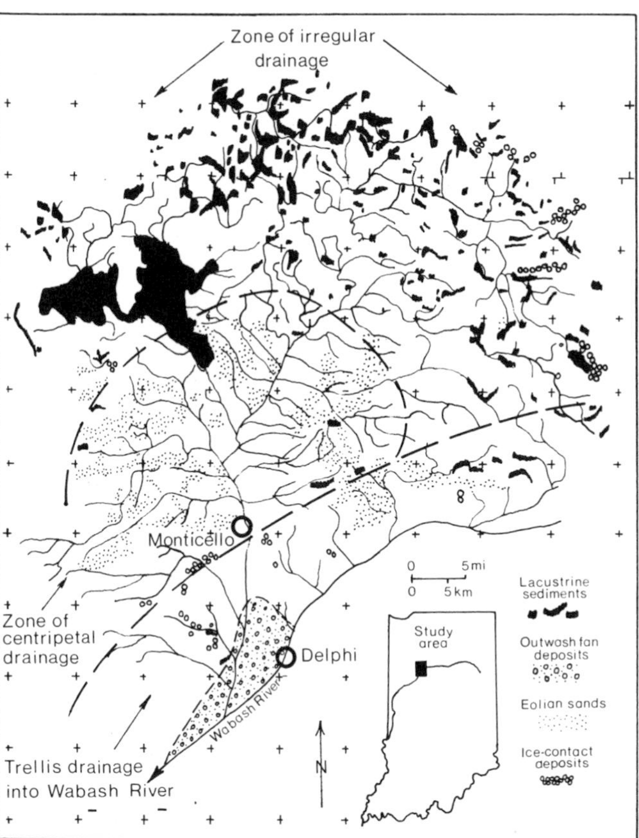

Figure 6. Map of the Tippecanoe Fan area showing drainage features and elements of the surficial geology.

cial Lake Alsek. Sediments in the valley consist of cobble or boulder gravels in lunate dunes locally exceeding 5 m in height and 200 m in wavelength (Clague and Rampton, 1982). Such bedforms could produce cross sets similar in shape and scale to those in the Wabash Valley.

Flow structure. Most jökulhlaups decline from peak discharge in a matter of hours, but they may reach peak discharge in periods ranging from hours to days (Ashley and others, 1985). We initially interpreted that the jökulhlaup in the Wabash Valley had reached peak flow rather quickly because of the sudden appearance of very coarse sediments overlying a boulder lag in the valley-fill sequence. Such a flow would have produced a sharply peaked hydrograph, consistent with an event that lasted no more than a few days.

Two exposures, however, open only for a short time during the latter part of our field investigations, provided evidence that the flow peaked somewhat more slowly, perhaps in a period of days rather than hours. One of the exposures was just west of Lafayette and consisted of a narrow excavation that was elongate parallel to the axis of a trough-shaped cut-and-fill structure. A transverse section of the trough was exposed on the narrow face of the excavation, and the limbs of the trough were exposed on the two long sides. The trough was set in braided-stream alluvium, typical of that found in the alluvial fill of the Wabash Valley, and it was filled with sediments divisible into three units on the basis of grain size, structures, and presence of erosional bounding surfaces (Fig. 8).

The basal unit in the axis of the trough consisted of medium- to coarse-grained sand in climbing ripples that graded upward into plane beds (Fig. 9a). Updip and on the limbs of the trough, the ripples and planar beds passed into high-angle planar foresets; together these sediments appear to represent the bottomsets and avalanche face of a two-dimensional dune that migrated down the axis of the trough.

In the trough axis, the middle unit consists of sand and fine pebbles in festoon trough sets (Fig. 9b). On the limbs of the trough, the trough sets pass first updip into medium-scale, amalgamated, planar crossbeds and then into a single set of large-scale planar crossbeds. A few of the medium-scale planar beds climb subcritically (Fig. 9c). As in the basal unit, the smaller-scale trough and planar sets of the middle unit appear to represent the bottomsets and toesets of the avalanche face of a two-dimensional dune that formed the large-scale planar foresets. The sets of the middle unit, however, are each proportionately larger than the corresponding sets of the basal unit, and the sediments are coarser grained (Fig. 8).

The uppermost unit consists of coarse pebbles with some cobbles in a single trough set similar in size to the jökulhlaup deposits elsewhere in the valley (Fig. 8). Each of the three units probably represents the deposits of separate sets of bedforms that carried progressively coarser sediment and that progressively increased in size and complexity through time. The upward-coarsening-and-thickening sequence suggests that the flow

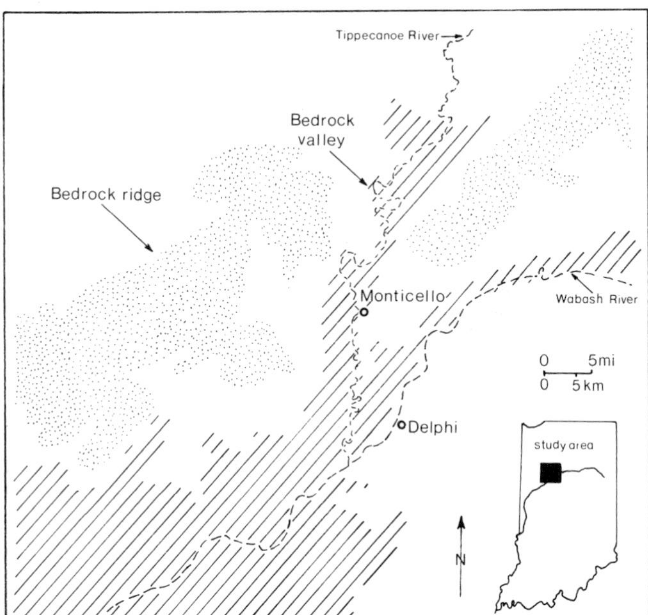

Figure 7. Bedrock topography of north-central Indiana showing a northeast-to-southwest-trending bedrock high north of the Wabash River and a north-to-south-trending valley that cuts through the high.

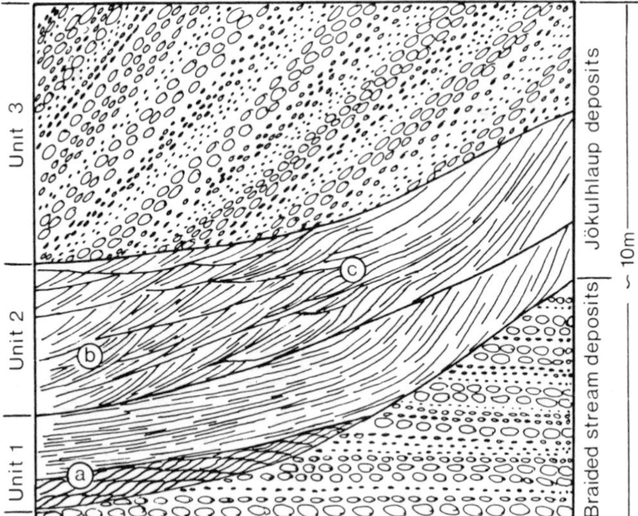

Figure 8. Diagrammatic representation of the sediments at the base of the jökulhlaup sequence showing lateral and vertical changes in bedding structures, grain size, and bed thickness; letters refer to features shown in photographs of Figure 9.

increased in depth and velocity more slowly than had originally been interpreted. Given the time necessary to mobilize the substrate into bedforms and aggrade the channel floor to produce the upward-coarsening-and-thickening sequence seen in outcrop, it is more reasonable to assume that peak flow was reached in a period of days rather than hours.

The occurrence of crossbedded bottomsets and toesets as lateral equivalents of larger-scale planar foresets in the basal two units also reveals something about the nature of the flow during the jökulhlaup. Such structures can be derived in several ways, but for most of them either the type of structures produced, their orientation with respect to expected flow, or the sequence that would have accumulated is inconsistent with the outcrop evidence.

Regressive ripples, for instance, can form under reverse currents produced in the lee of bedforms where flow separation occurs. Plane beds can form at the same time as bottomsets where the suspended load begins to settle out from the base of the separated flow (Fig. 10). These two bedding types are found in the basal unit of the trough, but under conditions of separated flow, an advancing bedform would produce an inverse sequence from that found in outcrop (Fig. 10a). In addition, the plane beds in the basal unit were not formed by fallout from suspension but by fast-flowing traction currents, as indicated by the presence of scour pockets around outsized clasts, and both climbing ripples of the basal unit and trough sets of the middle unit were formed by currents coincident in orientation to the mean downstream flow. Also, it is improbable that the trough crossbeds in the bottomsets of the middle unit could have originated as regressive dunes. It would probably be impossible to maintain a reverse flow strong enough to maintain three-dimensional dunes of moderate scale in the lee of the primary bedform without invoking downstream flows sufficiently strong to destroy it.

On the other hand, if the pressure gradient produced by the expanding flow in the lee of the bedform was insufficient to produce flow separation, simple flow divergence might take place instead (Fig. 10b). Where this occurs, tractive flow would be maintained down the lee face, but it would slow gradually as flow expansion continued. Plane beds would form on the foresets of the larger bedform where flow was most rapid, two- or three-dimensional dunes would form at the toe, and ripples (or lower-flow-regime plane beds in unit 2 where sediments are too coarse to permit ripples) would form the bottomsets (Fig. 10b).

Cross sets characteristically formed by such bedforms occur in the trough of the outcrop, and in the case of flow expansion, the expected orientation of the cross sets is coincident with that seen in outcrop. As in the case of simple flow separation, however, the sequence that would be produced by flow divergence over the crest of a migrating bedform is not consistent with that actually produced (Fig. 10b). Lower-flow-regime ripples or plane beds would be overlain by medium-scale trough sets or planar foresets in turn overlain by large-scale foresets. Instead, climbing ripples are overlain by upper-flow-regime plane beds in unit 1, and only trough sets occur in unit 2.

A third possibility considers the occurrence of a continuous turbidity flow down the lee face of the advancing bedforms (Fig. 10c). If the sediment was available, a very fast flow would be heavily charged with sand at its base where shear-induced turbulence was great. At the point of flow separation at the crest of a bedform, however, shear would cease and a reduction in

Figure 9. Photograph of basal deposits of the jökulhlaup exposed at Meduse Aggregate pit west of West Lafayette: (a) climbing ripples and plane beds of unit 1; (b) trough sets changing laterally to high-angle foresets of unit 2; (c) medium-scale climbing foresets of unit 2.

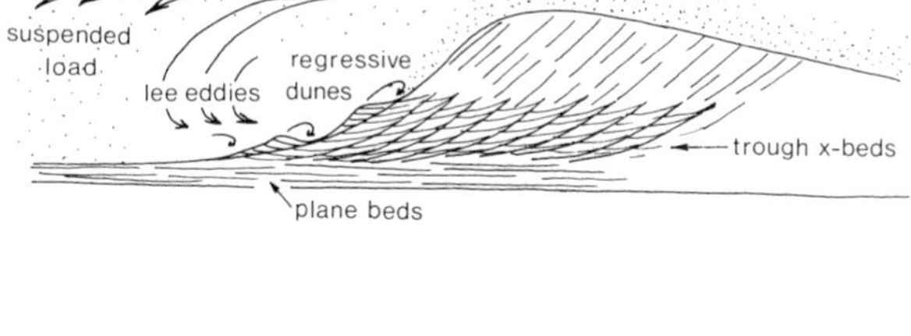

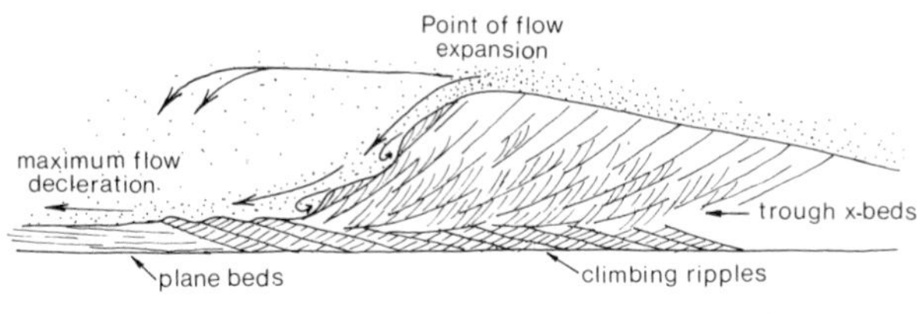

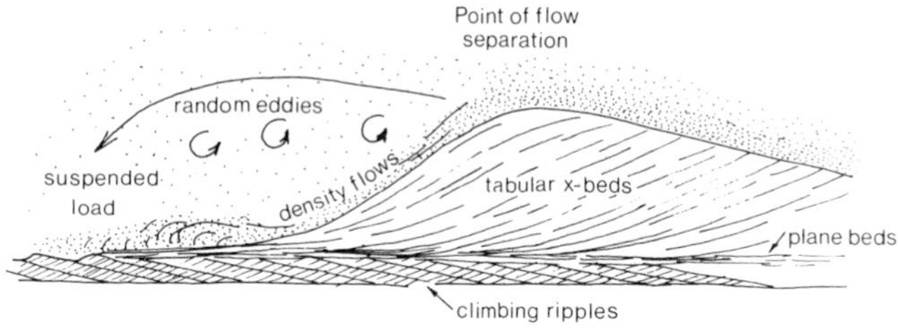

Figure 10. Potential mechanisms for formation of bedding sequence seen at the base of the jökulhlaup deposits: (a) flow separation with strongly developed reverse circulation in the lee of the bedform; (b) flow divergence over the crest of the bedform; (c) continuous density currents driven by sand released from transport at the crest of the bedform that accelerate down the lee of the bedform and begin to decelerate at the toe.

turbulence and eddy viscosity would result. Sand would begin to settle rapidly out of suspension, but if enough sand was available, it could initiate a high-density turbidity flow down the lee face of the bedform (Fig. 10c). Flow velocity would initially be fast down the steep avalanche face, but it would begin to decelerate at the toe of the bedform, and maximum deceleration would occur along the bottomset.

Where sand is the dominant sediment in transport, such a process might deposit plane beds at the toe of the bedform, where flow deceleration began, and climbing ripples along the bottomsets where maximum deceleration occurred. A migrating bedform would deposit a sequence of climbing ripples at the base grading upward into plane beds and high-angle foresets at the top. This is, in fact, the sequence seen in unit 1 in outcrop (Fig. 8).

This interpretation suggests that flow velocities were generally sufficient to keep much of the sand of the sediment load in suspension. This is consistent with the fact that sediments in the megasets normally contain less than 10 percent sand, whereas sediments in the underlying braided alluvium contain much greater amounts.

A similar sequence was later exposed in an excavation at Williamsport about 25 km downstream from the first, and these two remain the only exposures of their kind observed in the valley. Elsewhere, the jökulhlaup deposits consist entirely of amalgamated sets of very large scale trough crossbeds, and in both places where they do occur, the coarsening-upward and thickening-upward sequences are overlain by large-scale trough sets. Apparently, at some point during the history of the flow, velocities became sufficient to maintain sand in suspension beyond the point of flow separation and to produce sufficient turbulence at the point of reattachment to erode the previously deposited sediments.

Flow apparently declined from flood stage more rapidly than it increased to it. Because grain sizes of sediments at the top are not significantly finer than those lower in the flood sequence, shear stress and flow velocity probably remained relatively constant. Set thickness does decrease upward, however, indicating that flow depth and sediment flux decreased as the channel aggraded, and this factor may have helped to maintain constant velocities even as discharge declined.

Only at Williamsport is there any indication that late-stage reworking of sediments occurred. There, a channel was cut in crossbedded cobble gravels and was then filled with finer-grained sediments in smaller-scale trough cross sets. This locality, however, occurs behind a bedrock buttress that may have modified the flow during peak discharge to produce the channeling.

Maumee Torrent

Products. In addition to the jökulhlaup, a second extreme flow is recorded in the Wabash Valley. The origin of the floodwaters and their erosive effects in the Wabash Valley have long been recognized. Gilbert (1871, 1873) noted beach ridges on the Fort Wayne Moraine that he attributed to a predecessor of Lake Erie, which emptied through a notch in the moraine at Fort Wayne. Scovell (1899), McBeth (1901, 1902), Leverett (1902), Malott (1922), Shrock (1928), Fidlar (1936, 1948), Wayne and Thornbury (1951), Thornbury and Deane (1955), and Thornbury (1958) all recognized the erosive nature of the Maumee Torrent where it cut through previously deposited outwash in the valley.

Erosion was not confined to the valley. It also occurred on upland surfaces adjacent to the valley where the torrent must have overflowed the valley margins. At Perrysville, for example, the valley was narrow and apparently could not accommodate the flood. The torrent overflowed the valley walls, crossed a drainage divide separating the Wabash Valley and the valley of the Vermilion River, and cut troughs in the upland till surface before entering the tributary valley and rejoining the main flow (Fig. 11a).

Although the effects of the flood were mainly erosional, lee-eddy bars, expansion bars, and pendant bars, analogous to those in the Channeled Scablands described by Baker (1973) and Baker and Bunker (1985), can be found in some of the upstream reaches. The bars are as much as 1.5 km long and 0.5 km wide and tend toward a streamlined shape that is elongate parallel to the river.

These bars have been interpreted as remnants of the original valley train outwash left after passage of the Maumee Flood (see Wayne and Thornbury, 1951; Thornbury and Deane, 1955; Thornbury, 1958), and in many places their geomorphic expression is consistent with this interpretation. In other places, however, it is difficult to reconcile their shape and placement in the valley with an origin as erosional remnants. At Richvalley, for instance, a string of bars, broken only by narrow gaps, extends across a reentrant in the valley wall (Fig. 12). It is difficult to conceive how these bars could have been left intact while erosion was occurring between them and the valley wall. It is more reasonable to believe that the bars formed from lee eddies in the slack-water zone formed in a preexisting reentrant in the valley wall.

The composition and internal structure of the bars also indicate that they are not intact remnants of previously deposited outwash. In places these bars form a thin veneer over bedrock, but in other places they overlie lacustrine clays, outwash sand and gravels, and till of pre-Wisconsinan age. These sediments are strikingly dissimilar to those in the bars that consist of unsorted, poorly organized, cobble-boulder gravels (Fig. 11c). The bars are also markedly different from the outwash sediments exposed in the terraces lining the valley walls. Therefore, the occurrence, composition, and internal organization of the bars suggest that they were deposited in a trough cut into subjacent and laterally adjacent sediments rather than that they are erosional remnants of these materials. Their textural characteristics and lack of internal organization also indicate that they were deposited by a flow substantially greater in magnitude than the flows that deposited the underlying and laterally adjacent sediments.

Interpretation. The bars are probably the result of deposi-

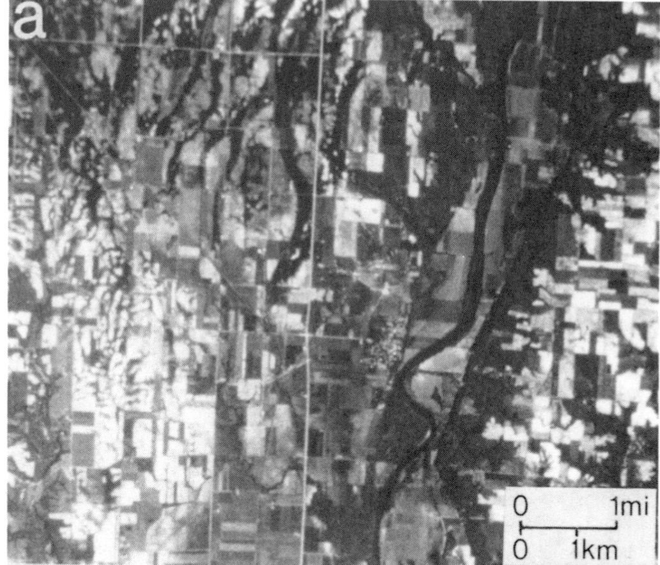

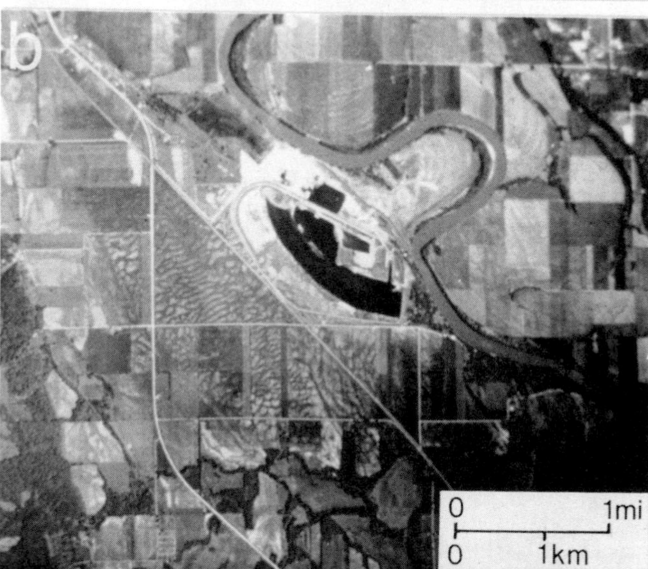

tion during the waning stages of the Maumee Torrent. Peak flows eroded a trough in the pre-flood alluvium in the valley and left a boulder lag on the erosional surface. As the water level in glacial Lake Maumee dropped, the hydraulic head declined, however, and discharge decreased. Concomitantly, as lateral erosion widened the channel, flow velocities slowed until a threshold was reached and deposition was initiated.

Deposition was localized at places where the valley widened and the flow decelerated, or at sharp turns in the valley and in front of tributaries where flow separation occurred and slack-water conditions were established. Deposition also occurred at obstructions in the valley floor where the flow was locally modified. Most of the obstructions are bedrock outcrops of Silurian reefs called klintar (singular form is "klint") (Fig. 13). Flow decelerated rapidly at the upstream face of the outcrops, and intense turbulence was produced. Erosion was localized at the upstream face and along the lateral margins where trains of turbulent eddies occurred, and deposition took place in the lee of the structure.

The combination of erosion and deposition around these klintar produced erosional troughs at their upstream face and along their lateral margins and also produced gravel-rich depositional ridges in the lee of the structure extending downstream from either side. Holocene deposition and erosion have modified these features to some extent, but remnants of the depositional ridges and erosional troughs can still be seen around some of the klintar (Fig. 14).

Because depositional features associated with the klintar are large, however, and because topographic patterns are subdued and masked by vegetation, it is often difficult to unambiguously identify the various features around a given klint. Soils, however, reflect the nature of the underlying parent material and its drainage characteristics, so that the distribution of soil types is a sensitive indicator of these structures in the valley floor. Klintar, themselves, show on soils maps as circular to semicircular areas consisting of exposed rock or as areas where thin soils are developed on steep slopes. Depressions at the upstream face and along lateral margins are underlain by poorly drained soils, and soils indicative of moderate to steep slopes mark the depositional ridges in the lee of the structures. More poorly drained soils occur in the central depression enclosed by the ridges.

Another purely depositional event occurred near Cayuga, where a train of megaripples was deposited on the valley floor (Fig. 11b). Preserved amplitudes of the megaripples exceed 1 m, and wavelengths exceed 50 m. The megaripples consist of a

Figure 11. Features of the Maumee flood in the Wabash Valley: (a) air photo of fluted topography on an upland surface near Perrysville; (b) air photo of a train of sand waves on the valley floor near Cayuga; (c) outcrop photo of internal structures of an expansion bar near Peru, Indiana, showing unorganized cobble-boulder gravels at the base and a well-organized fining-upward sequence at the top.

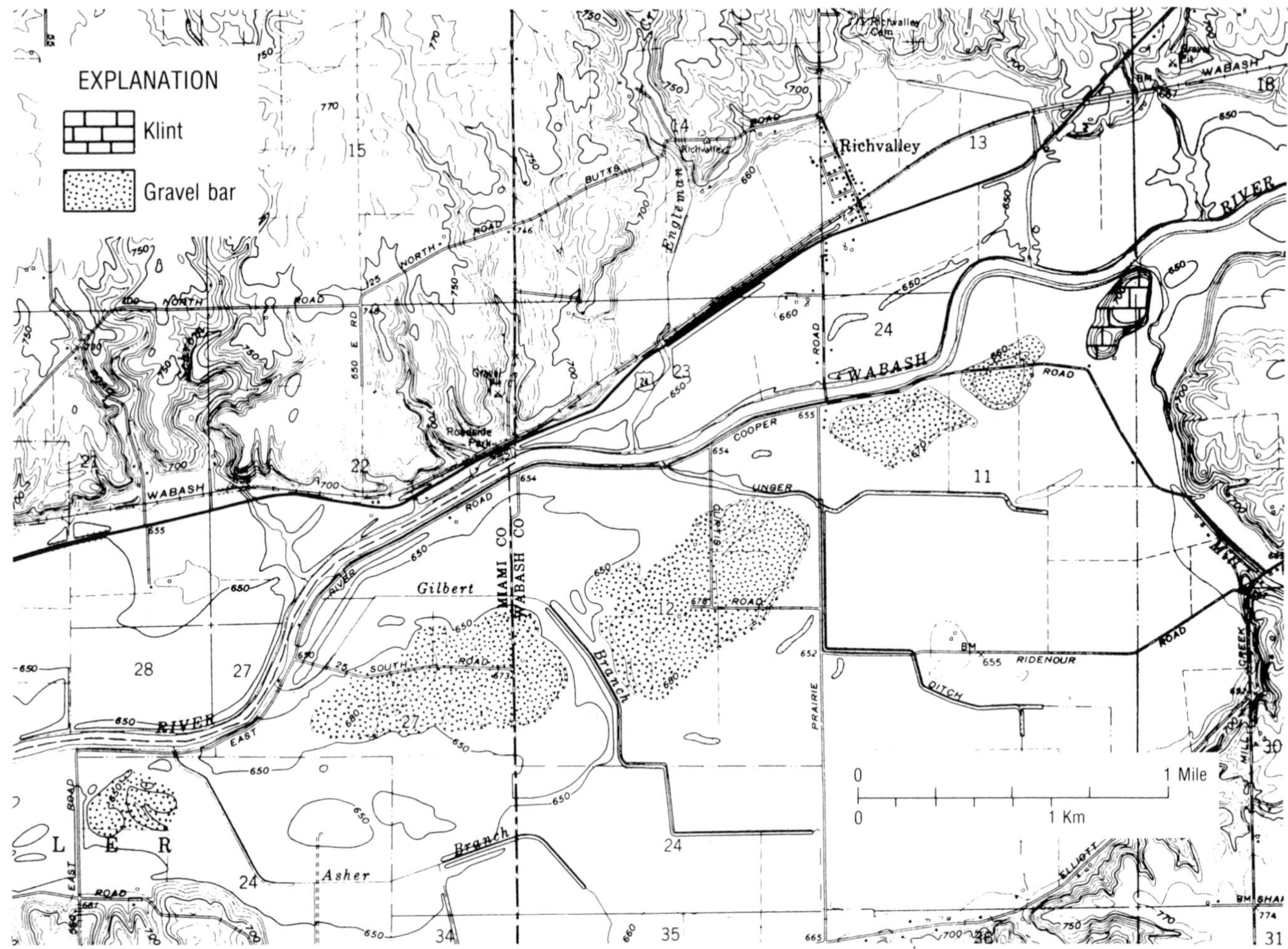

Figure 12. Pendant bars formed during the Maumee flood near Richvalley, Indiana (after Fraser and others, 1983).

veneer of silt (loess and/or modern overbank alluvium) over cores of sand and gravel, and except for a possible reduction in amplitude, the ripples appear to be relatively unmodified. The slopes of the present river and the Pleistocene valley floor are nearly coincident, and because the grain size of the megaripples is relatively fine, they probably formed during very late-stage flow.

Flow structure. Evidence within the valley suggests that the Maumee flood was initially more powerful, peaked more quickly, and declined more slowly than the jökulhlaup. The base of the Maumee Torrent deposits is seen in many places, but nowhere is there evidence of a gradually rising stage. Of course, this may only be a function of preservation, but the extreme dichotomy between flood and pre-flood deposits suggests that fines were selectively eroded during early stages of the flood, leaving a coarse residuum that was deposited rather rapidly once a threshold was attained.

Well-developed fining-upward sequences and the presence of channels in the uppermost parts of bar deposits indicate also that there was a relatively pronounced stage of waning flow capable of modifying bar surfaces and depositing fining-upward sequences. Surfaces of some of these bars display evidence of late-stage reworking during waning flow. In a few places, channels were cut into bars and filled with crossbedded sand and gravel. More commonly, the uppermost part of the bar deposits consists of well-organized, horizontally bedded sand and gravel in pronounced fining-upward sequences (Fig. 11c).

Differences in the structure of the flow may be attributed to differences in the source of the flow. Water released during the Maumee flood was ponded before the flood and was immediately available at the point of outflow. Flow was most intense during initial stages because the pressure head was greatest when the lake stood at its highest elevation, but as the lake drained, the hydraulic head was reduced and the flow waned gradually. This is consistent with flood hydrographs for catastrophic flows from failed dams. Jarrett and Costa (1985), for example, report that peak flow from a failed dam in Colorado occurred within 30 minutes of initiation and that flow gradually waned from maximum in a period of hours. During the jökulhlaup, on the other

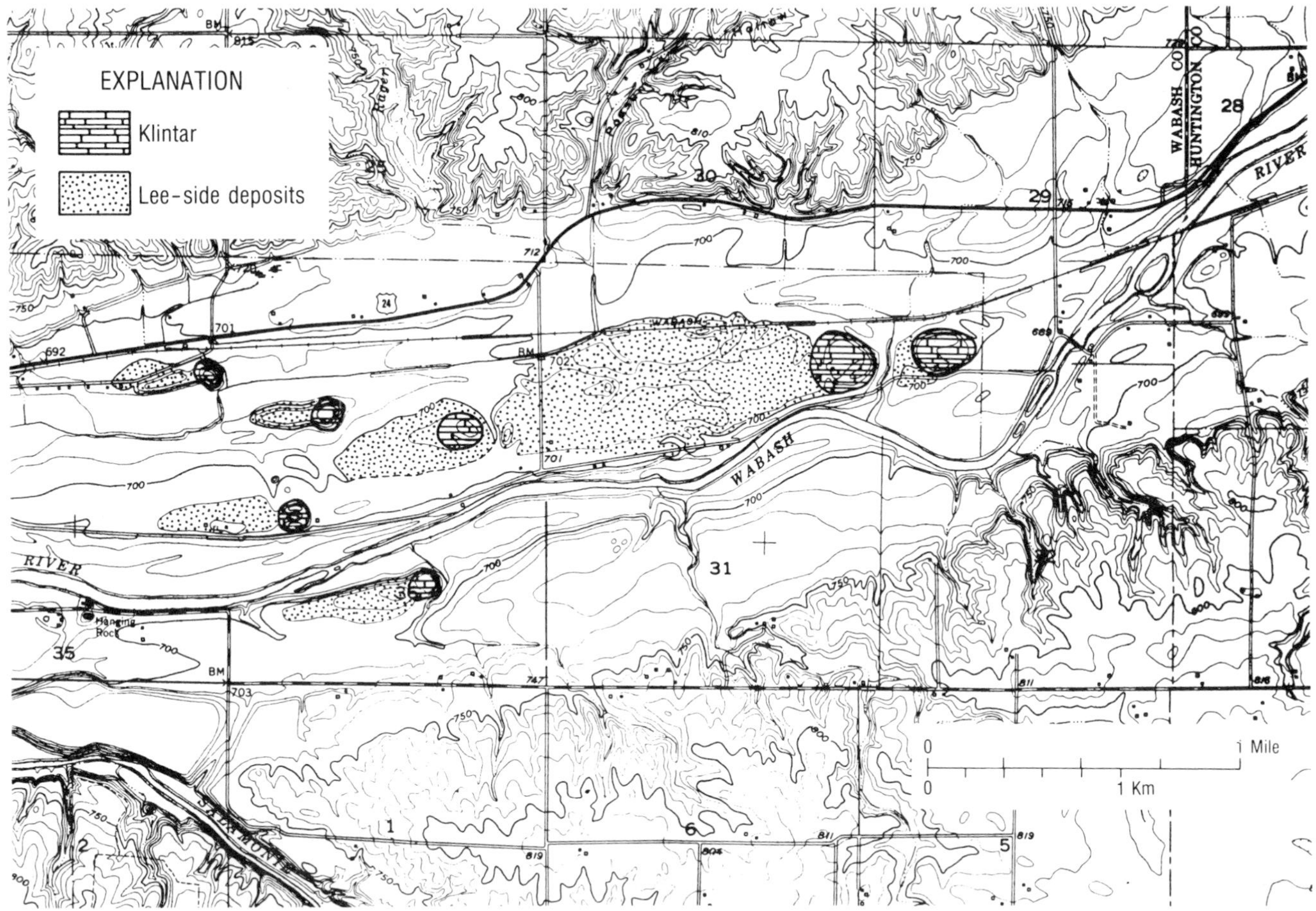

Figure 13. Deposits of the Maumee flood formed in the lee of klintar on the valley floor near Wabash, Indiana.

hand, flow peaked more gradually because the water was stored in intra-ice voids connected by tortuous passageways and was not immediately available at the outlet point.

DISCUSSION

Geomorphic and sedimentologic features within the Wabash Valley point to the occurrence of at least two floods of extreme magnitude during the late Wisconsinan. The first occurred when meltwaters, stored in disintegrating ice north of the valley, were catastrophically released. Before entering the valley and being redirected downstream, these floodwaters eroded and then backfilled deep troughs in the adjacent uplands. In the valley, gradually rising stage deposited a thickening- and coarsening-upward sequence that is capped in most places by cobble gravels in very large-scale crossbeds.

The second occurred when meltwaters, backed up between the Fort Wayne Moraine and an ice front in central Ohio, burst through a low point in the moraine. This flood cut a trough through preexisting glacigenic sediments in the valley and left their remnants as terraces lining the valley walls. Although the flood was primarily erosional, lee eddy bars, expansion bars, and pendant bars in the upper reaches of the valley are evidence of localized deposition, probably during the latter stages of the flood. The bars are very coarse grained and poorly organized except in their uppermost parts where thin, fining-upward sequences of well-defined plane beds were probably deposited by a rapidly waning flow.

When postulating the occurrence of flows of extreme magnitude, one is normally tempted to assess their importance as sedimentary agents and their potential for recurrence. Jökulhlaups occur with relative frequency and are part of the normal hydraulic regime of temperate glaciers in Iceland. Glacial hydraulics and topography may also combine in valley glaciers to produce frequent floods. Lake George, for instance, drained yearly down the Knik Valley until 1966 (Fahnestock and Bradley, 1973), and several floods occurred in the Alsek Valley (Clague and Rampton, 1982).

The jökulhlaup in the Wabash Valley was an extreme event that may have been the result of a one-time draining of the Huron–Erie Lobe north of the Wabash River, but other flows of large magnitude may also have interrupted normal processes of ice drainage elsewhere. Ashley and others (1985) pointed out that the potential for storage of meltwater in temperate-zone Pleistocene ice sheets was probably high, and the abundance of lacus-

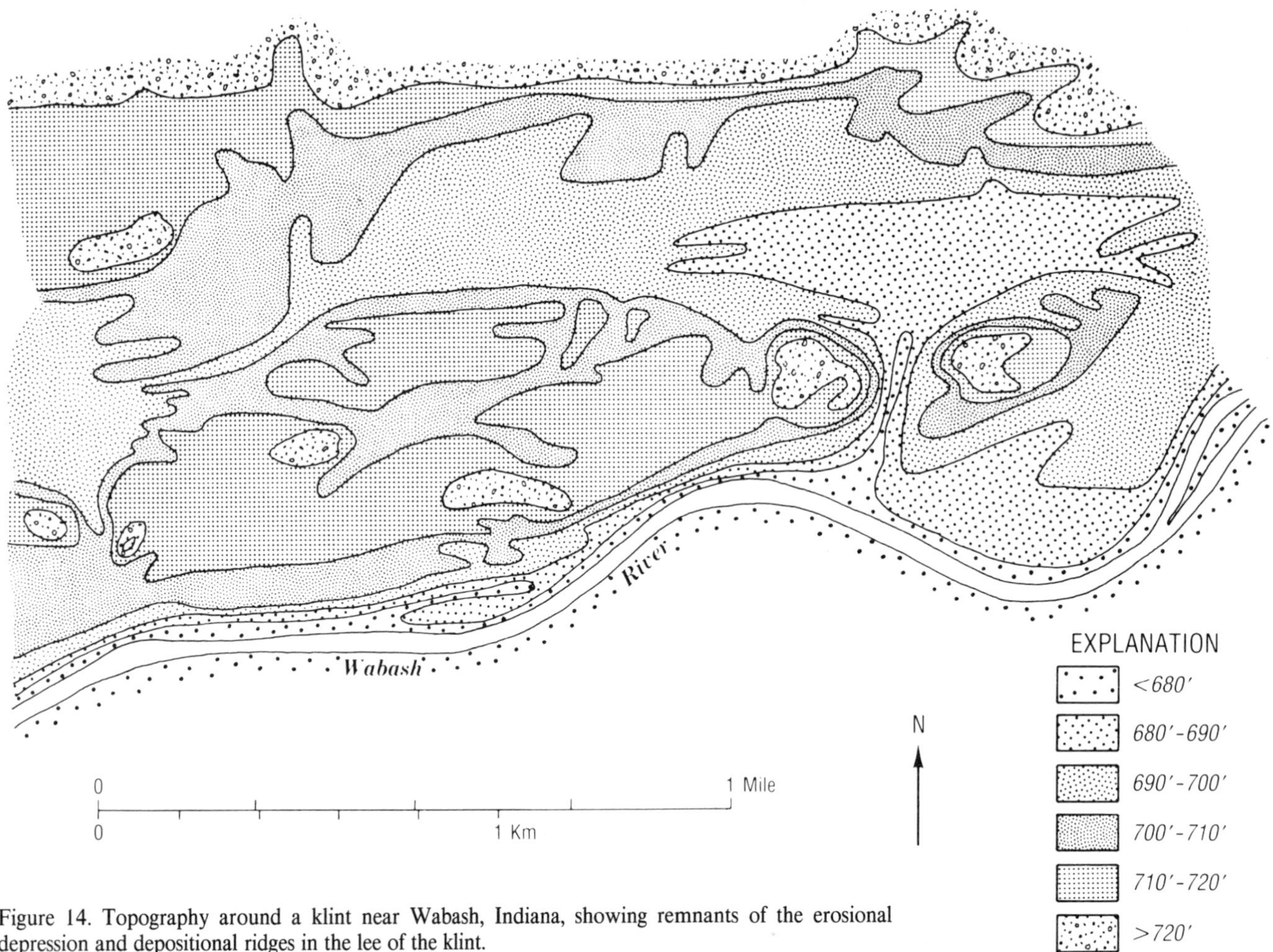

Figure 14. Topography around a klint near Wabash, Indiana, showing remnants of the erosional depression and depositional ridges in the lee of the klint.

trine sediments deposited in intra-ice lakes in Indiana suggests that stored meltwater was abundant. A reexamination of valley-train and outwash-fan sediments may reveal evidence of more frequent occurrences of large-magnitude flows.

The factors involved in producing the Maumee Torrent are somewhat different. The jökulhlaup was dependent on the confluence of factors that could conceivably recur as meltwater was cyclically stored and then released. The Maumee Torrent, on the other hand, was dependent mainly on factors that cannot predictably occur in any ice sheet or recur within a given ice sheet.

In the case of the Maumee Torrent, ice advanced out of a large basin and then retreated rapidly to a position where a large moraine was built. The ice then retreated and left a basin in which meltwater collected until the moraine was breached. Similar situations may also have occurred at the distal ends of the other Great Lakes basins. There is evidence, for example, of high velocity flows from Pleistocene precursors to Lake Michigan (Willman, 1971).

Smaller flows may have occurred outside of bedrock basins where retreat moraines trapped lakes of large extent but shallow depth. Intermorainic lakes such as Lake Douglas (Fraser and Steinmetz, 1971), and Lakes Wauponsee, Watseka, and Pontiac (Moore, 1981) are numerous in Illinois, and there is evidence of catastrophic drainage from some of them.

Ice sheets may also interact with other elements of the local topography to produce large proglacial lakes. Meltwaters were impounded by ice in several river valleys in North Dakota and Saskatchewan, and large proglacial lakes were formed. Once drainage was initiated from several of these lakes, large-magnitude floods resulted (Kehew, 1982).

Large volumes of water are the primary requisite for catastrophic flows, and such volumes are always available from melting ice sheets. The conditions necessary to impound meltwater also seem to occur frequently, so that catastrophic flows from glacial lakes associated with continental glaciers may be relatively common events.

REFERENCES CITED

Ashley, G. M., Shaw, J., and Smith, N. D., 1985, Glacial sedimentary environments: Society of Economic Paleontologists and Mineralogists Short Course Lecture Notes 16, 246 p.

Baker, V. R., 1973, Paleohydrology and sedimentology of Lake Missoula flooding in eastern Washington: Geological Society of America Special Paper 144, 79 p.

Baker, V. R., and Bunker, R. C., 1985, Cataclysmic late Pleistocene flooding from glacial Lake Missoula; A review: Quaternary Science Review, v. 4, p. 1–41.

Ballard, W. T., 1985, Sedimentology and controls on drainage development in the glaciofluvial White River system, northeast-central Indiana [M.S. thesis]: Bloomington, Indiana University, 70 p.

Bleuer, N. K., Melhorn, W., and Fraser, G. S., 1982, Geomorphology and glacial history of the Great Bend area of the Wabash Valley, Indiana: Geological Society of America North-Central Section 16th Annual Meeting Field Trip Guidebook, 63 p.

Bradley, W. C., Fahnestock, R. K., and Rowekamp, E. T., 1972, Coarse sediment transport by flood flows on Knik River, Alaska: Geological Society of America Bulletin, v. 83, p. 1261–1284.

Clague, J. J., and Rampton, V. N., 1982, Neoglacial Lake Alsek: Canadian Journal of Earth Sciences, v. 19, p. 94–117.

Fahnestock, R. K., and Bradley, W. C., 1973, Knik and Matanuska Rivers, Alaska; A contrast in braiding, *in* Morisawa, M., ed., Fluvial geomorphology: Proceedings, 4th Geomorphology symposium, Binghamton, State University of New York, p. 220–250.

Fidlar, M. M., 1936, Features of the valley floor of the Wabash River near Vincennes, Indiana: Indiana Academy of Science Proceedings 1935, v. 45, p. 175–182.

—— , 1948, Physiography of the lower Wabash Valley: Indiana Division of Geology Bulletin 2, 112 p.

Fraser, G. S., 1982, The Independence Fan, *in* Bleuer, N. K., and others, eds., Geomorphology and glacial history of the Great Bend area of the Wabash Valley, Indiana: Geological Society of America North-Central Section 16th Annual Meeting Field Trip Guidebook, p. 35–41.

—— , 1983, Sand and gravel resources of the Wabash Valley; Causes of variability, *in* Proceedings, 18th forum on geology of industrial minerals: Indiana Geological Survey Occasional Paper 37, p. 27–42.

Fraser, G. S., and Fishbaugh, D. A., 1980, Sedimentary structures of the late Wisconsinan terraces along the Wabash River, *in* Fraser, G. S., ed., 10th annual field conference, Bloomington, Indiana: Society of Economic Paleontologists and Mineralogists Great Lakes Section Field Trip Guidebook, p. 59–78.

—— , 1986, Sedimentology of the Coal Creek Fan, Fountain County, Indiana: Indiana Geological Survey Occasional Paper 48, 7 p.

Fraser, G. S., and Steinmetz, D. A., 1971, Deltaic sedimentation in glacial Lake Douglas: Illinois Geological Survey Circular 466, 12 p.

Fraser, G. S., Bleuer, N. K., and Smith, N. D., 1983, History of Pleistocene alluviation of the middle and upper Wabash Valley, *in* Shaver, R. H., and Sunderman, J. A., eds., Field trips in midwestern geology, Bloomington, Indiana: Geological Society of America, Indiana Geological Survey, and Indiana University Department of Geology, v. 1, p. 197–224.

Gilbert, G. K., 1871, On certain glacial and post-glacial phenomena of the Maumee Valley: American Journal of Science, v. 1, p. 339–345.

—— , 1873, Reports on the surface geology of the Maumee Valley, and on the geology of Williams, Fulton, and Lucas Counties and West Sister Island: Ohio Geological Survey Report 1, part 1, p. 535–590.

Jarrett, R. D., and Costa, J. E., 1985, Hydrology, geomorphology, and dam-break modeling of the July 15, 1982, Lawn Lake Dam and Cascade Lake Dam failures, Larimer County, Colorado: U.S. Geological Survey Open-File Report 84-612, 109 p.

Kehew, A. E., 1982, Catastrophic flood hypothesis for the origin of the Souris spillway, Saskatchewan and North Dakota: Geological Society of America Bulletin, v. 93, p. 1051–1058.

Leverett, F., 1902, Glacial formations and drainage features of the Erie and Ohio basins: U.S. Geological Survey Monograph 41, 802 p.

Malott, C. A., 1922, The physiography of Indiana, *in* Handbook of Indiana geology: Indiana Department of Conservation Publication 21, part 2, p. 59–256.

McBeth, W. A., 1901, The development of the Wabash drainage system and the recession of the ice sheet in Indiana: Indiana Academy of Science Proceedings 1900, p. 184–192.

—— , 1902, Wabash River terraces in Tippecanoe County, Indiana: Indiana Academy of Science Proceedings 1901, p. 237–243.

Moore, D. W., 1981, Stratigraphy of till and lake beds of late Wisconsinan age in Iroquois and neighboring counties, Illinois [Ph.D. thesis]: Urbana, University of Illinois, 118 p.

Scovell, J. T., 1899, Terraces of the lower Wabash: Indiana Academy of Science Proceedings 1898, p. 274–277.

Shrock, R. R., 1928, Some interesting physiographic features of the upper Wabash drainage basin in Indiana: Indiana Academy of Science Proceedings 1927, p. 125–139.

Stone, K. H., 1963, Alaskan ice-dammed lakes: Association of American Geographers Annals, v. 53, p. 332–349.

Theakstone, W. H., 1978, The 1977 drainage of the Austre Okstinbreen ice-dammed lake; Its cause and consequences: Norsk Geografiska Tidsskrift, v. 32, p. 159–171.

Thorarinsson, S., 1953, Some new aspects of the Grimsvotn problem: Journal of Glaciology, v. 2, p. 267–274.

Thornbury, W. D., 1958, The geomorphic history of the upper Wabash Valley: American Journal of Science, v. 256, p. 449–469.

Thornbury, W. D., and Deane, H. L., 1955, The geology of Miami County, Indiana: Indiana Geological Survey Bulletin 8, 49 p.

Wayne, W. J., and Thornbury, W. D., 1951, Glacial geology of Wabash County, Indiana: Indiana Geological Survey Bulletin 5, 39 p.

Willman, H. B., 1971, Summary of the geology of the Chicago area: Illinois Geological Survey Circular 460, 77 p.

MANUSCRIPT ACCEPTED BY THE SOCIETY APRIL 4, 1988
PUBLISHED WITH PERMISSION OF THE INDIANA STATE GEOLOGIST

Printed in U.S.A.

Geological Society of America
Special Paper 229
1988

Deposits of a middle Tertiary convulsive geologic event, San Emigdio Range, southern California

Peter G. DeCelles, *Department of Geological Sciences, University of Rochester, Rochester, New York 14627*

ABSTRACT

The upper Eocene through lower Miocene rocks of the San Emigdio Range in southern California present a remarkably well-exposed, east-west cross section of a portion of the sedimentary fill of the southern San Joaquin basin. During middle Tertiary time, a complex of west-northwestward prograding fan-deltas deposited the nonmarine Tecuya Formation and interfingering shallow marine sandstones of the Pleito, San Emigdio, and Temblor Formations. A 50-m-thick, 1.6-km-wide wedge of granitic breccia was deposited near the base of the Tecuya in the eastern (proximal) study area. Several kilometers to the west, a zone of large-scale soft-sediment deformation occurs in stratigraphically equivalent, interfingering nearshore and nonmarine deposits. Approximately 6 km to the west of this disturbed zone, in probably equivalent but relatively deep-water deposits of the lower Pleito Formation, there occurs a 15-m-thick, submarine, mass-flow unit. All three of these units apparently were produced by the depositional and deformational effects of a large earthquake or series of earthquakes.

Major differences in the large-scale lithologic (facies) packaging of two thick (several hundred meters each) depositional sequences below and above the disturbed zones indicate that the architecture of the basin margin, and thus the related depositional systems, changed drastically after the postulated seismic event(s). The previously extant, stepped basin margin (shelf, slope, basin) was destroyed as the shelf edge foundered. The system of shelf fan-deltas, which occupied the basin margin, was replaced by a system of slope fan-deltas. Thus, although the disturbed zones are volumetrically minor in comparison to the sandwiching depositional sequences, they provide key insights into the long-term evolution of the basin margin.

INTRODUCTION

The upper Eocene to lower Miocene rocks of the San Emigdio Range in southern California (Fig. 1) comprise an east-west cross-section of the sedimentary fill of the southern San Joaquin basin. As in many other parts of California (Nilsen, 1984), the Oligocene portion of the stratigraphic section in the San Emigdio Range is represented in part by a thick wedge of coarse, nonmarine detritus and volcanic rocks, in this case the Tecuya Formation (Nilsen and others, 1973; DeCelles, 1986, 1988). The Tecuya consists of southeasterly derived alluvial fan deposits, which interfinger to the west and northwest with shallow marine deposits of the upper Eocene to lower Miocene, upper San Emigdio, Pleito, and lower Temblor Formations. Collectively, these deposits formed in a system of fan-deltas (DeCelles, 1988).

A 50-m-thick lens of very coarse, angular, granitic breccia occurs at or near the base of the Tecuya in the eastern (proximal) study area. At approximately the same stratigraphic level 6 to 7 km to the west, in interfingering nearshore and fluvial deposits, several large penecontemporaneous deformation structures are present. This disturbed zone, in turn, can be approximately correlated with a unit that crops out 6 to 7 km farther west and consists of a mixture of chaotically deformed sandstone, mass-flow conglomerate, and granitic breccia. The stratigraphic and paleogeographic contexts of the latter unit suggest that it was deposited in relatively deep water, probably on the slope.

Several lines of evidence suggest that the three contrasting disturbed zones were produced by seismic events. Although precise physical correlation of the three units is not possible, some evidence (discussed in a later section) strongly suggests that they were produced by the same seismic event. Notwithstanding the potential problems of exact correlation, the sedimentologic results

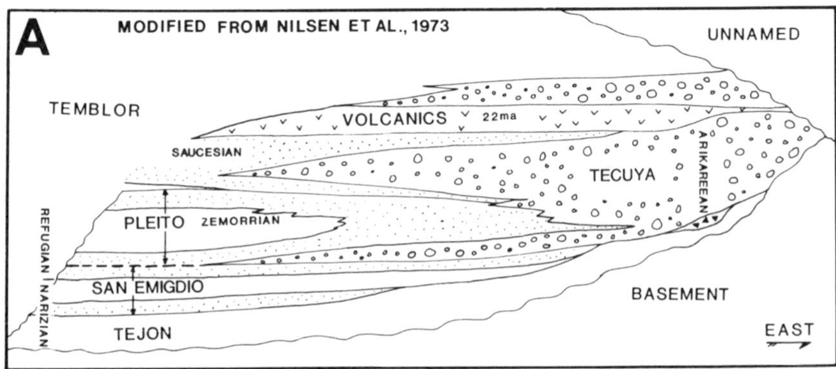

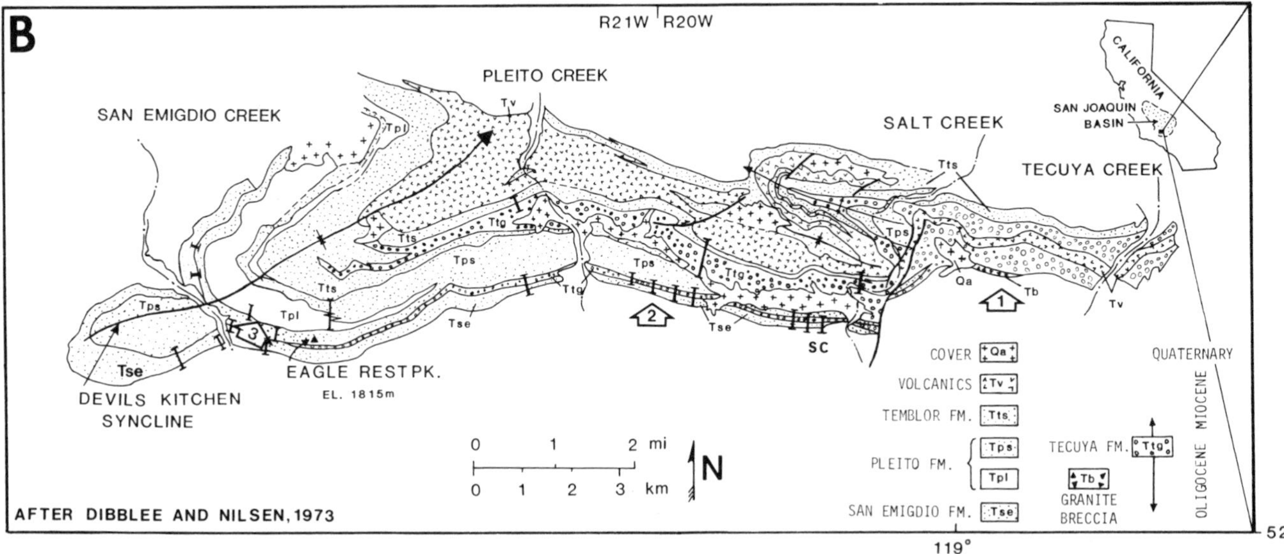

Figure 1. (A) Ages (benthic foraminiferal and mammalian stages of California; K-Ar radiometric date from Turner, 1970) and generalized stratigraphic relationships of upper Eocene to lower Miocene rocks of the San Emigdio Range. Stippled pattern represents shallow-marine sandstone and conglomerate; open circle pattern represents nonmarine conglomerate; and solid triangle pattern represents Tecuya granitic breccia. (B) Geologic map of a portion of the San Emigdio Range. Numbered arrows point to the locations of: (1) the Tecuya granitic breccia, (2) Graham Ridge disturbed zone, and (3) Devil's Kitchen disturbed zone. Bars represent measured sections. SC = Salt Creek sections.

of a seismic event, or series of events, can be observed in three contrasting environments: nonmarine, shallow marine, and relatively deep marine. Moreover, profound changes in large-scale sedimentary packaging, reflecting changes in depositional systems, occurred after the formation of the disturbed zones, suggesting that the disturbance(s) may have signaled the onset of tectonic foundering of the basin margin.

TECUYA DISTURBED ZONE

Description

Hoots (1930) mapped a lens of granitic rock at the base of the Tecuya Formation between Tecuya Creek and Salt Creek (Fig. 1) and interpreted it as a faulted "sliver" of Mesozoic basement rock. However, he acknowledged that the Tecuya rests in depositional contact upon Eocene rocks in a gulch just east of Tecuya Creek, with no evidence for faulting. In contrast, Dibblee and Nilsen (1973) depicted the granitic lens in normal depositional contact with the overlying Tecuya and the underlying Reed Canyon Siltstone Member of the Eocene Tejon Formation. Nilsen and others (1973) considered the granitic lens to be a sedimentary breccia, probably the result of a large landslide that heralded deposition of the Tecuya in the vicinity of Salt Creek.

The granitic lens is roughly 1.6 km long, with a maximum thickness of approximately 50 m. Poor exposure hinders rigorous mapping and description of the breccia. The rock is medium- to coarse-grained and ranges in composition from granitic to granodioritic. It is pervasively fractured into angular blocks and fragments.

Interpretation

The lack of evidence for faulting and the presence of local sandstone-filled cracks and thin (<0.02 m) layers of arkosic sandstone near the top of the breccia unit indicate that it is, indeed, sedimentary. However, the extreme coarseness and angularity of the clasts suggest that they were not subjected to even the slightest amount of reworking and/or abrasion after initial deposition. The breccia could have been deposited by either a rockslide or a sturzstrom (in the sense of Hsü, 1975). According to Hsü (1975, p. 130), a sturzstrom is defined as "a stream of very rapidly moving debris derived from the disintegration of a fallen rock mass of very large size; the speed of a sturzstrom often exceeds 100 km/hr, and its volume is commonly greater than 1×10^6 m^3." Sturzstroms flow with the clasts being held aloft by dispersive pressures exerted by interstitial mud, air, or dust (Hsü, 1975, 1978). Consequently, deposits of sturzstroms typically contain abundant pulverized rock fragments and dust (Hsü, 1975). The Tecuya breccia, on the other hand, contains little, if any, fine-grained material, suggesting that during emplacement, the rock mass may have behaved as a sliding, coherent block (i.e., a rockslide), rather than a flowing stream of loose rocks with interstitial dust or mud.

Paleogeographic reconstruction (DeCelles, 1987) indicates that the southern shoreline of the San Joaquin basin was oriented roughly northeast-southwest during the middle Tertiary. Paleomagnetic data (McWilliams and Li, 1985; Plescia and Calderone, 1986) indicate that the middle Tertiary rocks of the San Emigdio and Tehachapi Ranges were rotated several tens of degrees in a clockwise sense sometime between the Late Cretaceous and early Miocene, which suggests that the shoreline may have been oriented more northerly. The shoreline reconstruction of DeCelles (1987), however, is primarily based on paleocurrent data from a single outcrop belt, which trends east-west and has not changed in orientation with respect to the paleocurrent indicators (i.e., cross strata) contained therein. Therefore, the angle between the outcrop belt (which contains the breccia lens) and paleoshoreline is still 45° to 60°, regardless of tectonic rotations in the horizontal plane. With this in mind, if the breccia was originally square-shaped in plan view, its volume may have been $50\text{m} \times (1{,}600 \text{ m} \sin 45°)^2 = 6.4 \times 10^7 \text{ m}^3$. If the breccia was originally circular in plan view, its volume may have exceeded 1×10^8 m^3. Rockslide and sturzstrom deposits, however, typically are lobate or digitate in plan view (Voight and Pariseau, 1978), so these estimates may be too low by a substantial factor. The size of the Tecuya breccia falls approximately within the middle range of a list of known rockslide and sturzstrom deposits compiled by Hsü (1975). For comparison, the well-known Madison Canyon rockslide in southwestern Montana involved roughly 2×10^7 m^3 of material (Hadley, 1964).

Rockslides and sturzstroms can be triggered by a variety of mechanisms acting alone or in concert. Intrinsic rock properties, such as composition, bedding, foliation, weathered zones, and structural weaknesses (e.g., faults and joints) can promote massive failure (Hadley, 1964; Voight, 1978). Many episodes of mass movement are initiated by a combination of extended periods of heavy precipitation and favorably oriented zones of weakness in the rock mass (Voight and Pariseau, 1978). Probably the most common cause, however, of major rockslides and sturzstroms is seismic activity. Thousands of rockslides were unleashed during the Great Alaska earthquake (M = 8.5) and its aftershocks of 1964 (Long, 1968). The Madison Canyon rockslide occurred in response to the Hebgen Lake earthquake (M = 7.1) of 1959 (Hadley, 1964), and numerous other rockslides have been directly or indirectly caused by earthquakes (Voight, 1978).

An earthquake mechanism is favored as the cause of the Tecuya rockslide breccia for several reasons: (1) the breccia is composed of massive, unfoliated, granitic rock, which probably lacked major through-going zones of weakness resulting from compositional layering or structural fabric; (2) although local zones of altered rock are present in the breccia, they are neither pervasive nor necessarily "primary" (i.e., they could have formed after deposition of the breccia, or they could be modern); and (3) the San Joaquin basin has been located in a tectonically active setting throughout its history, and the close association with the breccia of abundant, coarse, alluvial-fan deposits suggests that the study area was tectonically active during deposition of the middle Tertiary stratigraphic section.

GRAHAM RIDGE DISTURBED ZONE

Stratigraphic and paleogeographic context

Approximately 6.5 km along strike to the west, a zone of large-scale deformation occurs in complexly interbedded Tecuya nonmarine and Pleito nearshore deposits on Graham Ridge (Figs. 1 and 2). The interfingering of marine and nonmarine rocks resulted from local oscillations in the position of the paleoshoreline. In addition to the interfingering relationships, the outcrop is further complicated by a large (>250 m wide and 19 m thick) body of marine Pleito sandstone that is encased within a 30-m-thick unit of nonmarine Tecuya conglomerate (Fig. 3). The marine sandstone body consists of a single progradational sequence of inner-shelf, surf-zone, and longshore-trough facies (measured sections 2 and 3, Fig. 5; see DeCelles, 1987 for descriptions of the nearshore facies). The progradational sequence begins and ends with conglomeratic transgressive lags (facies 1 in Fig. 5), which developed as the nearshore sediment prism migrated landward during transgressions (e.g., Swift, 1968). The marine sandstone body probably was deposited within an entrenched channel on the surface of a fan-delta lobe during and immediately after a marine transgression. In other words, the sandstone body represents a marine-backfilled, entrenched alluvial-fan channel.

Description

Several deformation structures are exposed on the south face of Graham Ridge (Figs. 3, 4, and 5). The margins of the back-

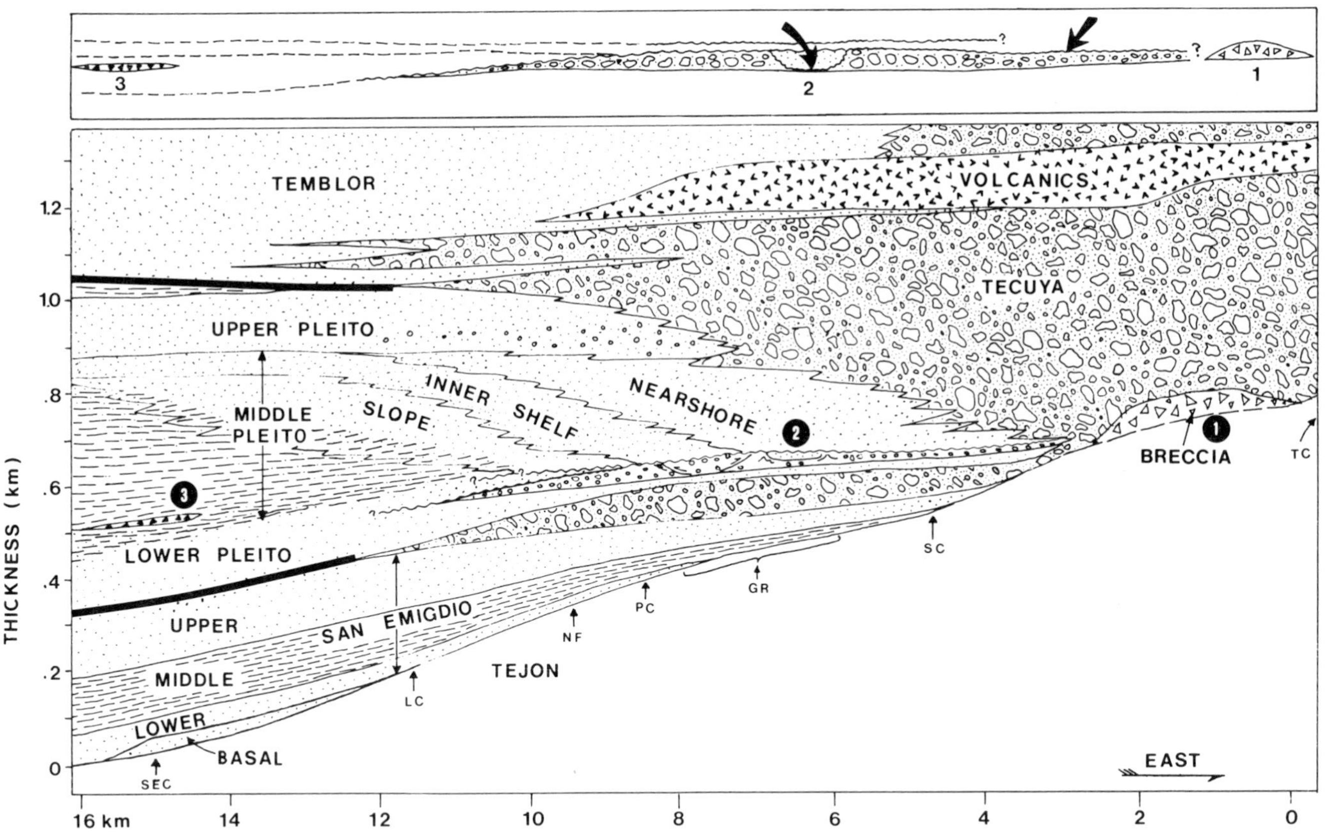

Figure 2. Schematic diagram depicting large-scale facies relationships in the middle Tertiary strata and geographic and stratigraphic positions of : (1) Tecuya granitic breccia, (2) Graham Ridge disturbed zone, and (3) Devil's Kitchen disturbed zone. Generalized depositional subsystems are labeled in the Pleito Formation. Locations are abbreviated as follows: SEC = San Emigdio Canyon, LC = Lost Creek Canyon, NF = Neason's Flat, PC = Pleito Creek Canyon, GR = Graham Ridge, SC = Salt Creek, TC = Tecuya Creek. Contacts between the San Emigdio, Pleito, and Temblor are accentuated with heavy lines on the left (western) side of diagram. Upper panel shows a schematic close-up diagram of important stratigraphic relationships between the three disturbed zones (1, 2, 3) and three transgressive lags (wiggly lines). Conglomerate lags highlighted by arrows are shown in Figure 5.

filled channel have been deformed by a combination of settling and lateral movement. On the eastern flank of the channel, beds in the underlying interbedded Tecuya and Pleito Formations form a nearly symmetrical, 20-m-high buckle structure (Fig. 4A; Fig. 5, structure A). Near the top of the structure, the beds are vertical to slightly overturned, but they can be traced downward into stratigraphically concordant beds on both sides of the structure. The core of the structure is occupied by nonmarine conglomerate. No detachment plane was observed, but the geometry of the structure indicates that it formed by a combination of lateral movement and vertical settling. On the opposite (western) side of the same backfilled channel, beds of the channel fill and subjacent marine sandstones are contorted and internal sedimentary structures have been destroyed, probably by dewatering and lateral shearing.

Approximately 200 m to the west, the nonmarine conglomerates that encase the backfilled channel sandstone are interrupted by a 4-m-thick unit of sandstone, which is inclined approximately 50° from the horizontal (Fig. 4B; Fig. 5, structure B). This unit clearly is not in its original depositional position. The overlying beds are undeformed and continuous across the top of the inclined unit. The inclined unit consists of longshore-trough and beach-foreshore facies, which can be matched with equivalent in situ facies that crop out approximately 100 m to the west (structure B, Fig. 5). The in situ facies have been deformed by lateral forces that stretched and tore some of the beds. It appears that the enigmatically oriented unit (Figs. 4 and 5, structure B) was torn from the in situ facies and transported roughly 100 m eastward along a horizontal zone of detachment.

Approximately 200 m to the west of structure B (Fig. 3), equivalent strata are cut by at least three minor, west-verging reverse faults (Fig. 4C). The eastern end of this outcrop displays beds that have been upturned toward the west. Palinspastic restoration of the faulted beds indicates that they were displaced approximately 30 m to the west.

Several smaller-scale deformation structures occur in the same stratigraphic zone farther to the west. The entire zone is underlain by undeformed, interfingering shallow marine (Pleito)

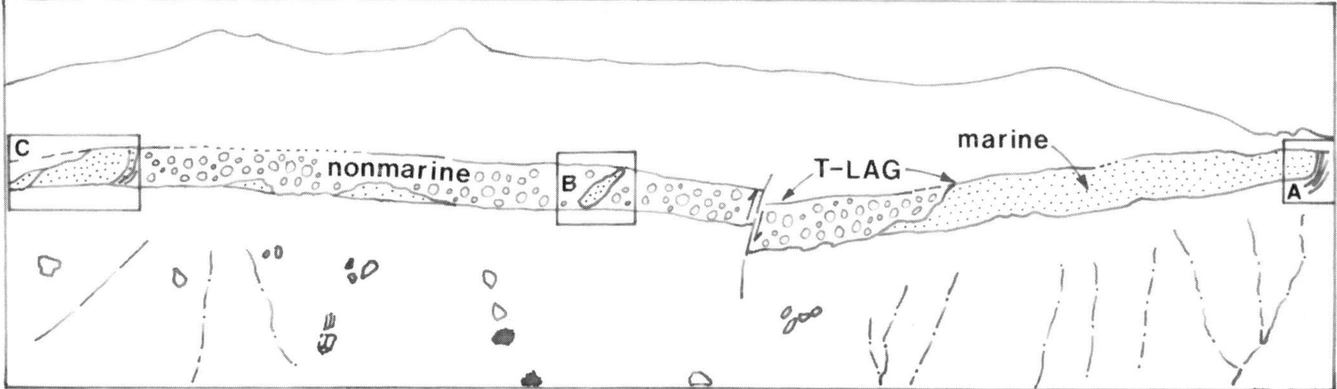

Figure 3. Oblique-aerial photomosaic (top) and tracing (bottom) of the Graham Ridge disturbed zone. East is toward the right. Lettered boxes (A, B, C) on tracing show locations of structures shown in close-up photographs in Figure 4. White arrow on photomosaic points to structure B (see Fig. 4B). Width of view is approximately 800 m.

sandstone and nonmarine (Tecuya) conglomerate. The disturbed zone is overlain by an undeformed transgressive lag, which is in turn overlain by concordant beds of inner-shelf and nearshore sandstone of the Pleito Formation (Figs. 3 and 5).

Interpretation

Several lines of evidence suggest that the disturbed zone exposed on Graham Ridge was formed soon after deposition of the involved beds, rather than by later tectonic processes (i.e., folding and faulting). First, the disturbed zone is more than 25 m thick and is abruptly truncated above and below. If the zone were a bedding-parallel fault zone, deformation would be expected to die out gradually upward (Wojtal and Mitra, 1986). Movement of the somewhat jumbled beds in the disturbed zone would have caused disruption of the overlying material. Instead, an undeformed transgressive lag truncates the top of the disturbed zone (lag 2, Fig. 5). Also, if the disturbed zone were a fault zone, its relatively great thickness would suggest the existence of a major low-angle to horizontal fault, evidence for which is lacking in laterally equivalent beds.

Second, the directions of lateral movement on the various faults in the disturbed zone (both inferred and observed) are, in several cases, opposite to each other and incompatible with movement along a single horizontal fault zone. For example, the faults in structure C (Fig. 4C) indicate movement of several tens of meters to the west, whereas the inclined bed (Fig. 4B) has been displaced at least 100 m to the east.

Finally, no evidence for metamorphism, shearing, or brittle failure, on a microscopic or mesoscopic scale, is present in the lithologies involved in the deformation. The sandstone and conglomerates obviously were not cemented when the deformation occurred. Dislocations of the clasts occurred in a quasi-liquefied state, as indicated by stretched and distorted layers and the lack of microscopic evidence for grain and/or cement deformation.

Figure 6 depicts a hypothetical sequence of events that could have produced the anomalous configuration of sedimentary units on Graham Ridge. A major, probably seismic, disturbance apparently caused intense lateral movement in the sedimentary apron at the toe of the subaerial fan-delta. The amount of lateral displacement of sedimentary units exhibited on Graham Ridge is several tens of meters (Fig. 3). However, given the orientation of the outcrops along Graham Ridge (east-west) with respect to relative paleoshoreline (northeast-southwest), it is not possible to determine from surface data whether any downslope (i.e., northwestward) movement occurred in the sedimentary apron. Conceivably, the oppositely verging reverse faults (Fig. 4B and 4C) represent lateral ramps of a slide zone that facilitated signifi-

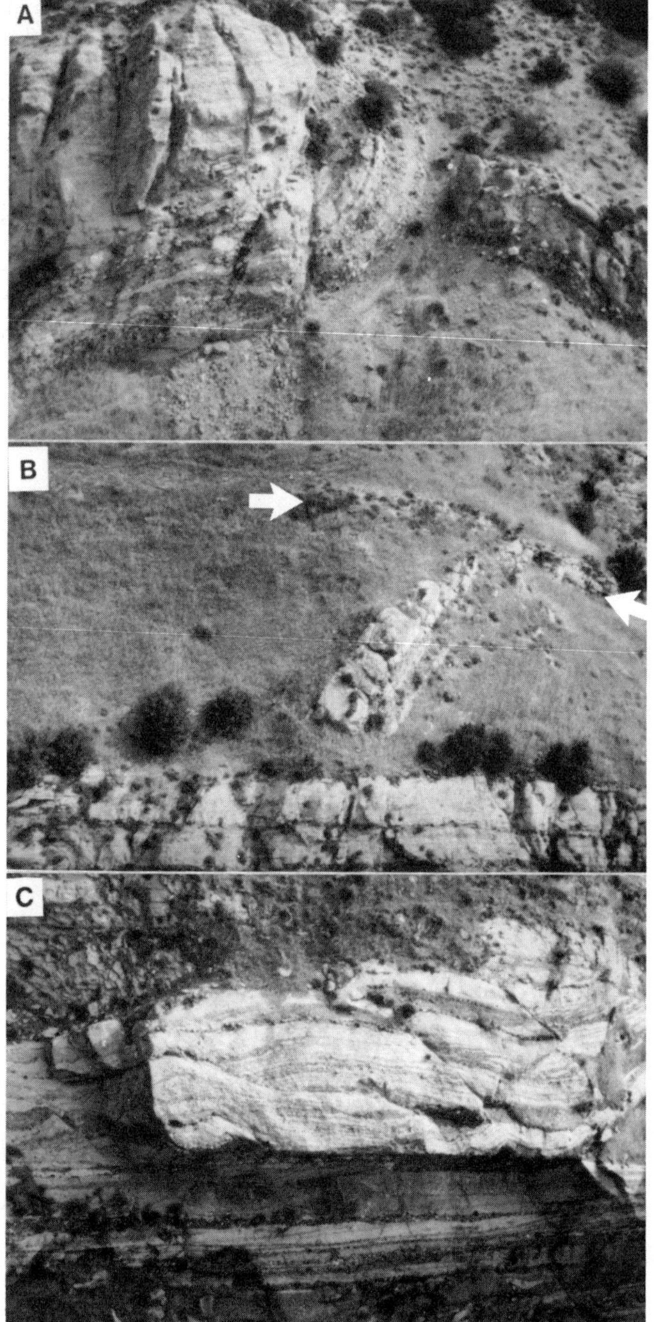

Figure 4. Close-up photographs of structures A, B, and C in Figure 3. East is to the right. (A) Buckle structure exposed at the eastern end of marine-backfilled channel deposit on Graham Ridge. Covered area below the apex of the structure is underlain by nonmarine conglomerate. Beds involved in the structure are interfingering nonmarine conglomerate and nearshore sandstone and conglomerate. See structure A, Figure 5, for lithofacies map. (B) Obliquely oriented beds of nearshore sandstone and conglomeratic sandstone, encased laterally by nonmarine conglomerate (mostly covered by grassy slope). Underlying beds are nearshore sandstone (below bushes). Arrows show position of overlying, undeformed, transgressive lag. Maximum thickness of obliquely oriented unit is 4 m (C) Thrust-faulted nearshore strata on Graham Ridge. Note upturned beds at eastern (right) end of outcrop. Deformed zone (above the shadow line) is 25 m thick.

cant downslope ("into the outcrop") mass-movement during the seismic event.

The marine-backfilled channel within the nonmarine conglomerate suggests that relative sea level had already risen at least 20 m prior to the disturbance (Fig. 6C). The topographic profile of the sedimentary surface probably was disrupted above the deformation structures (Fig. 6D). Uplift accompanying the disturbance may have actually raised the sediment prism above sea level, but evidence for subsequent subaerial exposure and/or sedimentation is not preserved. Most likely, the disturbance was followed by a rapid and major marine transgression, during which the overlying transgressive lag and several hundred meters of shallow marine Pleito sandstone were deposited.

DEVIL'S KITCHEN DISTURBED ZONE

Stratigraphic and paleogeographic contexts

Approximately 7.5 km to the west of Graham Ridge, in a narrow canyon that cuts along strike on the south limb of the Devil's Kitchen syncline (location 3, Fig. 1), a unit in the lower Pleito Formation consists of a chaotic association of massive conglomeratic sandstone, granitic and granodioritic breccia, and deformed sandstone blocks. This disturbed zone lies above dark silty shales and thin (<50 cm) sandstone beds of the lower Pleito sandstone (Fig. 7). Several sandstone beds below the deformed unit comprise Bouma $T_{b,c}$ and $T_{a\text{-}c}$ sequences (Fig. 7), suggesting that they were deposited by turbidity currents, probably on the slope or basin plain. Below these fine-grained deposits lie bioturbated, outer-shelf, fine-grained sandstones (Nilsen and others, 1973; DeCelles, 1987). The lithologic sequence reflects generally upward-deepening conditions, from outer shelf to slope, during deposition of the lower Pleito Formation in the Devil's Kitchen area.

In the Devil's Kitchen area, the disturbed zone is overlain by alternating dark silty shales (dominant) and thin (generally less than 70 cm) beds of fine-grained sandstone. Benthic foraminifera reported by DeLise (1967) from these dark shales include *Cyclamina incisa,* a middle to lower-middle bathyal (1,500 to 2,000 m deep) species (Bandy and Arnal, 1969; Ingle, 1980). Dibblee (1961) also reported relatively deep-water foraminifera, including *Siphogenerina nodifera,* collected from roughly equivalent shales of the middle Pleito Formation. In subsurface sections to the west of San Emigdio Canyon, equivalent strata are dominated by mudstone containing upper to middle bathyal foraminifera (Lagoe, 1986).

The thin sandstone beds consist of tabular, upward-fining (very fine-grained sandstone to siltstone) units, generally 30 to 60 cm thick. The base of each sandstone bed is erosional and overlain in turn by a 5 to 15-cm-thick zone of massive sandstone, a zone of delicate horizontal lamination, and a zone of dewatering structures and soft sediment deformation. The units usually are capped by a 10 to 20-cm-thick bed of bioturbated silty sandstone or siltstone. These beds resemble Ricci Lucchi's (1975) turbidite

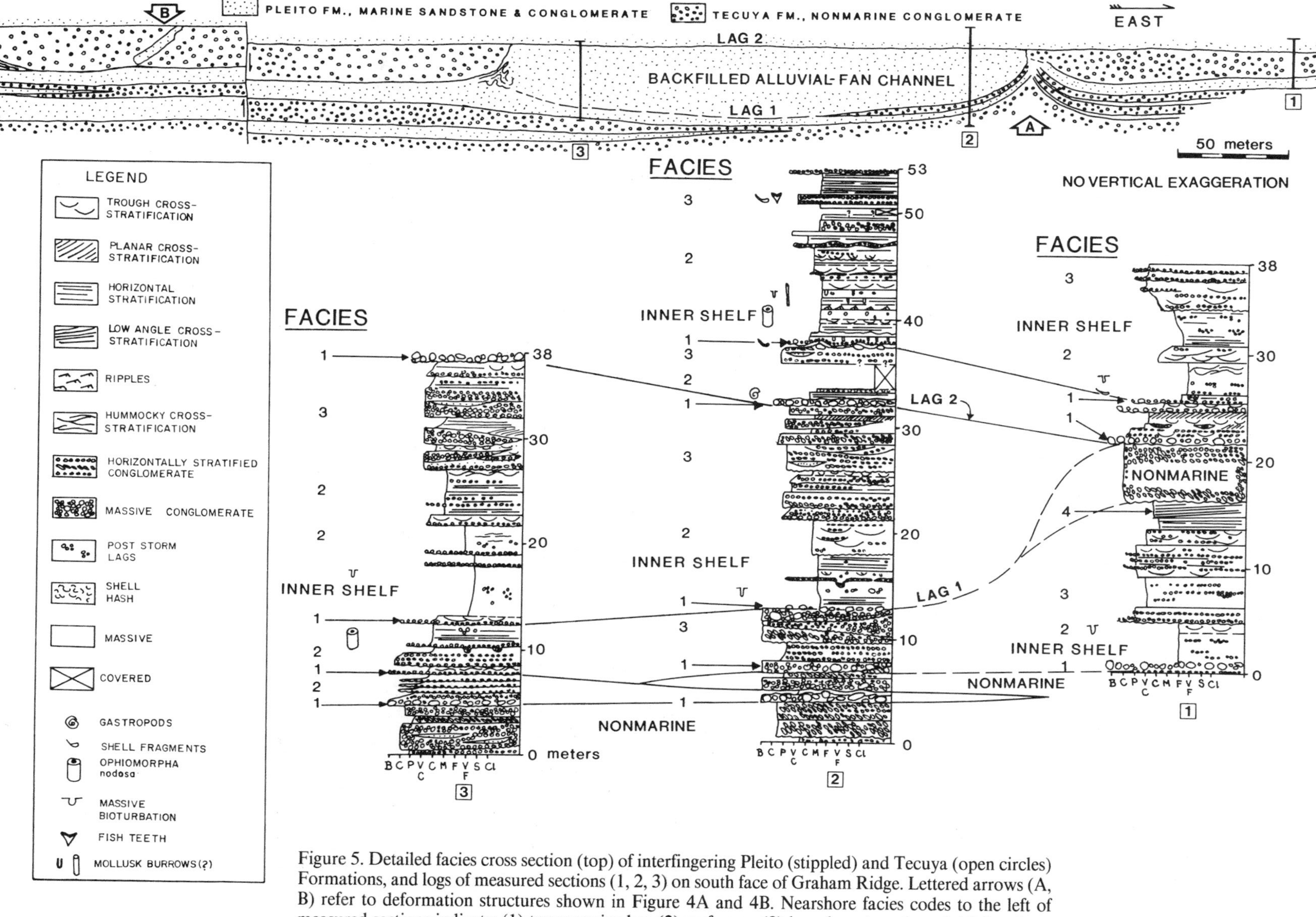

Figure 5. Detailed facies cross section (top) of interfingering Pleito (stippled) and Tecuya (open circles) Formations, and logs of measured sections (1, 2, 3) on south face of Graham Ridge. Lettered arrows (A, B) refer to deformation structures shown in Figure 4A and 4B. Nearshore facies codes to the left of measured sections indicate: (1) transgressive lag, (2) surf-zone, (3) longshore-trough, and (4) foreshore deposits.

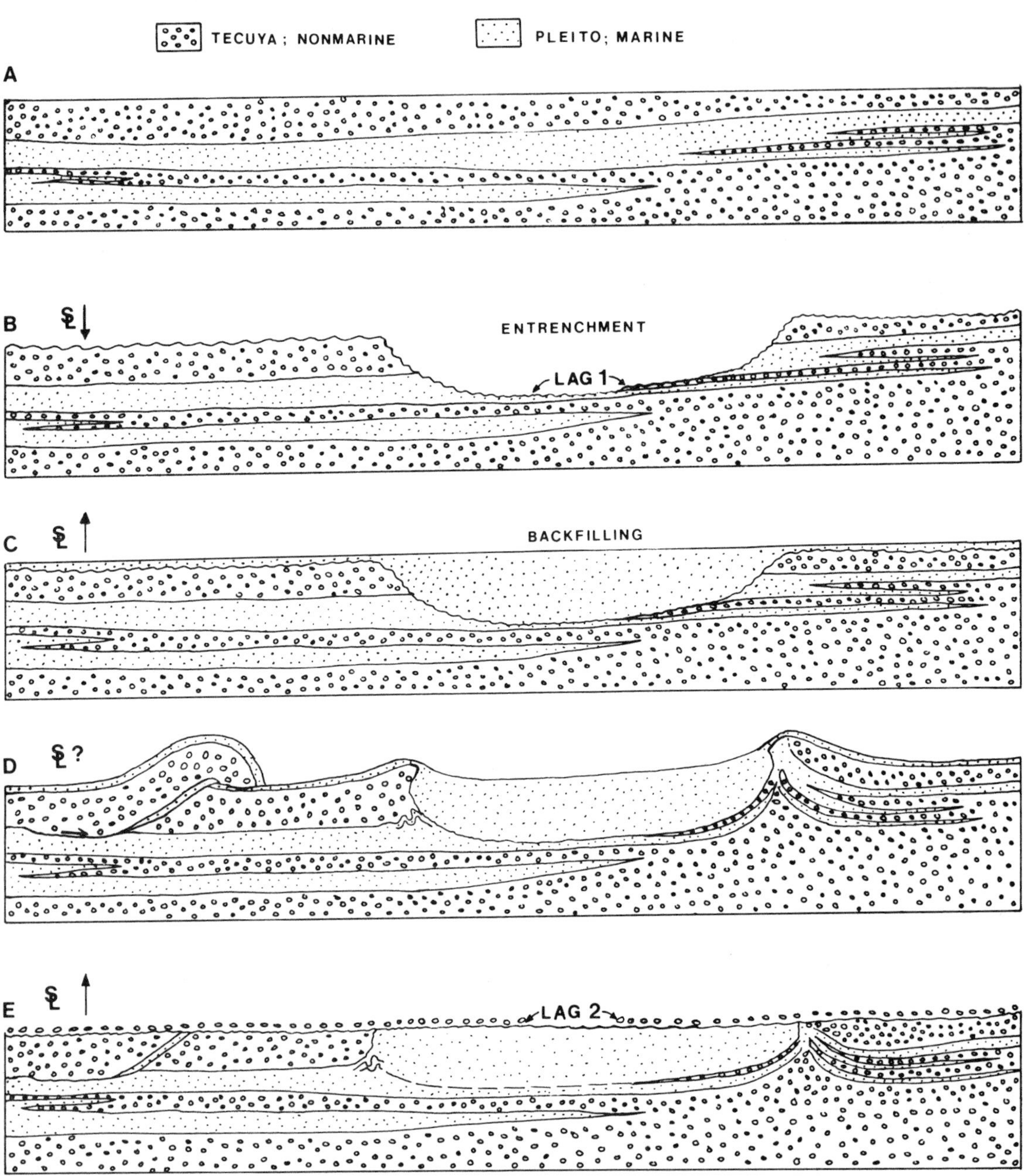

Figure 6. Schematic diagram depicting the postulated sequence of events that produced the configuration of facies and deformation structures on the south face of Graham Ridge. (A) Deposition of interfingering nonmarine and nearshore conglomerate and sandstone; (B) relative lowering of sea level and entrenchment of fan-delta channel, producing lag 1; (C) marine transgression and backfilling of entrenched channel; (D) seismic event(s), causing deformation in sediment prism; (E) continued rise of sea level, producing a transgressive lag (lag 2) on top of planed-off deformed sediment prism (compare with Figs. 3 and 5). The width of the backfilled channel is approximately 250 m.

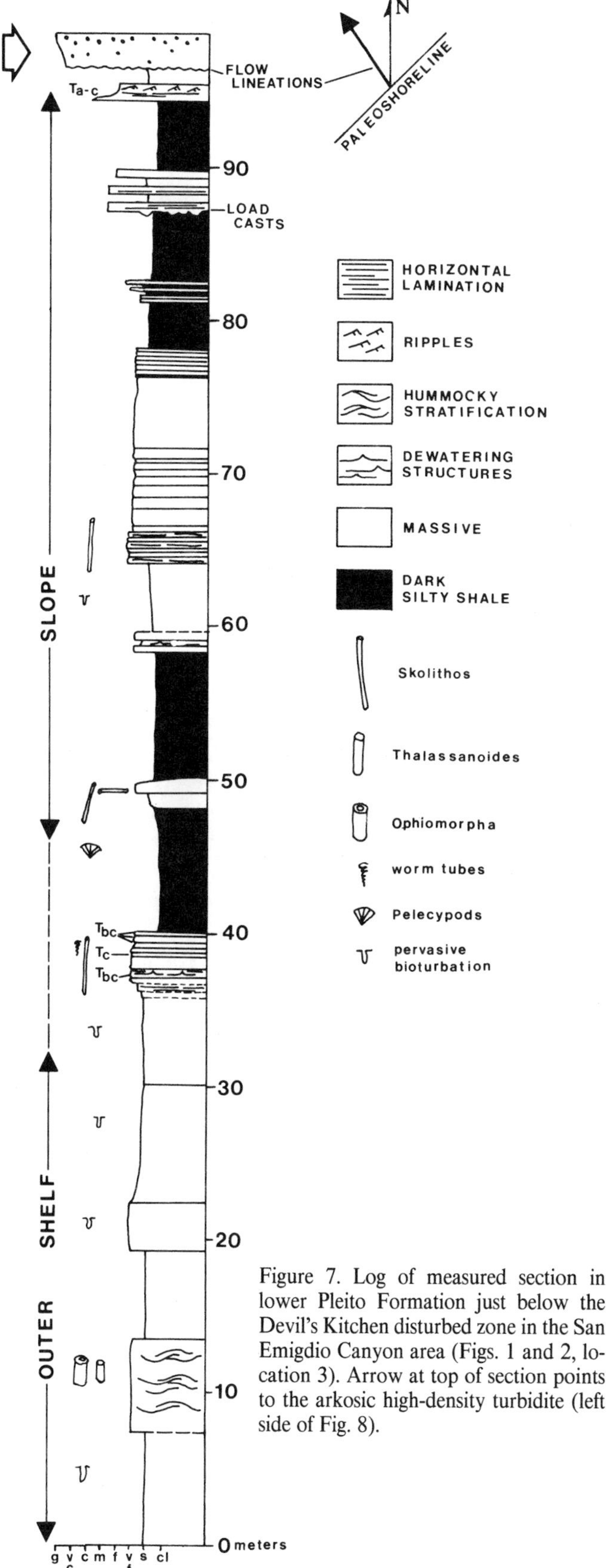

Figure 7. Log of measured section in lower Pleito Formation just below the Devil's Kitchen disturbed zone in the San Emigdio Canyon area (Figs. 1 and 2, location 3). Arrow at top of section points to the arkosic high-density turbidite (left side of Fig. 8).

facies B and C and may have been deposited by turbidity currents in relatively deep water. Alternatively, these beds may be distal storm deposits, which were formed on the outer shelf or upper slope.

Description

The disturbed zone in the Devil's Kitchen area is as much as 15 m thick and can be mapped over a minimum lateral distance of 860 m (Fig. 8). It comprises four lithologic components. The bulk of the unit consists of massive or laminated, fine-grained, tan or gray sandstone (Fig. 9A). Bedding within the sandstone is chaotically deformed, with vertical beds, folds, and pervasively contorted layers. Large blocks of sandstone beds, several meters thick and more than 10 m wide, are juxtaposed with each other or separated by zones of intense deformation where bedding cannot be recognized. Where stratification is preserved, the dominant sedimentary structures are horizontal and subhorizontal lamination and possible hummocky cross-stratification. It is noteworthy that the Pleito outer-shelf sandstones that occur approximately 60 m down section are characterized by the same textures and sedimentary structures (Fig. 7).

The second lithologic component in the deformed zone is massive, matrix-supported, conglomeratic, fine-grained sandstone (Fig. 9A and 9B). Clasts within the sandstone matrix range from 1 to 50 cm in size (Fig. 9B). Most clasts are volcanic or metamorphic and are very well rounded, but some, particularly the granitic ones, are subangular to angular (Fig. 9B). Sorting is poor and textural grading is absent, but locally the clasts exhibit a preferred subvertical orientation (Fig. 9C). The sandstone matrix, which constitutes the majority of the deposit, is structureless but well-sorted. The lithology as a whole lacks stratification and occurs in irregular masses with high-angle sides, rounded bases, and flat tops (Fig. 9A and 9D). The largest unit observed is more than 8 m thick and approximately 5 m wide. Although these lithologic bodies appear to have channelized geometry, evidence of erosion (e.g., truncation of beds, rip-ups, etc.) is lacking along the margins of the encasing, massive, deformed sandstone. Instead, the conglomeratic sandstone appears to have been intruded into and/or deformed along with the encasing deformed sandstone (e.g., Fig. 9D).

The third lithologic component of the deformed unit constitutes a bed, generally less than 1 m thick, of very angular granitic to granodioritic clasts. At the bottom and top of the bed, well-rounded metamorphic and volcanic clasts are mixed with the angular granitic detritus. The unit exhibits both matrix and clast support. The angular granitic clasts are very poorly sorted and range in size from less than 1 cm to more than 10 cm. Matrix is fine-grained, gray or brown sandstone. In some places it is difficult to determine with certainty the boundaries of individual clasts because they are so densely concentrated within the bed (Fig. 9E). Locally, the well-rounded pebbles in the upper part of the unit are normally graded, but grading is absent in the very angular, granitic clasts. No stratification was observed. The bed of

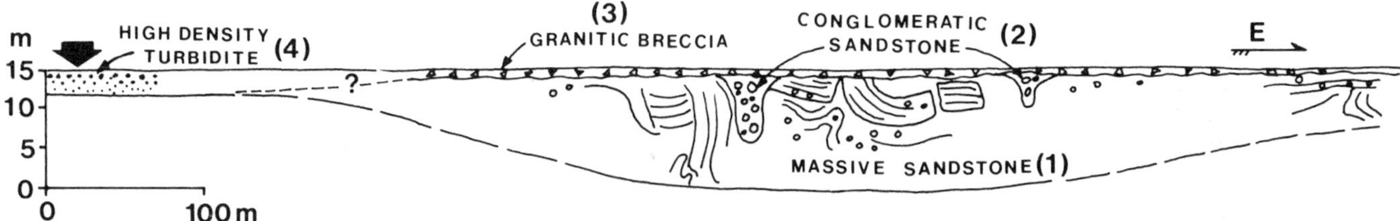

Figure 8. Schematic cross section of the Devil's Kitchen disturbed zone in the lower Pleito Formation exposed in a narrow canyon in the south limb of the Devil's Kitchen syncline (Fig. 1, location 3). Arrow on left side of diagram depicts position of measured section shown in Figure 7. Numerals in parentheses refer to various components discussed in text. Note vertical exaggeration.

angular granitic clasts is undeformed, consistently occurs at the top of the entire disturbed zone in the Devil's Kitchen area (Figs. 8 and 9A), and can be correlated at several localities over a lateral distance of approximately 850 m. The angular granitic clasts are lithologically identical to the clasts in the granitic breccia at or near the base of the Tecuya Formation.

The fourth lithologic component of the Devil's Kitchen disturbed zone consists of a 2 to 3-m-thick bed of massive, arkosic, very coarse-grained to conglomeratic sandstone (Fig. 8). The larger angular clasts are composed of granitic material that is similar to the granitic breccia in the Tecuya. Clasts are very angular to subrounded. Rip-ups of the underlying, very fine-grained sandstone occur in the lower few centimeters of the bed. Sorting is poor. The upper one-third of the unit locally is inversely graded. With the exceptions of basal load structures and sole lineations, sedimentary structures are absent. The restored orientation of the sole lineations is N34°W (Fig. 7), approximately perpendicular to the paleoshoreline postulated by DeCelles (1987).

Interpretation

The disturbed zone in the lower Pleito Formation in the Devil's Kitchen area is interpreted as the deposit of a submarine mass movement. It incorporates material derived from three different subenvironments on the southern margin of the San Joaquin basin, including the shelf, nearshore, and mountainous source area. The massive, deformed sandstone probably is a large segment of the shelf that slid downslope and came to rest somewhere on the slope. The well-rounded cobbles and pebbles of the conglomeratic sandstone were derived from the inner-shelf and/or nearshore systems. The granitic breccia and massive, very coarse-grained to conglomeratic sandstone bed represent the deposits of a subaqueous sturzstrom and related high-density turbidity flow, probably the distal equivalents of a subaerial rockslide. The extreme angularity of the clasts in the Devil's Kitchen breccia indicates that they must have rapidly bypassed the nearshore system, where they would have been rounded by high-energy processes had they spent any time in the longshore-trough or surf zone.

A hypothetical sequence of events during emplacement of the disturbed zone is illustrated in Figure 10. The superposition of the various components suggests that the shelf edge detached and slid downslope prior to arrival of the distal subaqueous rockslide and accompanying high-density turbidity flow. The general lack of deformation in the breccia bed suggests that it was deposited on top of the highly deformed sandstone after the latter had been deformed, and presumably near the end of its downslope movement. It is, therefore, unlikely that sudden deposition of the breccia bed on the outer shelf caused the mass movement. Local mixing of the lower part of the breccia with the upper part of the underlying massive sandstone and/or conglomeratic sandstone implies that minor readjustments may have occurred after the entire zone was deposited. The presence of occasional angular granitic boulders in the massive conglomeratic sandstone suggests that some large clasts preceded, or "outran," the main part of the sturzstrom. Some clasts in the conglomeratic sandstone experienced rotation into subvertical orientations (Fig. 9C), probably owing to internal shearing and/or dewatering of the sliding mass. The normally graded, rounded pebbles in the upper several centimeters of the breccia bed probably were deposited by the tail of the high-density turbidity flow that followed the sturzstrom. An-

Figure 9. Photographs of the Devil's Kitchen disturbed zone in the lower Pleito Formation (location 3 in Figs. 1 and 2). (A) Massive, deformed, fine-grained sandstone (left of dashed line above Jacob staff); massive conglomeratic sandstone (right of and above dashed line); and granitic breccia (above arrows). Area within white-lined box in lower right is shown in B. Jacob staff is 1.5 m long. (B) Closeup of massive conglomeratic sandstone, showing large angular to subangular granite boulder and a well-rounded metamorphic pebble (below scale). (C) Subvertical orientation of long axes of clasts in massive conglomeratic sandstone. Horizontal is parallel to base of photograph. (D) Mixed angular granite and well-rounded metamorphic and volcanic clasts in the upper portion of the disturbed zone. Note the downward taper of the body of coarse-grained material and massive deformed sandstone on either side. (E) Closely packed, very angular clasts in the granitic breccia at the top of the disturbed zone. Note two, well-rounded, metamorphic clasts adjacent to scale. Scale in B, C, D, and E is 20 cm long.

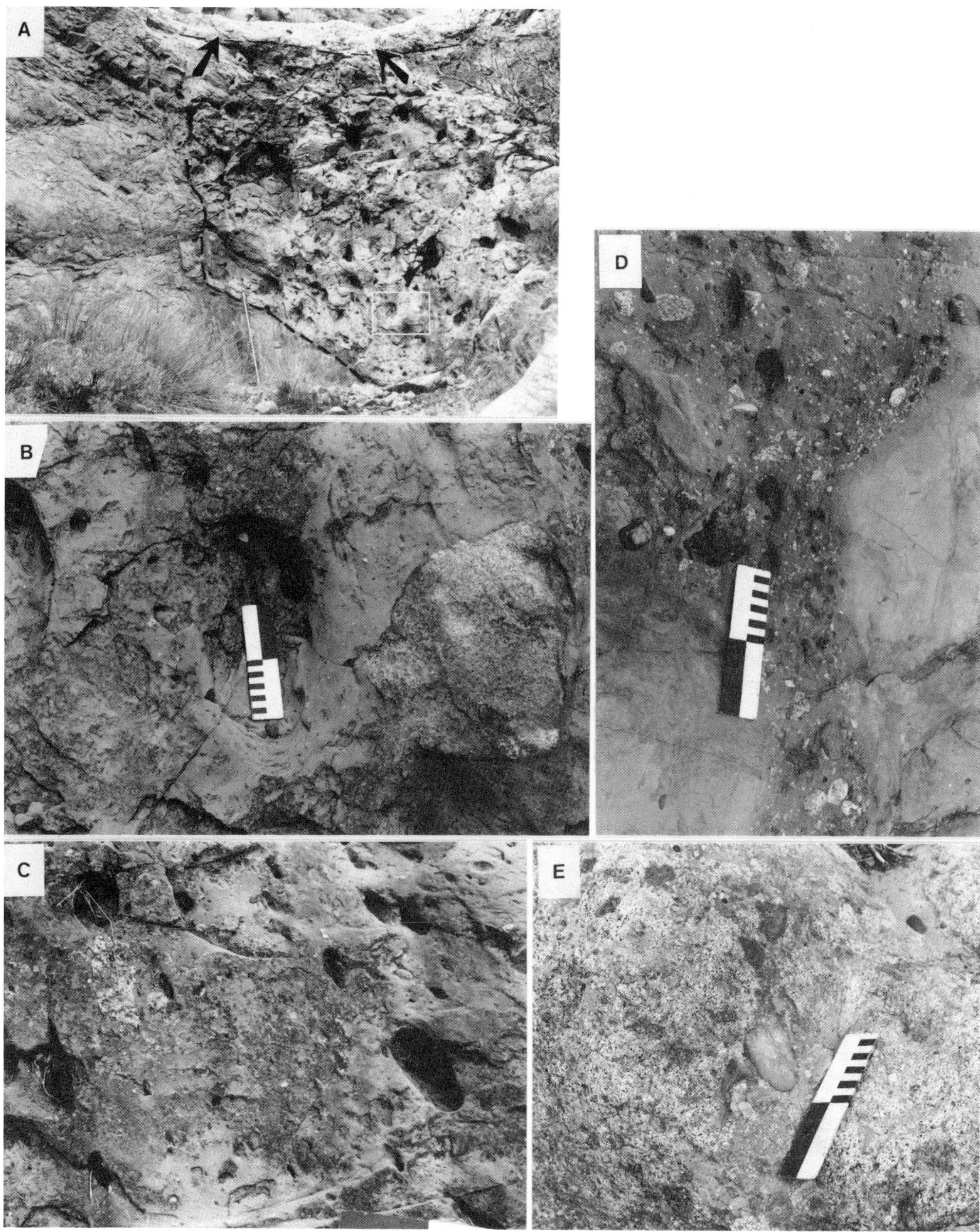
A
B
C
D
E

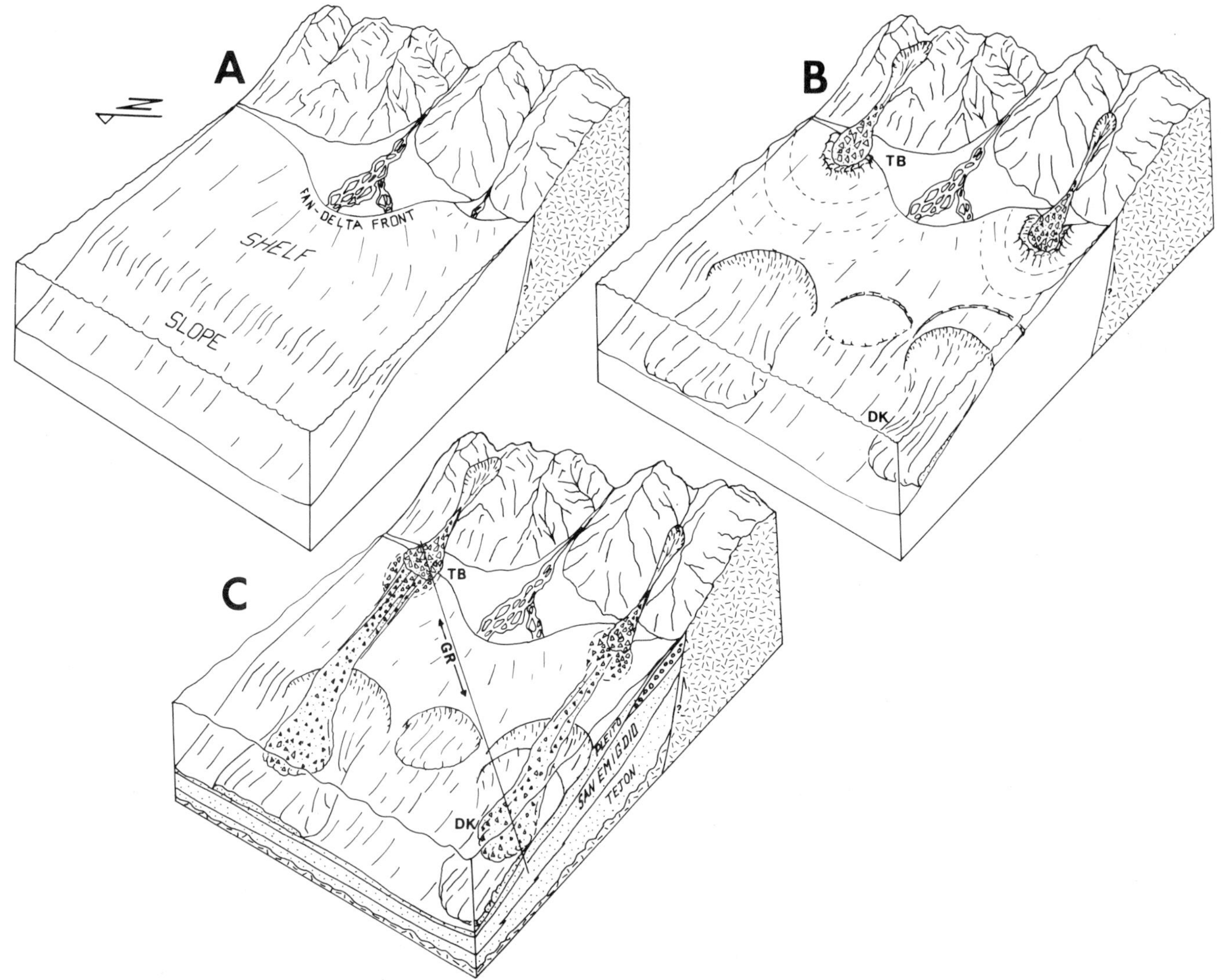

Figure 10. Schematic sequential block diagrams illustrating the paleogeographic reconstruction of the southern margin of the San Joaquin basin and development of the disturbed zones in the middle Tertiary strata. TB = Tecuya breccia; DK = Devil's Kitchen disturbed zone; GR = Graham Ridge disturbed zone. Diagonal line in C shows the orientation of the present outcrop belt with respect to the paleoshoreline and the three disturbed zones. Distance between DK and TB is approximately 15 km. Not to scale vertically.

gular debris that was deposited on the inner shelf (or what may have remained thereof) was subsequently abraded and transported shoreward by landward-asymmetric orbital currents that were driven by shoaling waves (e.g., Komar, 1976).

The two main causes of submarine mass movements are: (1) the development of anomalously high pore pressures owing to rapid rates of sediment accumulation (most common in deltaic systems); and (2) seismic disturbances (Moore, 1978). The former mechanism is unlikely in this case because the mass movement primarily involved Pleito shelf deposits, which lack evidence for rapid rates of sedimentation (e.g., dewatering structures and thick, structureless beds). On the other hand, the granitic breccia-bed atop the Devil's Kitchen disturbed zone indicates that crystalline bedrock must have been dislodged by the disturbance. The similarity in composition and angularity of the Tecuya and Devil's Kitchen breccias suggest that they were derived suddenly from the same source area. By all indications, that source area consisted of a basement uplift that lay to the east-southeast (Nilsen and others, 1973; DeCelles, 1986, 1988). No evidence has been found for the existence of a submarine bedrock exposure in the vicinity during late Eocene–early Oligocene time. Given that no causal link exists between deposition of the Devils's Kitchen breccia and generation of the submarine mass flow, I propose that both were triggered by a seismic event.

DISCUSSION

The three separate disturbed zones occur at approximately the same stratigraphic level (Fig. 2), but direct physical correlation is impossible because of locally poor exposure and structural complications (e.g., the north-south trending fault in the eastern part of the study area, Fig. 1). Nevertheless, some evidence suggests that all three were produced by the same event or several related events.

The two tongues of nonmarine Tecuya conglomerate that are involved in the Graham Ridge disturbed zone appear to postdate the Tecuya breccia (Fig. 2). However, these two conglomerate tongues attain maximum thicknesses (88 and 30 m, respectively) in the Graham Ridge area. Both tongues pinch out less than 5 km to the west, and thin significantly toward the east (Fig. 2). They are not traceable east of the Salt Creek area. In addition, the grain sizes of the conglomerates decrease to both the east and west of Graham Ridge. At Salt Creek (section SC, Figs. 1 and 2), both intervals contain at least 30 percent mudstone, whereas both are composed almost entirely of pebble-boulder conglomerate at Graham Ridge. Clast imbrications indicate west-northwestward paleocurrent directions (DeCelles, 1987, 1988). I interpret these lower Tecuya conglomerate tongues as subaerial fan-delta lobes that are exposed in outcrop sections oriented nearly transverse to the alluvial-fan axes. The material that they incorporate was derived from the east-southeast prior to deposition of the Tecuya breccia and the overlying main body of the Tecuya farther to the east. The Tecuya above the breccia zone east of Salt Creek is wedge-shaped thickening and coarsening dramatically eastward (Fig. 2). This, together with sedimentological and provenance data (DeCelles, 1988), indicates that the main body of the Tecuya, which is greater than 700 m thick between Tecuya and Salt Creeks, was derived from a more easterly source. The outcrop belt is oriented roughly parallel to the axes of the alluvial fans which deposited the main mass of the Tecuya. It is thus plausible that deposition of the Tecuya breccia roughly coincided in time with deformation of the previously deposited, interfingering nearshore and alluvial-fan deposits exposed on Graham Ridge. The breccia was probably deposited on the flank of a pre-existing fan-delta lobe, which is exposed in nearly transverse cross section along Graham Ridge.

The wedge of nonmarine Tecuya conglomerate that lies directly below the Graham Ridge disturbed zone pinches out approximately 4.5 km to the west on the south face of Eagle Rest Peak, where it is stratigraphically separated from the Devil's Kitchen disturbed zone by approximately 120 m of overlying lower Pleito outer-shelf and slope deposits (Figs. 1 and 2). The westward increase in stratigraphic separation suggests that the Devil's Kitchen disturbed zone is younger than the Graham Ridge disturbed zone. However, as discussed above, the disturbed zone on Graham Ridge occurs within interbedded nearshore and nonmarine deposits, which have relatively little potential for preservation during transgressions (Swift, 1968; Komar, 1976). Tens of meters of outer-shelf or slope deposits in the lower Pleito near Eagle Rest Peak may be represented by only a 50-cm-thick transgressive lag in nearshore deposits on Graham Ridge.

If the disturbance that deformed the interbedded Pleito and Tecuya on Graham Ridge also produced the lens of granitic breccia at the base of the Tecuya 6.5 km to the east, abundant, angular, granitic boulders ought to be present in the Pleito just above the deformed zone on Graham Ridge. Angular granitic clasts are absent right above the deformed zone, but their absence probably is a result of the high-energy nearshore processes that dominated the depositional system along the toe of the subaerial fan-delta system. Theoretical reconstructions of wave heights in the southern San Joaquin basin suggest that storm waves may have exceeded 5 m in height (DeCelles, 1987). Thus, any detritus that was deposited in the nearshore system was reworked and quickly rounded by the action of powerful wave-deriven currents. Moreover, the disturbed zone along Graham Ridge is truncated by a transgressive lag (Fig. 5), which suggests that any rockslide material that may have been deposited in the nearshore zone was reworked during the ensuing transgression. The transgressive lag contains abundant, very coarse (>1 m in diameter) granite boulders, which may be the well-rounded remnants of the breccia. Alternatively, the rockslide that produced the Tecuya breccia may have been localized to the area east of present-day Salt Creek.

The lithologic similarity between the Tecuya and Devil's Kitchen breccias suggests that they were produced by at least closely related events. Sole lineations on the base of the arkosic high-density turbidite bed in the Devil's Kitchen area indicate flow direction was N34°W. If the flow traveled roughly perpendicular to paleoshoreline, its source would have been located approximately 6 km to the southwest of the Tecuya granitic breccia. Thus, unless the Tecuya granitic breccia extended for approximately 6 km along paleoshoreline, the Devil's Kitchen breccia probably does not make up its distal equivalent. Conceivably, the earthquake that triggered the Tecuya rockslide also could have generated another rockslide alongshore to the southwest.

The postulated sequence of events that produced the three disturbed zones involves three stages (Fig. 10). Uplift along the east-southeast margin of the San Joaquin basin generated fan-deltas that prograded to the north-northwest (Fig. 10A). Sometime during the early Oligocene, a major earthquake (or series of earthquakes) launched a number of large rockslides along the rugged shoreline (Fig. 10B). One rockslide deposited roughly 6×10^7 m^3 of granitic debris on or near the northeastern flank of a fan-delta lobe, which was located in the vicinity of present-day Graham Ridge, while a second rockslide was unleashed a short distance (6 km) alongshore to the southwest (Fig. 10B). The prism of sediment along the front of the fan-delta was violently shaken, producing several large-scale deformation structures (Fig. 6D). Simultaneously, approximately 2 to 3 km offshore, on the outer shelf, large portions of the shelf-edge slid downslope, producing the Devil's Kitchen disturbed zone (Fig. 10B). Eventually (probably within minutes), the distal end of the second sub-

Figure 11. The south limb of the Devil's Kitchen syncline, viewed toward the east, in the San Emigdio Canyon area. The highest point on the skyline (upper right) is Eagle Rest Peak. Note the lateral persistence of facies in the stratigraphic members of the San Emigdio Formation. In contrast, turbidites of the middle Pleito exposed on the slope above the Devil's Kitchen disturbed zone (white arrow, lower left) can be traced along strike toward the skyline (eastward) into light gray sandstone and siltstone slope deposits. These beds can, in turn, be traced beyond the skyline into inner-shelf and ultimately nearshore and nonmarine deposits. Topographic relief shown in photograph is approximately 800 m.

aerial rockslide was deposited on top of the Devil's Kitchen disturbed zone (Fig. 10C).

If the three disturbed zones were not produced by a single seismic event, their stratigraphic localization suggests that they were generated over a geologically brief period by a series of related seismic events. Regardless of the number of disturbances, the disturbed zones are significant in their own rights because they developed in three contrasting environments, probably in response to similar, if not the same, convulsive geologic events.

ACCOMPANYING CHANGES IN DEPOSITIONAL SYSTEMS

The upper Eocene to lower Miocene rocks of the San Emigdio Range consist of two complete depositional sequences (in the sense of Mitchum and others, 1977), referred to here as the San Emigdio and Pleito sequences (Fig. 11). In the vicinity of San Emigdio Creek, the San Emigdio sequence (305 m) consists of three informal stratigraphic members, each of which constitutes a separate facies (DeCelles, 1988; Figs. 2 and 11). These are: (1) a lower, retrogradational, deltaic sandstone facies; (2) a middle, fine-grained, basin-plain facies; and (3) an upper, shelf-sandstone facies. The upper facies can be traced eastward into nearshore deposits, which crop out almost continuously from the southeast face of Eagle Rest Peak to Salt Creek. The lower two facies in the San Emigdio sequence pinch out toward the east, in an onlapping relationship, against the underlying Tejon Formation (Fig. 2). The three members of the San Emigdio exhibit remarkable lateral continuity of facies. The lower member consists, throughout its lateral extent, of deltaic deposits. Similarly, the middle and upper members comprise only deep- and shallow-marine deposits respectively (DeCelles, 1988).

The Pleito sequence (680 m) also contains three stratigraphic facies in the San Emigdio Canyon area (DeCelles, 1988; Figs. 2 and 11). These are: (1) a lower, shelf-sandstone facies, which is similar to the upper San Emigdio sandstone facies; (2) a middle, fine-grained, slope facies; and (3) an upper, coarse-grained, deltaic facies. The lower and upper Pleito facies are laterally persistent, consisting entirely of shallow-marine deposits. However, the thick, middle, fine-grained facies is replaced to the east by progressively coarser-grained, shallower-water sandstones (Fig. 11; DeCelles, 1988). Ultimately, all three Pleito facies interfinger eastward with alluvial-fan deposits of the Tecuya (Fig. 2). The Devil's Kitchen disturbed zone lies at the base of the middle Pleito fine-grained facies (Fig. 2).

The Tecuya Formation, below its middle volcanic unit, comprises two major wedges of alluvial-fan deposits (Fig. 2). The lower wedge consists of the two relatively thin tongues of conglomerate, which crop out on Graham Ridge and pinch out several km to the west. As discussed above and in the foregoing section, these tongues of conglomerate occur within the Graham Ridge disturbed zone, and interfinger to the west with widespread nearshore and shelf deposits, which can be traced to the west beyond the nose of the Devil's Kitchen syncline (Nilsen and others, 1973; Fig. 1). This complex of interfingering alluvial-fan and shallow-marine facies thus composes the deposits of a system of shelf fan-deltas (Wescott and Ethridge, 1980).

In contrast to the lower Tecuya wedge, the upper wedge (not to be confused with the informal upper member of the Tecuya, which is stratigraphically above the volcanic unit; Fig. 1A) is more than 300 m thick and interfingers to the west with a narrow (<3 km wide) zone of nearshore and shelf deposits, which, in turn, give way to fine-grained, channelized, slope deposits (Fig. 2). This entire facies assemblage embodies the deposits of a system of slope fan-deltas (DeCelles, 1988). According to Wescott and Ethridge (1980), deposits of slope fan-deltas are distinguished from those of shelf fan-deltas by their generally greater thicknesses and close associations with fine-grained, deeper water deposits. Slope fan-deltas typically form where a fan-delta progrades directly onto a submarine slope, whereas shelf fan-deltas prograde across continental or island shelves.

The two contrasting types of fan-delta deposits are separated by the disturbed zones (Fig. 2). Most likely, the seismic disturbance (or series of disturbances) that produced the disturbed zones was followed by major prolonged uplift and erosion of the source area. Moreover, because the shelf edge was apparently resedimented downslope in the form of the Devil's Kitchen disturbed zone, the subaerial fan was able to prograde almost directly to the top of the slope. In other words, the depositional system changed from a shelf fan-delta to a slope fan-delta because the shelf edge slid basinward during the postulated seismic events.

Furthermore, differences in large-scale sedimentary packaging above and below the disturbed zones (Fig. 11) demonstrate profound changes in the degree of influence on sedimentation exerted by changing relative sea level. In the middle Pleito, above the Devil's Kitchen disturbed zone, relatively deep-water deposits can be traced up-dip (eastward) into contemporaneous shallow-water and nonmarine deposits. Deposition occurred simultaneously across the entire subaqueous basin margin, regardless of sea level, apparently because the basin margin lacked a stepped shelf-slope-basin profile.

In contrast, sedimentary packaging below the disturbed zone, in the lower Pleito and underlying San Emigdio Formation, indicates that the locus of deposition was governed by the relative position of sea level with respect to a stepped basin margin. As suggested by Vail and others (1977) and Mutti (1985), deposition on a stepped basin margin does not occur simultaneously on the shelf, slope, and basin floor (except in areas with exceptionally high rates of sediment input). Rather, during high sea-level stands, most sediment is trapped on the shelf, while the basin is relatively sediment-starved. During low stands, the shelf is exposed and sediment is bypassed directly to the basin. The resultant packaging of large-scale sedimentary units displays stratigraphic segregation of the deposits of shelf, slope, and basin.

This type of segregation is exhibited by the lower Pleito and San Emigdio Formations, with their stratigraphically persistent facies units. The lower San Emigdio sandstone is deltaic and probably was deposited during a low stand (DeCelles, 1988). The middle San Emigdio comprises bathyal shales (DeLise, 1967), which were deposited during a period of rising relative sea level. The upper San Emigdio and lower Pleito sandstones were deposited on a prograding shallow shelf during a high stand in relative sea level (DeCelles, 1988). These four stratigraphic units are environmentally consistent throughout their respective lateral extents. That is, relatively deep-water deposits cannot be traced into contemporaneous shallow-water deposits, as can be done in the middle Pleito Formation, above the disturbed zones.

CONCLUSIONS

In late Eocene or early Oligocene time, a 50-m-thick wedge of granitic breccia was deposited on the flank of a shelf fan-delta lobe on the southern margin of the San Joaquin basin, probably by a major rockslide. The rockslide occurred in response to an earthquake or series of earthquakes. The adjacent prism of previously deposited sediment, comprising interfingering alluvial-fan and nearshore deposits, was violently shaken by the same or a related earthquake. Massive deformation of the sediment prism resulted. Farther seaward, portions of the shelf edge detached and slid downslope, eventually coming to rest on the slope in bathyal water depths. A high-density turbidity current and subaqueous sturzstrom, probably the distal equivalents of a second subaerial rockslide, deposited a bed of granitic breccia on top of the resedimented shelf-edge deposits.

Profound long-term changes in the geometry of the basin margin ensued, as the basin foundered and the source area was elevated. The shelf fan-deltas that had occupied the basin margin were replaced by a system of slope fan-deltas. The presence of the disturbed zones indicates that convulsive geologic events heralded major reorganization of the basin margin. Thus, although the disturbed zones are volumetrically minor depositional units in comparison with the sandwiching depositional sequences (both of which are several hundred meters thick), they are, nevertheless, a key element in the record of a major change in the architecture of the basin margin.

ACKNOWLEDGMENTS

Funding was provided by contributors to the San Joaquin Basin Project at Stanford University, including AMOCO, ARCO, BP Alaska, Champlin, Chevron, Cities Service, Conoco, EXXON, Getty, Gulf, Husky, Marathon, Mobil, Occidental, Phillips, Santa Fe Energy, Shell, SOHIO, Superior, Tenneco, Texaco, and Union. I thank Tenneco and the proprietors of the Tejon and San Emigdio Ranches for access to the study area, which lies on their property. I am grateful to S. A. Graham, my postdoctoral advisor at Stanford, for his encouragement and support throughout this study. The quality of this paper was greatly enhanced through critical reviews by Rick Stanley, Ed Clifton, and Tor Nilsen.

REFERENCES CITED

Bandy, O. L., and Arnal, R. E., 1969, Middle Tertiary basin development, San Joaquin Valley, California: Geological Society of America Bulletin, v. 80, p. 783–820.

DeCelles, P. G., 1986, Middle Tertiary depositional systems of the San Emigdio Range, southern California, *in* DeCelles, P. G., ed., Middle Tertiary depositional systems of the San Emigdio Range, southern California: Society of Economic Paleontologists and Mineralogists Pacific Section Field Trip Guidebook 47, 32 p.

—— , 1987, Variable preservation of middle Tertiary, coarse-grained, nearshore to outer-shelf storm deposits in southern California: Journal of Sedimentary Petrology, v. 57, p. 250–264.

—— , 1988, Middle Cenozoic depositional, tectonic, and sea-level history of the southern San Joaquin basin, California: American Association of Petroleum Geologists Bulletin (in press).

DeLise, K. C., 1967, Biostratigraphy of the San Emigdio Formation, Kern County, California: California University Publications in Geological Science, v. 68, 66 p.

Dibblee, T. W., Jr., 1961, Geologic structure of the San Emigdio Mountains, Kern County, California: Society of Economic Paleontologists and Mineralogists, Society of Exploration Geophysicists, and American Association of Petroleum Geologists, Pacific Sections, San Joaquin Geological Society Spring Field Trip Guidebook, 1961, p. 2–6.

Dibblee, T. W., Jr., and Nilsen, T. H., 1973, Geologic map of San Emigdio and western Tehachapi Mountains, Kern County, California: American Association of Petroleum Geologists Field Trip Guidebook 2, scale 1:62,500.

Hadley, J. B., 1964, Landslides and related phenomena accompanying the Hebgen Lake earthquake of August 17, 1959: U.S. Geological Survey Professional Paper 435, p. 107–138.

Hoots, H. W., 1930, Geology and oil resources along the southern border of the San Joaquin Valley, California: U.S. Geological Survey Bulletin 812–D, p. 243–332.

Hsü, K. J., 1975, Catastrophic debris streams (sturzstroms) generated by rockfalls: Geological Society of America Bulletin, v. 86, p. 129–140.

—— , 1978, Albert Heim; Observations on landslides and relevance to modern interpretations, *in,* Voight, B., ed., Rockslides and avalanches; 1, Natural phenomena: Developments in Geotechnical Engineering, v. 14A, p. 71–93.

Ingle, J. C., 1980, Cenozoic paleobathymetry and depositional history of selected sequences within the southern California continental borderland: Cushman Foundation Special Publication no. 19, p. 163–195.

Komar, P. D., 1976, Beach processes and sedimentation: Englewood Cliffs, New Jersey, Prentice Hall, 417 p.

Lagoe, M. B., 1986, Stratigraphic nomenclature and time-rock relationships in the Paleogene rocks of the San Emigdio Mountains, *in* Davis, T. L., and Namson, J. S., eds., Geologic transect across the western Transverse Ranges: Society of Economic Paleontologists and Mineralogists Pacific Section Fieldtrip Guidebook 48, p. 11–21.

Long, E., 1968, Earth slides and related phenomena, *in,* The Great Alaska earthquake of 1964; Part B, Engineering: National Academy of Science Publication 1603, p. 644–773.

McWilliams, M., and Li, Y., 1985, Oroclinal rotation of the southern Sierra Nevada batholith: Science, v. 230, p. 172–175.

Mitchum, R. M., Vail, P. R., and Thompson, S., III, 1977, Seismic stratigraphy and global changes of sea level; Part 2, The depositional sequence as a basic unit for stratigraphic analysis, *in,* Payton, C. E., ed., Seismic stratigraphy; Application to hydrocarbon exploration: American Association of Petroleum Geologists Memoir 26, p. 53–62.

Moore, D. G., 1978, Submarine slides, *in,* Voight, B., ed., Rockslides and avalanches; 1, Natural phenomena: Developments in Geotechnical Engineering, v. 14, p. 563–604.

Mutti, E., 1985, Turbidite systems and their relations to depositional sequences, *in* Zuffa, G. G., ed., Provenance of arenites: Dordrecht, Holland, D. Reidel Publishing Company, p. 65–93.

Nilsen, T. H., 1984, Oligocene tectonics and sedimentation, California: Sedimentary Geology, v. 38, p. 305–336.

Nilsen, T. H., Dibblee, T. W., Jr., and Addicott, W. O., 1973, Lower and middle Tertiary stratigraphic units of the San Emigdio and western Tehachapi Mountains, California: U.S. Geological Survey Bulletin 1372–H, p. H1–H23.

Plescia, J. B., and Calderone, G. J., 1986, Paleomagnetic constraints on the timing and extent of rotation of the Tehachapi Mountains, California: Geological Society of America Abstracts with Programs, v. 18, p. 171.

Ricci Lucchi, F., 1975, Depositional cycles in two turbidite formations of the northern Appennines (Italy): Journal of Sedimentary Petrology, v. 45, p. 3–43.

Swift, D.J.P., 1968, Coastal erosion and transgressive stratigraphy: Journal of Geology, v. 76, p. 444–456.

Turner, D. L., 1970, Potassium-argon dating of Pacific Coast Miocene foraminiferal stages: Geological Society of America Special Paper 124, p. 91–129.

Vail, P. R., Mitchum, R. M., and Thompson, S., III, 1977, Seismic stratigraphy and global changes of sea level; Part 3, Relative changes of sea level from coastal onlap, *in* Payton, C. E., ed., American Association of Petroleum Geologists Memoir 26, p. 63–81.

Voight, B., 1978, Lower Gros Ventre slide, Wyoming, U.S.A., *in* Voight, B., ed., Rockslides and avalanches; 1, Natural phenomena: Developments in Geotechnical Engineering, v. 14A, p. 113–166.

Voight, B., and Pariseau, W. G., 1978, Rockslides and avalanches; An introduction, *in* Voight, B., ed., Rockslides and avalanches; 1, Natural phenomena: Developments in Geotechnical Engineering, v. 14A, p. 1–67.

Wescott, W. A., and Ethridge, F. G., 1980, Fan-delta sedimentology and tectonic setting; Yallahs fan delta, southeast Jamaica: American Association of Petroleum Geologists Bulletin, v. 64, p. 374–399.

Wojtal, S., and Mitra, G., 1986, Strain hardening and strain softening in fault zones from foreland thrusts: Geological Society of America Bulletin, v. 97, p. 674–687.

MANUSCRIPT ACCEPTED BY THE SOCIETY APRIL 4, 1988

Printed in U.S.A.

Geological Society of America
Special Paper 229
1988

Cretaceous/Tertiary boundary sediment

Kenneth J. Hsü, *Swiss Federal Institute of Technology, Zurich, Switzerland*

ABSTRACT

In many parts of the world a thin clay or marly unit marks the boundary between Cretaceous and Tertiary rocks. In marine sequences this boundary is defined by the first appearance of typically Paleocene marine plankton in the clay. In continental rocks, the boundary sediment yields the stratigraphically highest occurrence of a Cretaceous assemblage of fossil pollen. Detailed analyses of the marine boundary sediment at Caravaca, Spain, permit a three-fold subdivision: the lowest is apparently a fallout deposit of impact ejecta, preserved as a 0.5-cm lamina of red clay. The main subdivision is a black or dark gray clay or marl, containing reworked extraterrestrial debris, laid down in an oxygen-deficient environment. The uppermost boundary clay is lighter gray in color, transitional in lithology to the overlying Paleocene sediments, which were deposited after the recovery from the terminal Cretaceous convulsive event. The boundary clay unit on land, represented by a section in Raton Basin, New Mexico, consists of a lower white clay, which is apparently a fallout deposit, and an upper carbonaceous shale. Boundary sections elsewhere are similar to those sections. The sedimentology of the boundary sediment records the convulsive environmental changes at/after a terminal Cretaceous event.

INTRODUCTION

Sediments deposited by convulsive events commonly accumulated at a very high sedimentation rate, as discussed in various chapters of this volume. The possibility should not be overlooked, however, that they could also have accumulated at an abnormally slow rate. The Cretaceous/Tertiary boundary clay is a case to be considered. The clay is commonly found in pelagic carbonates in marine environments and in clastic sequences when deposited on continents. The normal sedimentation rate of pelagic carbonate is of the order of centimeters per thousand years. Alvarez and others (1980) considered the clay to be fallout debris after an impact explosion, implying a very rapid deposition rate of centimeters per year (Wolbach and others, 1985). The clay can, however, also be considered insoluble residue of a dissolution event, deposited at a rate of millimeters per thousand years (Hsü, 1986). This chapter reviews the sedimentology of the boundary clay at several localities and discusses the evidence pertaining to its origin, rate of sedimentation, and depositional environment.

BOUNDARY IN MARINE SEQUENCES ON LAND

Stevns Klint, Denmark

The boundary clay at the Stevns Klint locality, Denmark, is especially easy to recognize. Intercalated between the latest Maastrichtian chalk and the early Paleocene Cerithium Limestone is a dark gray clay unit called the Fish Clay (*fiskeler*). Christensen and others (1973) subdivided the thin unit (Fig. 1). Resting on the Upper Cretaceous white chalk (bed I) with transitional contact is a gray marl. This subdivision (bed II), 1 to 2 cm thick, is similar in lithology to the underlying marly interbed in the chalk and may represent a dissolution facies of the Cretaceous chalk. Bed III, some 2 cm thick, is a black clay, or more exactly, a silty marl, containing pyrite concretions. Bed IV, 3 to 5 cm thick, is a black to light gray, laminated marl. Bed V is a light gray marl, transitional to the overlying Cerithium Limestone. While the chalk and the Cerithium Limestone are intensely bioturbated, the Fish Clay shows little evidence of bioturbation. The clay contains few fossils except dinoflagellates and fish remains and owes its

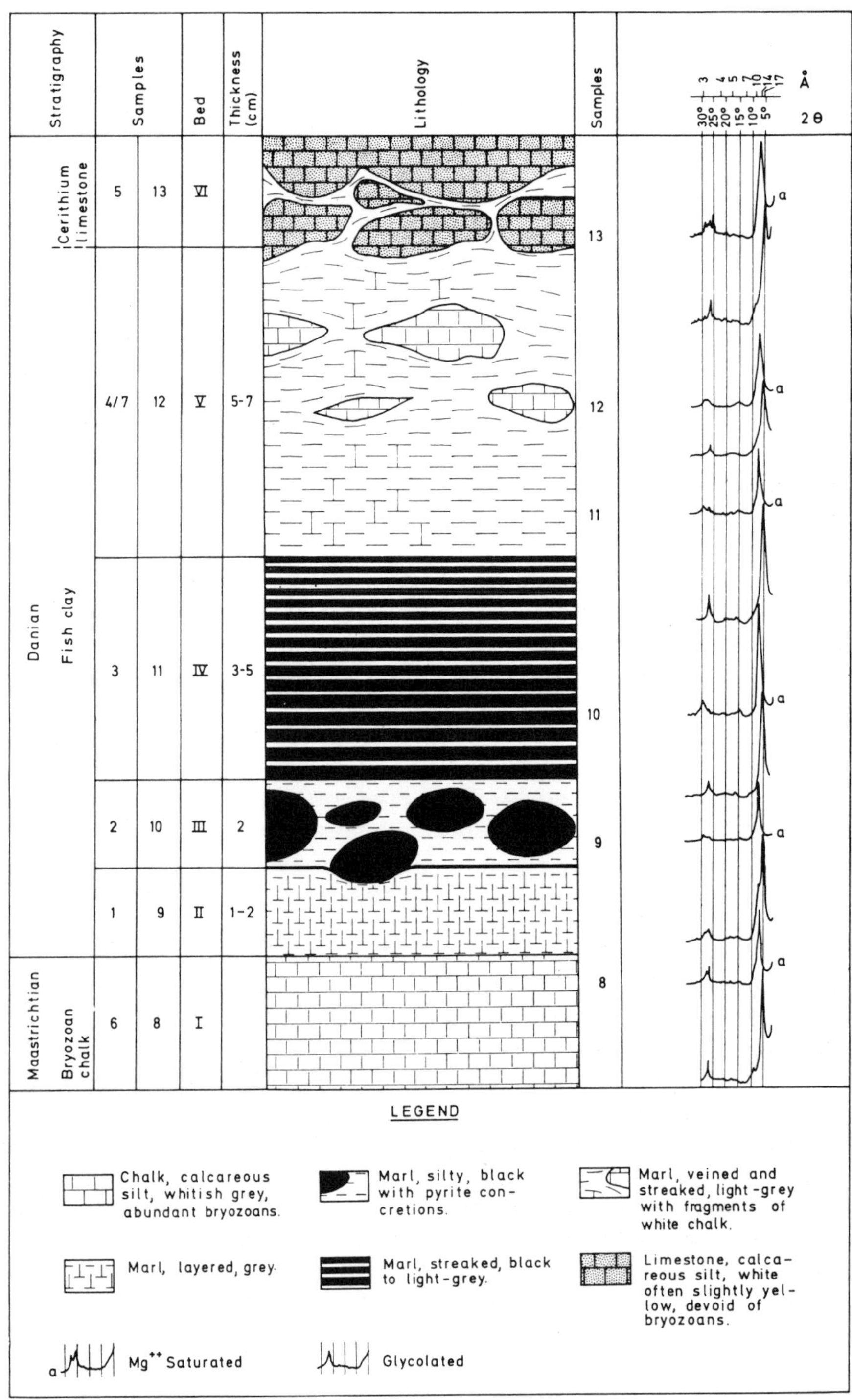

Figure 1. The Cretaceous/Tertiary boundary at Stevns Klint, Denmark. The K/T boundary in this sketch, which is reproduced from Christensen and others (1973), has been placed at the base of bed II. This bed was, however, very probably a chalk sediment of the last Cretaceous sea and was subjected to sea-floor dissolution at the beginning of Tertiary time. On the basis of event stratigraphy, the K/T boundary should be placed at the top of bed II. No fallout deposit has been positively identified at Stevns Klint. The iridium-rich clays and marls (beds III, IV, V) are probably redeposited fallout deposit.

dark color to the high organic carbon content. Amorphous carbon or soot has been identified in the clay (Wolbach and others, 1985).

The boundary clay is low in carbonate content, enriched in siderophile elements, and depleted in rare earths (Alvarez and others, 1980; Smit and ten Kate, 1982). Spherules resembling microtektite have been identified (Varecamp and Thomas, 1982). The bed III is particularly distinctive. This black clay has the highest insoluble residue content (more than 50 percent), the highest iridium and gold concentration, and consists almost exclusively of pure smectite. The pyrite occurrence and the undisturbed laminations of the black clay indicate deposition in an anoxic environment. The Fish Clay is restricted in distribution, limited to the deepest and more central part of a Cretaceous shelf-basin; it wedges out laterally (Christensen and others, 1973).

Gubbio and Petriccio, Italy

The boundary clay at Gubbio, Italy, is present between carbonate strata. The Cretaceous and Paleocene *scaglia rossa* of the Apennines are typically reddish brown, indurated marls of pelagic origin. The sequence at Gubbio is, however, mainly limestone that contains about 5 percent insoluble residue. The boundary clay contains more than 50 percent insoluble residue, and is about 1 cm thick. The lower half-centimeter layer is gray, and the upper half-centimeter layer is red. Alvarez and others (1980) analyzed the concentration of 28 trace elements in the insoluble-residue fraction of the boundary sequence at Gubbio. They found that 27 of the 28 elements (including numerous rare earth elements) show very similar patterns of random abundance-variation with time, but the only siderophile analyzed, iridium, increases by a factor of 30 (above the background values) in the boundary clay. They also reported that the iridium maximum is in the upper red clay. Spherules of magnetite and glauconite are present in the clay, and are especially abundant in the middle part of the 2.5-cm-thick boundary clay at Petriccio (Montanari and others, 1983; see Fig. 2).

The base of the boundary clay unit marks the extinction horizon of Cretaceous planktic foraminifera, which are diverse and represented by numerous individuals in the chalk. Only sparse remains of a small dwarf fauna, the *Globigerina eugubina* assemblage, are present in the boundary clay of Gubbio (Luterbacher and Premoli-Silva, 1962).

Caravaca, Spain

The boundary clay at Caravaca is ten times thicker than that at Gubbio, and its biostratigraphic record is one of the most complete on land. Smit and Romein (1985) recognized a basal red lamina, 0.5 cm thick. The clay has insoluble residue content of about 70 percent (Smit, 1982) and is characterized by an unusual enrichment of iridium and other siderophiles, and by a corresponding depletion of the rare earths (Smit and ten Kate, 1982; see Fig. 3). Spherules of sanidine, believed to be diagenetically altered microtektite glass, were first identified at Caravaca by Smit and Klaver (1981). Overlying the red lamina is a dark gray marl, grading upward to greenish gray, and its insoluble residue content varies between 40 and 60 percent. The siderophile elements are somewhat enriched, but the rare-earth elements are not much depleted in the marl. This sediment contains an indigenous fauna of planktic foraminifers belonging to Cretaceous taxa (Smit and Romein, 1985), and is biostratigraphically assigned to the *Guembelitria cretacea* zone. The gray marl grades upward into the marl of the *Globigerina eugubina* zone.

Isotope analysis of the calcite component in the sediments across the boundary revealed a sharp carbon-isotope anomaly at the boundary-clay horizon; the anomaly is a negative $\delta^{13}C$ perturbation of about 2 pro mil (Smit, 1982), similar to that found in deep-sea sequences to be described later.

Other marine sections on land

A boundary clay or marl, similar in lithology but variable in thickness, has been identified in many other outcrop localities of continuous marine sedimentation across the K/T transition in France, Germany, Austria, Tunisia, Israel, Pakistan, Soviet Union, North America, and New Zealand. The sediment at the base of the unit recording the first appearance of Tertiary microfaunas and nannofloras is not only characterized by their low calcite content (compared to underlying the overlying sediments), but also by siderophile enrichment and by carbon-isotope anomaly. Spherules or altered microtektites and shock quartz grains have also been recognized at several localities (Varecamp and Thomas, 1982; Bohor, personal communication, 1987).

BOUNDARY IN DEEP SEA

The sediment marking the boundary of a K/T transition at deep-sea sites is not invariably a clay, but a marl or a calcareous ooze at numerous sites. The presence of calcareous oozes and marls in the boundary sequences of DSDP sites at 3,000 or 4,000 m paleodepth indicates calcite-compensation-depth (CCD) did not simply rise at the end of the Cretaceous from 4 or 5 km depth to the photic zone (Supko and Perch-Nielsen, 1977).

DSDP Site 524, South Atlantic

The boundary clay at Site 524 in the South Atlantic, 29°29″S,3°31″E, is about 2 cm thick. The subjacent and suprajacent marl oozes are light reddish brown; the clay is similar in coloration, but much darker. The boundary is defined by the datum plane marking the first appearance of the Tertiary microfaunal (P-1a) and nannofossil (NP-1) assemblages. Fossil foraminifers and nannoplankton belonging to Cretaceous taxa are present in the boundary clay and the transitional zone above the K/T boundary.

The K/T boundary is recognizable visually (Hsü and others, 1984, inside cover photograph). Cretacous sediments are biotur-

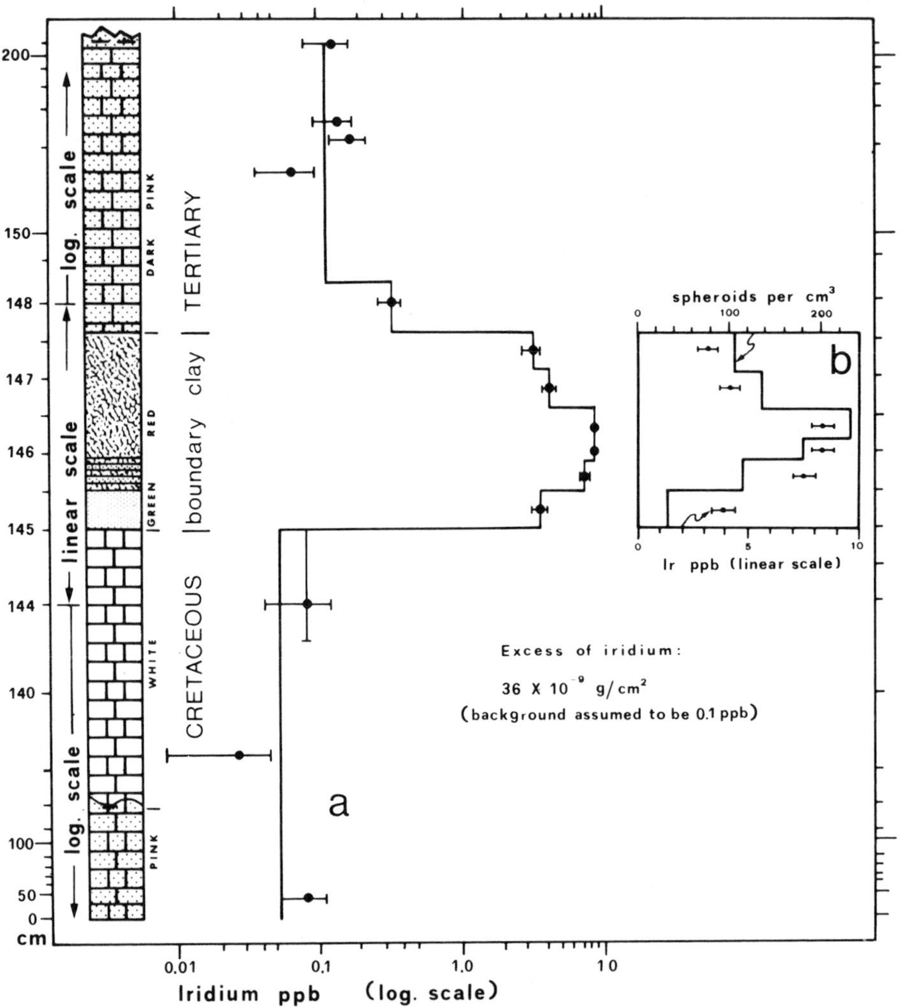

Figure 2. Boundary section at Petriccio, Italy (after Montanari and others, 1983). (a) Lithology iridium profile. (b) Spheroid abundance. Note that the iridium maximum is found in the red boundary clay, which was deposited after a short-lived phase of oxygen deficiency in coastal waters at the very beginning of Tertiary time.

bated and truncated by the contact. The boundary clay and the immediately overlying Paleocene marl and turbidite beds show little evidence of bioturbation. The geochemical anomalies across the K/T transition are illustrated in Figure 4.

The boundary clay has a maximum iridium enrichment and a minimum calcium-carbonate content. Small spherules have been found at the very base of the clay. Those at Site 524 are composed of glauconite; they, like the sanidine spherules at Caravaca, are believed to have been formed by submarine diagenesis of microtektite (Smit and Romein, 1985).

Deep-sea samples, being less indurated and less diagenetically altered, are particularly suitable for isotope analysis. The oxygen-isotope data suggest rapid temperature variations after the convulsive boundary event, and the oscillation pattern is not unlike that found by Shackleton and Opdyke (1973) for the late Quaternary. A maximum temperature was reached some 50,000 yr after the beginning of the Tertiary. The temperature then decreased and the oscillation dampened.

The most remarkable geochemical anomaly recognized at Site 524 is the carbon-isotope perturbation. A negative anomaly of $\delta^{13}C$ of about 2 to 3 pro mil in planktic fossils is found at the boundary horizon defined paleontologically and by the iridium spike. A second-peak anomaly was noted at the horizon of the temperature-maximum, some 50,000 yr later. The carbon-isotope anomaly at the boundary has been called Strangelove perturbation. The carbon-12 atoms were enriched in ocean surface-water because they were not utilized for building the soft tissues of marine planktons in an ocean that was not fertile (Hsü and McKenzie, 1985). The second anomaly, on the other hand, may reflect a more gradual change of the isotopic composition of the

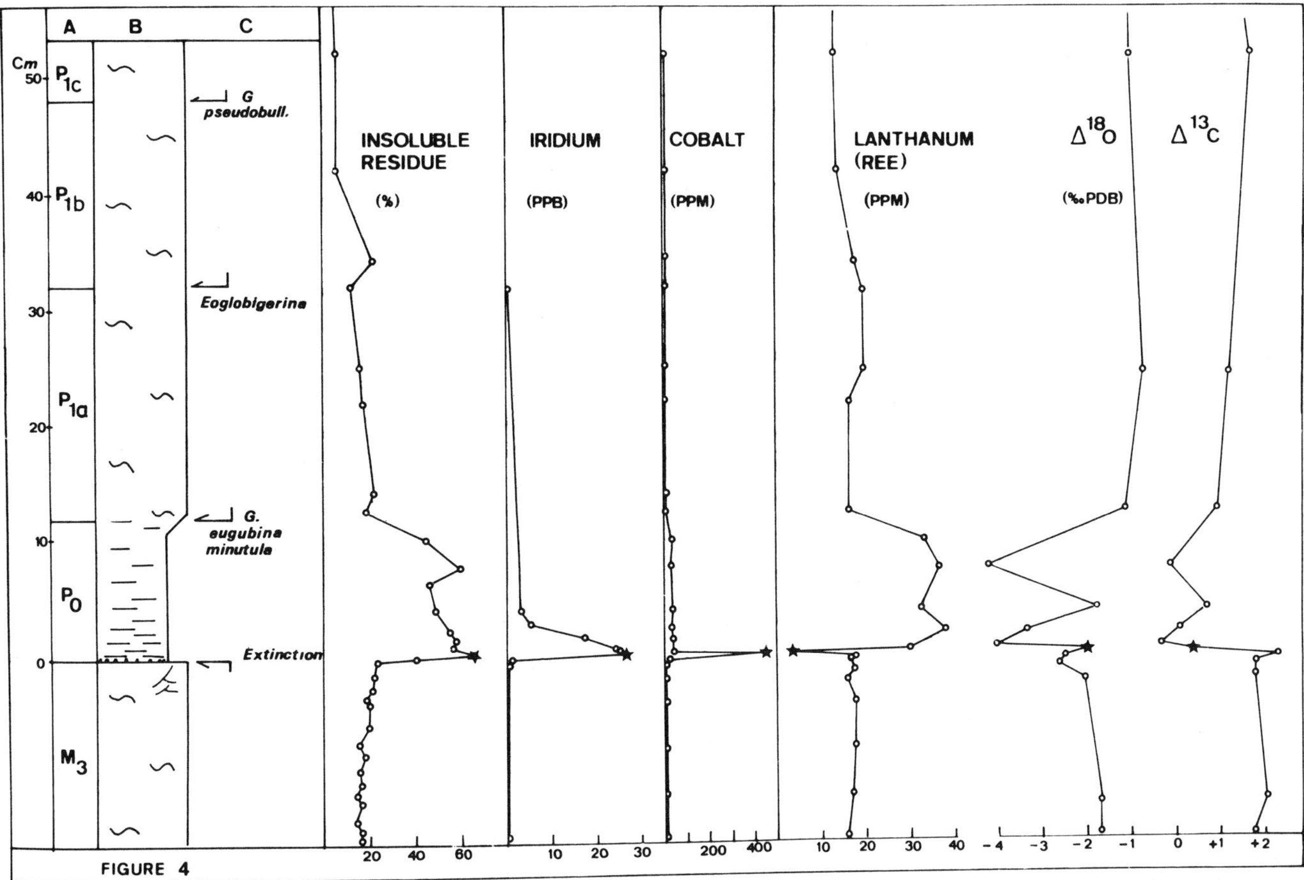

Figure 3. Geochemical anomalies of the boundary section at Caravaca, Spain (after Smit, 1982). (A) Biozonation. (B) Lithology: dash lines, clay; wavy lines, marl; dot, basal lamina of fallout deposit. (C) Foraminiferal events, including extinction horizons of typically Cretaceous taxa, and first appearance horizons of three Tertiary taxa defining the biozones. Note the double-peaked carbon-isotope anomaly at this section.

ocean during the first 50,000 yr of the Tertiary. This pattern of a two-peaked carbonate-isotope anomaly has been found in the most complete K/T sequences (Perch-Nielsen and others, 1982), in contrast to the pattern of one carbon-isotope perturbation that is characteristic of sequences where the boundary sediments have been disturbed or reworked.

Other localities in the Atlantic

The K/T boundary has been cored at several other South Atlantic sites. They differ from Site 524 in that the boundary sediment is more calcareous, probably because those sites had shallower paleodepth, being located on the Rio Grande Rise and Walvis Ridge or other submarine highs.

The boundary at Site 516 (30°17″S, 35°17″W) on the Rio Grande Rise, as defined by foraminiferal evidence, namely, the first appearance of P-1a assemblage, is marked by a change from the Cretaceous white chalk to the Paleocene dark marl (Hamilton, 1983, p. 950). The first appearance of typical Paleocene nannoplankton taxa and the base of a 20-cm-thick zone of iridium enrichment were noted at a horizon slightly deeper than the boundary defined by the microfaunas. The photograph of the cored boundary shows that the lower boundary of the dark marl is transitional (see Barker and others, 1983, inside cover photograph). The current structures and the evidence of bioturbation indicate that materials from the basal boundary clay, including oldest Tertiary nannofossils and iridium-rich detritus but not oldest Tertiary foraminifers, have been slightly reworked and mixed with the youngest Cretaceous sediment to form the sediment in the 7-cm interval between the apparent first appearance of the oldest Paleocene nannoplankton and that of first Paleocene microfauna. The negative shift of $\delta^{13}C$ and δO values started at the horizon marked by an iridium spike, and the isotope anomalies reached a maximum about 1 m above the boundary. The pattern

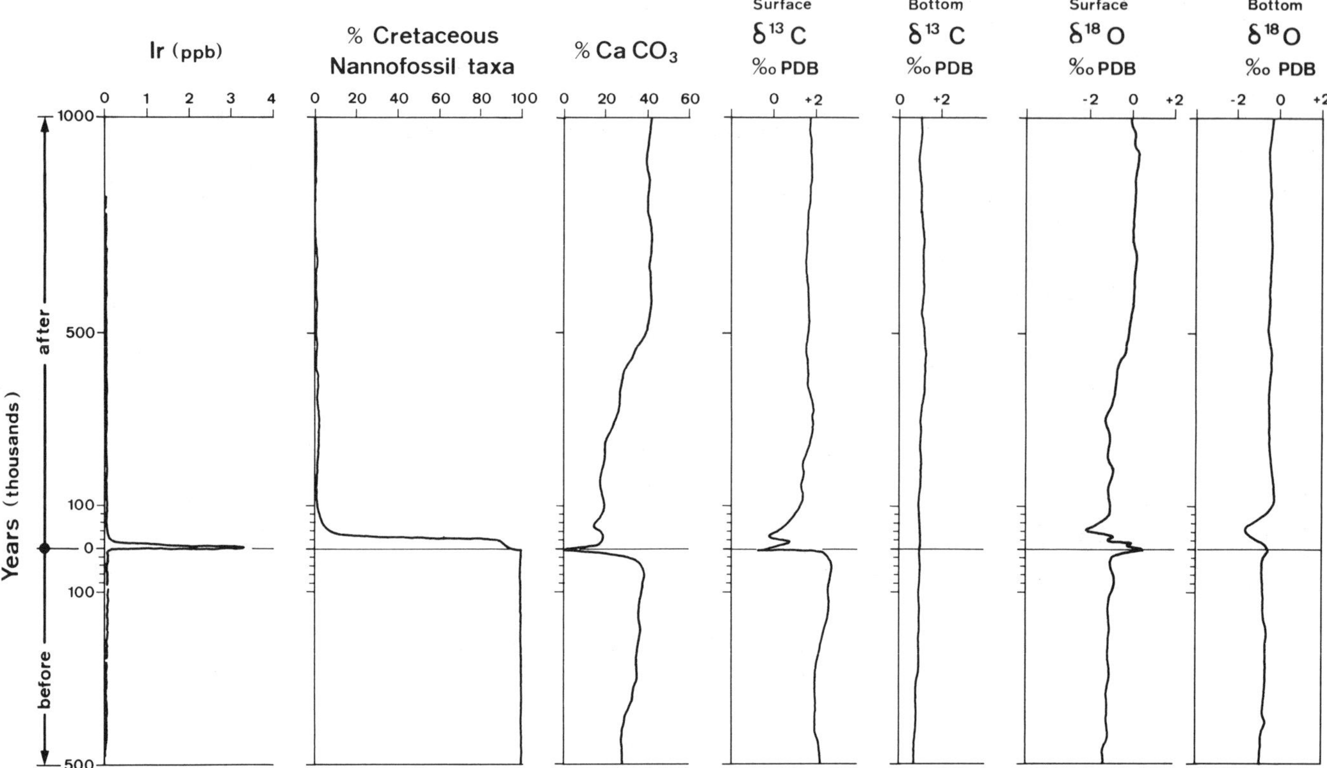

Figure 4. Geochemical anomalies of the boundary section at site 524, South Atlantic (after Hsü and McKenzie, 1985). The boundary sediment at site 524 is a marl except for the boundary clay that has 4 percent calcite. Note that considerable Cretaceous nannofossils, with Tertiary chemical signals, are found in sediments deposited during the first 30,000 yr of Tertiary time. Note also the Strangelove perturbance at this site.

is thus similar to that at Site 524, except the very sharp carbon-isotope perturbation at the base of the boundary there is missing at Site 516. The absence of the "Strangelove perturbation" at Site 516 may be attributed to the erosion and reworking of the boundary sediment deposited immediately after a convulsive terminal-Cretaceous event, a conclusion in agreement with the previously mentioned paleontological interpretation of the K/T boundary.

The K/T boundary at the Walvis Ridge sites has been determined by the first appearance of the oldest Paleocene nannoplankton. In fact, the boundary can be visually recognized at all sites because of a color change at the boundary. The calcite content of the boundary sediment is different at Walvis Ridge sites of various paleodepths. Carbonate content of 10 or 20 percent has been reported at or near the boundary (Moore and others, 1983). A sharp anomaly of the carbon-isotope value in planktic fossil has also been observed in the sediment immediately above the boundary (Shackleton and Hall, 1983).

The K/T boundary at other Atlantic DSDP sites has also been determined on the basis of the first appearance of the oldest Paleocene nannoplankton (NP-1). The boundary is marked by a transitional zone of color change from light gray upward to dark gray at Site 356 (28°17″N,41°05″W), a site drilled previously on the Rio Grande Rise (Supko and Perch-Nielsen, 1977). The boundary is obscure at the North Atlantic Site 384 (40°22″N,51°40″W) where the last Cretaceous and the first Paleocene sediments are separated by a transitional zone of bioturbated ooze (Thierstein, 1982). A negative carbon-isotope anomaly has been found at both of those Atlantic sites (Boersma and others, 1979). The presence of one, instead of two, perturbations at the boundary of those two sites suggests that the oldest sediment above the boundary, that of the Strangelove Ocean, has been reworked by current erosion, by dissolution, and/or by bioturbation. This conclusion is in agreement with that deduced on the basis of paleontological or sedimentological observations.

Pacific sites

The K/T boundary at Site 465 (33°49″N,178°55″E), on the Hess Rise in the Pacific Ocean, is defined by the first appearance datum of the oldest Paleocene nannoplankton (NP-1). The Cretaceous and Tertiary calcareous oozes cored at this site are all white or light gray in color; only the boundary sediment is black clay. Unfortunately this black sediment has been badly disturbed by drilling. (The black layer is estimated to have had a thickness of 1.5 cm.) The clay is almost entirely devoid of calcite, and is enriched in Ir, Au, Cr, Fe, Co, Ni, As, Sb, and Se. The trace-element chemistry and the discovery of the sanidine spherules in this layer suggest correlation to the basal lamina at Caravaca (Kyte, 1983), but the color here is black, not red. A negative carbon-isotope anomaly at the boundary has been identified as a Strangelove perturbation, because only the planktic, but not the benthic, fossils show the negative perturbation typical of an ocean depleted in plankton productivity (Boersma and Shackleton, 1981).

A K/T transition in an ooze sequence was investigated at Site 577 (32°27″N,157°44″E) on the Shatzky Rise in the west Pacific, and the cored boundary, sampled from three boreholes at this site, has not been disturbed by drilling. The boundary sediment is a light brown marl, slightly darker than the underlying Cretaceous ooze, but no black clay is present. The K/T contact has been determined by the datum plane marking the first appearance of typically Paleocene nannoplankton, although the presence of fossil tests belonging to Cretaceous taxa is not uncommon in the earliest Tertiary sediments (Monechi, 1985). Iridium anomaly, microtektite-like spherules, and carbon-isotope perturbation have all been found at the boundary, as predicted (Michel and others, 1985; Shackleton and others, 1985; Gerstel and others, 1986). However, the oldest Tertiary planktic foraminifera, *Globigerina eugubina,* with typically Tertiary geochemical signature, has been found in the sediments several centimeters below the boundary defined by nannoplankton and by spherules. Gerstel and others concluded that a paleoceanographic crisis started before the terminal Cretaceous impact event. This conclusion is probably not correct, because core photographs show clearly that the K/T contact at this site has been disturbed by bioturbation (Heath and others, 1985, inside cover photographs).

The K/T boundary in a red clay core (LL44 GPC-3) from the North Pacific abyssal plain has been recognized on the basis of icthyolith chronology. The boundary sediment is visually indistinguishable from the overlying and underlying reddish brown clay, but is characterized by an enrichment of the siderophiles and As and Sb sulphides (Kyte, 1983). The original boundary sediment was probably a thin lamina, but it has been smeared by bioturbation long after the deposition of the siderophiles.

BOUNDARY IN CONTINENTAL SEQUENCES ON LAND

The K/T boundary clay, 1 to 2 cm thick, intercalated in the continental facies of the Raton Basin, New Mexico and Colorado, commonly lies at the top of a carbonaceous shale unit below a thin coal bed (Tschudy and others, 1984). The boundary is defined by the extinction of numerous Cretaceous taxa. The clay, similar to tonstein, consists of nearly pure, well-crystallized kaolinite (Pillmore and others, 1984). Normal tonstein is an alteration product of volcanic glass, but the tonstein-like white clay of Raton Basin may have been altered by hot impact debris. Iridium and other siderophile elements are enriched in a 10-cm interval in the clay and in the overlying coal and shale (Fig. 4), and the maximum iridium concentration is several orders of magnitude higher than that of the background (Orth and others, 1982). Shock-quartz of impact origin has also been identified in the clay (Bohor and others, 1984). In addition to the extinction of numerous Cretaceous taxa, the level of maximum siderophile enrichment coincides approximately with the horizon marking the sudden appearance of an assemblage of fern spores with little or none of the pollen of the angiosperms, which had been flourishing (Tschudy and others, 1984). The high angiosperm–low fern spore flora reappeared, however, at a horizon only 15 cm above the boundary (Fig. 5). The presence of fusinite and macerated plant material immediately above the boundary clay has also been reported by Tschudy and others (1984).

The K/T boundary in south-central Saskatchewan is also marked by coincident anomalies in abundance of iridium and fern spores at the extinction level of a suite of Cretaceous pollen taxa (Nichols and others, 1986). The boundary near Morgan Creek marks the contact between a unit of terrigenous clastics containing dinosaur bones and a coal-bearing formation. A 2-cm-thick layer of brown to black carbonaceous claystone immediately below a coal bed yielded the youngest fossils of *Aquilapollenites* flora and greatest concentration of iridium (3 ppb). Tertiary samples yielded flora assemblages composed of species that survived the terminal Cretaceous event, and the typically Paleocene species first occur 5.5 m above the boundary. As in Raton Basin, a major shift in palynomorph abundance is found at the top of the boundary clay. This "fern-spore spike" is believed to reflect the initial recolonization of the terrestrial environment immediately after the devastation caused by a terminal Cretaceous catastrophe.

A similar clay has been found at the K/T boundary of other localities of terrestrial sedimentation in western North America (Pillmore and others, 1984; Smit and van der Kaars, 1984). The characteristic floral changes and the remarkable siderophile-enrichment signatures, first discovered in the Raton Basin, have also been noted at those sites (Pillmore and others, 1984).

SUMMARY OF SEDIMENTOLOGICAL AND GEOCHEMICAL EVIDENCE

The K/T boundary at all sections described in this chapter has been determined by paleontological criteria. The boundary in marine sequences is drawn at the horizon where the typically Paleocene nannoflora first appears, namely the bottom of the boundary clay; microfossils and nannofossils belonging to the Cretaceous taxa are not uncommon in the boundary clay above the contact. The paleontological boundary in marine sequences is an event-marker of mass mortality: the first Paleocene species

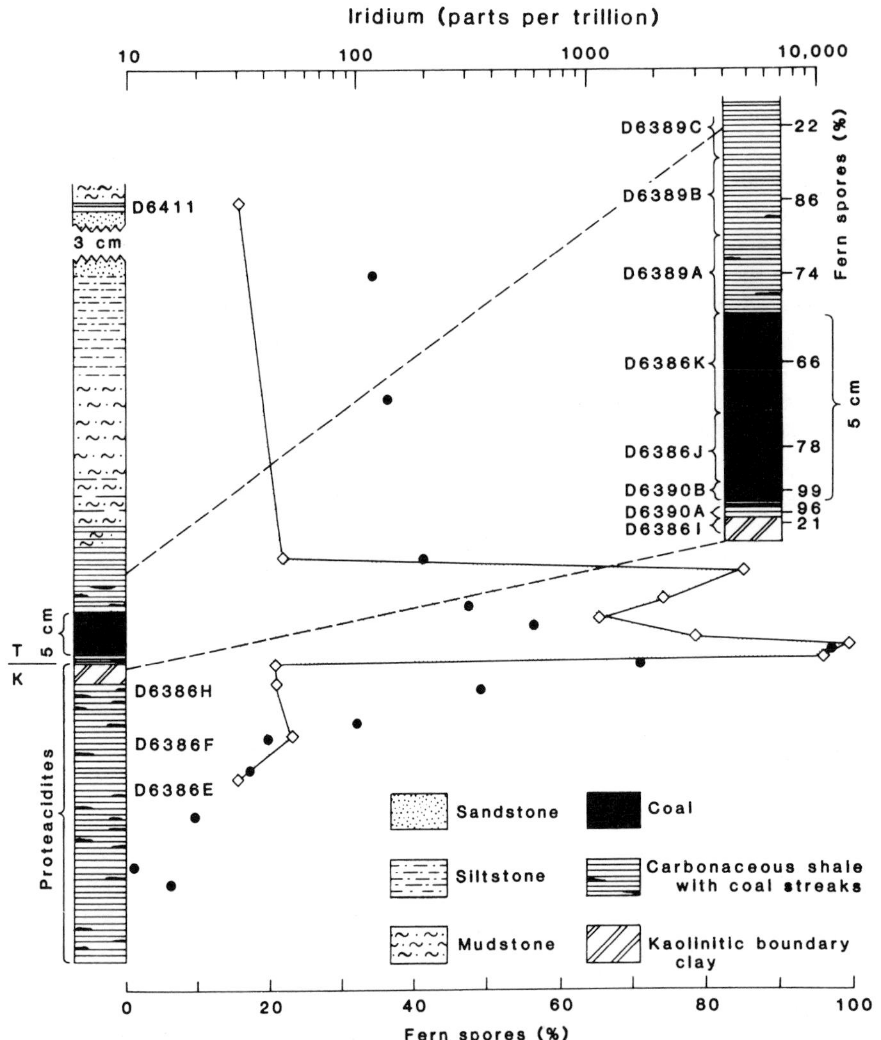

Figure 5. The K/T boundary section at Starkville North, Colorado (after Tschudy and others, 1984). The large black dots show the variation in iridium concentration, whereas the solid line and diamonds show the fern-spore percentages. The inset shows detail of the boundary interval. The USGS sample-locality numbers are shown adjacent to the columns. Note the iridium enrichment of the boundary clay, and that the fern-spore spike is positioned at the top of the clay. The relation suggests that recolonization of the deforested land by fern growth came after settling of the stratospheric fallout of impact debris.

existed prior to the catastrophe, but their numbers were relatively insignificant until the more numerous Cretaceous taxa had been decimated. The boundary in terrestrial consequences is, on the other hand, drawn at the top of the boundary clay—a horizon of the highest occurrence of the typically Cretaceous plant taxa; the last pollen grains of the old world settled down together with impact debris.

Remarkable is the presence of sedimentological and geochemical anomalies revealed by the sediments immediately above or below the boundary. The calcium-carbonate content in the boundary sediment of marine sequences is invariably less than the immediately subjacent and suprajacent sediments. Whereas the boundary sediments deposited in open oceans at paleodepth of a few thousand meters are marls or calcareous ooze, those deposited in shallow coastal waters and now found in sections on land are almost devoid of calcium carbonate. The boundary sediment on land is similar to tonstein, which is altered volcanic glass.

An enrichment of siderophiles, especially iridium, in the boundary sediment is typical of all sequences, marine or terrestrial, on land and under the sea. The relation of the peak anomaly to the paleontologically defined K/T boundary is, however, not exactly the same at all localities. Resedimentation or bioturbation may have played a role in a slight redistribution of the siderophiles. Also present in the boundary sediment of many localities are spherules of various lithology (altered microtekite) and shock-quartz grains. Depletion of the rare-earth elements has been noted in some boundary sediment, notably in the thin lamina considered by some to be the fallout deposit of stratospheric dust.

Where a marine sequence is proved by paleontological investigations to be more or less continuously deposited, a negative anomaly of carbon-isotope, with a carbon-12 enrichment of 2 to 3 pro mi, is registered by the planktic fraction of boundary sediment. A two-peaked anomaly is noted at localities where the section is most complete, and a broad peak at localities where the boundary has been slightly disturbed by reworking or bioturbation.

The most remarkable anomaly in the boundary sediment of terrestrial sequences of western North America is the so-called fern-spore spike, which occurs at the top of the boundary clay, signifying the recolonization by ferns in areas deforested by the terminal Cretaceous event. The tonstein-like boundary clay on land has been assumed to be a fallout deposit. Since the settling time for stratospheric dust is at most a few years, the time difference between the bottom and top of the boundary clay is very short, and it makes little practical difference if the boundary should be drawn at the top or bottom of this deposit. When it was found out, however, that the iridium anomaly and the concentration of micro-tektite are present at the top, but not at the bottom of the boundary clay (Bohor, personal communication, 1987), the interpretation that the tonstein-like sediment is a fallout debris, of either impact or volcanic origin, seems unlikely. I suggest, therefore, that the presently drawn boundary, defined by the iridium-rich horizon near the top of the boundary clay, is the bottom of the fallout deposit. The clay itself may have been originally an uppermost Cretaceous marsh clay, which was altered to tonstein when it was buried under a thin blanket of very hot fallout debris. Considering that the interior of a mushroom cloud above an impact site should have had a temperature of 10^5°C, a very high temperature of the debris after their settling could be assumed. The marsh clay turned white when its organic content was oxidized, and its mineralogy changed by thermal metamorphism.

GENESIS OF THE BOUNDARY CLAY

The iridium anomaly and the isotope signals in the K/T boundary clay were the basis for postulating a large-body impact at the end of the Cretaceous (Alvarez and others, 1980; Hsü, 1980; Smit and Herzogen, 1980). An alternative hypothesis of catastrophic volcanism has been postulated (McLean, 1985; Officer and Drake, 1985). The Deccan trap registered a short and intensive period of volcanism in the million years or so after the end of the Cretaceous (McLean, 1985), but the volcanism may have been triggered by a meteorite-hit in the Indian Ocean (Hartnady, 1986). Since the hypothesis of catastrophic volcanism fails to explain the siderophile-enrichment and the occurrence of shock-quartz in the boundary clay, the meteorite-impact idea is a more viable working hypothesis, even though the consequences of environmental devastation derived from those two alternative causes are very similar (McLean, 1985).

The carbon-isotope anomaly provides evidence for the great reduction of biomass in the oceans after the terminal Cretaceous event (Hsü and McKenzie, 1985), and a scenario for environmental devastation after the terminal Cretaceous event has been proposed in a recent paper (Hsü, 1986). This paper describes the stratigraphy and sedimentology of the K/T boundary sediment, and uses a deductive approach to test the theory of the impact hypothesis.

The ejecta debris of a large bolide-impact should have settled within a year or two (O'Keefe and Ahrens, 1982). It has been postulated that the boundary clay at Stevns Klint and elsewhere represents such a fallout deposit (Alvarez and others, 1980). This led in turn to an assumption of a catastrophic sedimentation rate (in pelagic realm) of centimeter(s) per year for the boundary clay, and implied a catastrophic flux of organic carbon ($10^5 \times$ normal) into the ocean (Wolbach and others, 1985). Although the geochemistry of the boundary clays supports the interpretation that impact debris are present in the boundary clay, the sedimentologic evidence suggests that the original fallout deposit, especially that in the marine realm, may have been disturbed by bioturbation, removed by dissolution and/or by current erosion, and the reworked impact debris were redeposited in many places as the boundary clay (Kyte and others, 1985; Hsü, 1986).

The best documented occurrence of a fallout deposit is the white boundary clay in Raton Basin. Not only is the sediment enriched in extraterrestrial component (ETC), but little terrigenous detritus is present so that the clay consists of almost pure kaolinite, which is the common alteration product of glassy detritus in the acid environment of coal swamps. Other debris of impact origin include microtektite and shock-quartz grains. This fallout sediment was probably deposited within a few years after the terminal Cretaceous event, because the fern-spore spike was found at the top of the clay. In view of the fact that the dominance of the fern vegetation lasted only a very brief time (Tschudy and others, 1984), the stratigraphic relationship indicates that the fallout deposit on land has been preserved as the white boundary clay, and has not been reworked during the millenia after the catastrophe.

The paleobotanic evidence of deforestation in western North America explains the presence of abundant carbon soot in the marine boundary clay. Like the fusinite in the coal and in the carbonaceous shale immediately above the boundary clay, the soot was apparently the product of widespread forest fires (Tschudy and others, 1984; Wolbach and others, 1985). The insignificant organic-carbon content in the fallout deposit suggests, however, that the carbon particles did not all go up to the stratosphere and come down together with the impact debris; much of the organic carbon may have been transported to, concentrated, and redeposited at their final resting places in swamps or on sea floor long after the fallout deposit was laid down.

The terminal Cretaceous event on land caused the extinction

of the *Aquilapollenites* flora in high-latitude regions of the northern hemisphere. Those plants were probably parasitic plants (Jarzen, 1982), and they may have died out when symbiotic trees temporarily disappeared during deforestation. There is, however, a persistence of numerous climatically sensitive taxa across the boundary, and this fact indicates that the climate returned quickly to the condition before the terminal Cretaceous catastrophe; angiosperm forests soon returned and displaced the fern growth (Nichols and others, 1986).

The record of the fallout deposit in the marine realms is not well preserved. Only the red lamina at the base of the boundary clay unit at Caravaca gives clear evidence of being a fallout deposit (Smit and Romein, 1985; Lindiger, personal communication, 1986). Even there the sediment has been disturbed by bioturbation and occurs as a broken lamina at the base of the boundary clay unit. The fallout at Caravaca consists of smectite, which is a common alteration of glass in marine environments. It is much enriched in iridium and other ETC, and is depleted in rare-earth elements; sanidine spherules are very common.

The sedimentology of the boundary sequences elsewhere records a history of sedimentation and resedimentation during the next millenia. A significant volume of the materials making up the boundary sediment in marine sequences has been derived from the reworking of the fallout detritus. The dark gray clay above the red laminar at Caravaca, the black clay at Stevns Klint, as well as the boundary sediment at several other sites may have derived part of their source materials from the eroded fallout deposit on the periphery of the basin (see Kyte and others, 1985). Therefore, iridium-rich boundary clay is not present everywhere, but is commonly found only in the deepest part of shelf basins.

The interpretation that impact debris were mixed with other oceanic sediments and deposited as pelagic or hemipelagic sediments on the bottom of a less-than-normally fertile ocean implies that the marine boundary clay (except the fallout lamina) is a normal pelagic sediment. The normal rate for the deposition of calcareous oozes, as estimated from many studies of deep-sea drilling cores, is of the order of centimeter or centimeters per thousand years. In an ocean of less-than-normal fertility, the insoluble residues constitute the bulk of the boundary clay, and such sediments are normally deposited at a very slow rate of millimeter(s) per thousand years.

What is distinctive about the Caravaca basal lamina is its light red color. Everywhere else, the boundary clay, including that above the red lamina at Caravaca, is gray, black, or a darker shade than the underlying Cretaceous sediments. It seems that the ocean was still oxidizing normally during the few years when the fallout deposit was accumulating. Later on, an unusual increase of biogenic carbon in the oceans led to temporary oxygen deficiency or even anoxia at many places. The previously deposited global fallout of impact debris was reworked and redeposited during this brief interval; preservation of the finely disseminated carbonate soot gives the sediment a dark coloration. Sulphide precipitation and other submarine diagenesis under reducing conditions took place during this time. In deeper Atlantic and Pacific sites, the organic carbon in sediment has been largely oxidized; the siderophile-rich boundary clays at those localities are thus light gray, yellowish brown, or reddish brown. Nevertheless, reducing bottom conditions may have prevailed for several millenia even in the normally oxygenated North Pacific, as indicated by the occurrence of pyrite and As and Sb sulphides at the K/T boundary of a red clay sequence there (Kyte, 1983; Brooks and others, 1984).

The destruction of biomass increased the carbon flux to the oceans, and the oxidation of this carbon generated biogenic carbon dioxide. This excess CO_2 added to the surface waters and to the intermediate waters of the oxygen-minimum zone, where the CO_2 concentration was already high, caused an undersaturation with respect to calcite at shallow depth (see computer modeling by Sundquist, 1986). Thus the boundary sediment deposited in coastal waters is commonly a clay (e.g., at Stevns Klint, Gubbio). However, the terminal Cretaceous CCD did not simply rise to the photic zone as postulated by Worsley (1974). At depth greater than 1,500 or 2,000 m (e.g., on Rio Grande Rise, Walvis Ridge), below the oxygen-minimum, the sea water was less corrosive than shallower waters; the earliest Tertiary sediment at those ocean sites is not a clay, but a marl or calcareous ooze. Only at still greater depth (abyssal North Pacific), below the true CCD of the open ocean, the boundary sediment is again a clay like the underlying and overlying red clay.

It has been noted that microfossils in the uppermost Maastrichtian sediments have been largely dissolved away at many DSDP sites, and the carbonate content of uppermost Maastrichtian sediment (e.g., bed II at Stevns Klint) is lower than that of the underlying limestones. This fact suggests that there was a short episode of extensive dissolution of bottom sediments immediately after a catastrophic event at the end of Cretaceous. Not only was the bottom water in many places more corrosive than usual, the decrease in productivity and the consequent decrease of calcareous fossils may have caused abnormal undersaturation of bottom waters and caused a widespread, if not global, sea-floor dissolution of previously deposited sediments. This calcite-dissolution episode was, however, very short-lived, lasting perhaps only a few hundred or a few thousand years, or the time of deposition of a thin lamina of almost calcite-free sediment at many places, because the early Paleocene nannofossils in calcite-bearing boundary clay are not much corroded (K. Perch-Nielsen, personal communication, 1985). The low calcareous content of boundary sediment is due, in part at least, to the low productivity of the calcareous planktons.

The oxygen-deficient conditions after the convulsive event lasted a little longer. At Gubbio the lower boundary clay, 0.5 cm thick, is green and overlain by a reddish brown clay, also 0.5 cm thick. The duration represented by the 0.5-cm green clay at Gubbio should be 2,500 yr if the sedimentation rate was 2 mm/thousand yr, and 5,000 yr if the rate was 1 mm/thousand yr. It seems that the anoxic event affecting the shelf realms of sedimentation may have lasted considerably less than 10,000 yr. This estimate is of the same order of magnitude as the one obtained by Kyte and

Wasson (1986) in their study of the As and Sb mineralization in the boundary clay of the deep North Pacific.

With the explosive radiation of the Tertiary planktons and the recovery of ocean fertility, conditions of normal pelagic sedimentation returned in less than 300,000 yr after the terminal Cretaceous event (Hsü and others, 1982; Gerstel and others, 1986). The Early Paleocene marine sediment, at a horizon a few meters above the boundary, is commonly calcareous ooze or limestone.

ACKNOWLEDGMENT

The study of the K/T boundary clay is only beginning. This is the main objective of Project 299 of the International Geological Correlation Project. New results of K/T boundary studies are published almost every month. Therefore, this progress report is likely to be somewhat out of date when it is published. In writing this review, I relied heavily on the work of many colleagues. I would like to acknowledge my indebtedness to Jan Smit, Frank Kyte, Walter Alvarez, Kathy Perch-Nielsen, Judith McKenzie, Chuck Pillmore, Richard Bromley, Isabella Premoli-Silva, Matthias Lindiger, and to many others who have discussed the K/T boundary sections investigated by them. Juergen Reinhardt, Chuck Pillmore, and Ed Clifton read a first draft of the manuscript, and I am grateful for their numerous suggestions, which led to improvement of the paper.

REFERENCES CITED

Alvarez, L. W., Alvarez, W., Asaro, F., and Michel, H., 1980, Extraterrestrial cause for the Cretaceous-Tertiary extinctions: Science, v. 208, p. 1095–1186.

Barker, P. F., Johnson, D. A., and others, 1983, Initial reports of the Deep Sea Drilling Project: Washington, D.C., U.S. Government Printing Office, v. 72, 1024 p.

Boersma, A., and Shackleton, N., 1981, Oxygen- and carbon-isotope variations and planktonic-foraminifer depth habitats, late Cretaceous to Paleocene, central Pacific, DSDP sites 463 and 465, *in* Initial reports of the Deep Sea Drilling Project: Washington, D.C., U.S. Government Printing Office, v. 62, p. 495–512.

Boersma, A., Shackleton, N., Hall, M., and Given, Q., 1979, Carbon- and oxygen-isotope records at DSDP 384 (North Atlantic) and some paleotemperatures and carbon-isotope variations in the South Atlantic, *in* Initial reports of the Deep Sea Drilling Project: Washington, D.C., U.S. Government Printing Office, v. 43, p. 695–718.

Bohor, B. F., Foord, E. E., Modreski, P. J., and Triplehorn, D. M., 1984, Mineralogic evidence for an impact event at the Cretaceous–Tertiary boundary: Science, v. 224, p. 867–869.

Brooks, R. R., Reeves, R. D., Yang, X. H., Ryan, D. E., Holzbecher, J., Collen, J., Neall, V. E., and Lee, J., 1984, Elemental anomalies at the Cretaceous-Tertiary boundary: Science, v. 226, p. 539–542.

Christensen, L., Fregerslev, S., Simonsen, A., and Thiede, J., 1973, Sedimentology and depositional environment of Lower Danian fish clay from Stevns Klint, Denmark: Bulletin of the Geological Society of Denmark, v. 22, p. 193–282.

Gerstel, J., Thunnel, R. C., Zhacos, J. C., and Arthur, M. A., 1986, The Cretaceous/Tertiary boundary event in the North Pacific; Planktonic foraminiferal results from Deep Sea Drilling Project site 577, Shatsky Rise: Paleoceanography, v. 1, p. 97–117.

Hamilton, N., 1983, Cretaceous/Tertiary boundary studies at Deep Sea Drilling Project site 516, Rio Grande Rise, South Atlantic; A synthesis, *in* Initial reports of the Deep Sea Drilling Project: Washington, D.C., U.S. Government Printing Office, v. 72, p. 949–952.

Hartnady, C.J.H., 1986, Amirante Basin, western Indian Ocean; Possible impact site of the Cretaceous/Tertiary extinction bolide?: Geology, v. 14, p. 423–426.

Heath, G. R., Burckle, L. H., and others, 1985, Initial reports of the Deep Sea Drilling Project: Washington, D.C., U.S. Government Printing Office, v. 86, 804 p.

Hsü, K. J., 1980, Terrestrial catastrophe caused by cometary impact at the end of Cretaceous: Nature, v. 285, p. 201–203.

——, 1986, Environmental changes in times of biotic crisis, *in* Jablonski, D., and Raup, D. M., eds., Patterns and processes in the history of life: Berlin, Springer-Verlag, p. 297–312.

Hsü, K. J., and McKenzie, J. A., 1985, A "Strangelove ocean" in the earliest Tertiary, *in* Sundquist, E. T., and Broecker, W. S., eds., The carbon cycle and atmospheric CO_2; Natural variations Archean to Present: American Geophysical Union Geophysical Monograph 32, p. 487–492.

Hsü, K. J., He, Q., McKenzie, J. A., Weissert, H., Perch-Nielsen, K., Oberhansli, H., Kelts, K., LaBrecque, J., Tauxe, L., Krahenbuhl, U., Peracival, S. F., Wright, R., Karpoff, A. M., Petrersen, N., Tucker, P., Poore, R. Z., Gombos, A. M., Pisciotto K., Carman, M. F., and Schriber, E., 1982, Mass mortality and its environmental and evolutionary consequences: Science, v. 216, p. 249–256.

Hsü, K. J., LaBrecque, J. L., and others, 1984, The initial reports of the Deep Sea Drilling Project: Washington, D.C., U.S. Government Printing Office, v. 73, 798 p.

Jarsen, D. M., 1982, *Aquilapollenites, in* Russell, D. A., and Rice, G., eds., Cretaceous/Tertiary extinction and possible terrestrial and extraterrestrial causes: Synllogeus, no. 39, p. 52.

Kyte, F. T., 1983, The Cretaceous–Tertiary boundary at two North Pacific sites: Geological Society of America Abstracts with Programs, v. 15, p. 620–621.

Kyte, F. T., and Wasson, J. T., 1986, The Cretaceous/Tertiary boundary clay GPC-3, an abyssal clay section: Geochimica et Cosmochimica Acta (in preparation).

Kyte, F. T., Smit, J., and Wasson, J. T., 1985, Siderophile interelement variations in the Cretaceous-Tertiary boundary sediments from Caravaca, Spain: Earth and Planetary Science Letters, v. 73, p. 183–195.

Luterbacher, H. P., and Premoli-Silva, I., 1962, Note preliminaire sur une du profil du Gubbio, Italie: Rivista Italiana Paleontologica e Stratigraphica, v. 68, p. 253–288.

McLean, D. M., 1985, Mantle degassing unification of the Trans-K-T geobiological record, *in* Hecht, M. K., and others, eds., Evolutionary biology, v. 19, p. 287–313.

Michel, H. V., Asaro, F., Alvarez, W., and Alvarez, L. W., 1985, Elemental profile of iridium and other elements near the Cretaceous/Tertiary boundary in hole 577B, *in* Initial reports of the Deep Sea Drilling Project: Washington, D.C., U.S. Government Printing Office, v. 86, p. 533–538.

Monechi, S., 1985, Campanian to Pleistocene calcareous nannofossil stratigraphy from the northwest Pacific Ocean, Deep Sea Drilling Project leg 86, *in* Initial reports of the Deep Sea Drilling Project: Washington, D.C., U.S. Government Printing Office, v. 86, p. 301–336.

Montanari, A., Hay, R. L., Alvarez, W., Asaro, F., Michel, H., and Alvarez, L., 1983, Spheroids at the Cretaceous-Tertiary boundary are altered impact droplets of basaltic composition: Geology, v. 11, p. 668–671.

Moore, T. C., Rabinowitz, P. D., and others, 1983, Initial reports of the Deep Sea Drilling Project: Washington, D.C., U.S. Government Printing Office, v. 74, 894 p.

Nichols, D.-J., Jarzen, D. M., Orth, C. J., and Oliver, P. Q., 1986, Palynological and iridium anomalies at Cretaceous–Tertiary boundary, south-central Saskatchewan: Science, v. 231, p. 714–717.

Officer, C. B., and Drake, C. L., 1985, Terminal Cretaceous environmental events: Science, v. 227, p. 1161–1167.

O'Keefe, J. D., and Ahrens, T. J., 1982, Impact mechanics of the Cretaceous–

Tertiary impact bolide: Nature, v. 298, p. 123–127.
Orth, C. J., Gilmore, J. S., Knight, J. D., Pillmore, C. L., Tschudy, R. H., and Fassett, J. E., 1982, Iridium abundance measurements across the Cretaceous/Tertiary boundary in the San Juan and Raton Basins of northern New Mexico: Geological Society of America Special Paper 190, p. 423–434.
Perch-Nielsen, K., McKenzie, J., and He, Q., 1982, Biostratigraphy and isotope stratigraphy and the "catastrophic" extinction of calcareous nannoplankton at the Cretaceous/Tertiary boundary: Geological Society of America Special Paper 190, p. 353–372.
Pillmore, C. L., Tschudy, R. H., Orth, C. J., Gilmore, J. S., and Knight, D'J. D., 1984, Geologic framework of nonmarine Cretaceous–Tertiary boundary sites, Raton Basin, New Mexico and Colorado: Science, v. 223, p. 1180–1183.
Shackleton, N., and Hall, M. A., 1983, Carbon isotope data from leg 74 sediment, *in* Initial reports of the Deep Sea Drilling Project: Washington, D.C., U.S. Government Printing Office, v. 74, p. 613–620.
Shackleton, N., and Opdyke, N. D., 1973, Oxygen isotope and paleomagnetic stratigraphy of equatorial Pacific core V28–39: Quaternary Research, v. 3, p. 39–55.
Shackleton, N., Hall, M. A., and Bleil, U., 1985, Carbon-isotope stratigraphy, site 577, *in* Initial reports of the Deep Sea Drilling Project: Washington, D.C., U.S. Government Printing Office, v. 86, p. 503–512.
Smit, J., 1982, Extinction and evolution of planktonic foraminifera at the Cretaceous/Tertiary boundary after a major impact: Geological Society of America Special Paper 190, p. 329–352.
Smit, J., and Herzogen, J., 1980, An extraterrestrial event at the Cretaceous Tertiary boundary: Nature, v. 285, p. 198–200.
Smit, J., and Klaver, G., 1981, Sanidine spherules at the Cretaceous–Tertiary boundary indicate a large impact event: Nature, v. 292, p. 47–49.
Smit, J., and Romein, A.J.T., 1985, A sequence of events across the Cretaceous–Tertiary boundary: Earth and Planetary Science Letters, v. 74, p. 155–170.
Smit, J., and ten Kate, W.G.H.Z., 1982, Trace element patterns at the Cretaceous–Tertiary boundary; Consequences of a large impact: Cretaceous Research, v. 3, p. 307–332.
Smit, J., and van der Kaars, S., 1984, Terminal Cretaceous extinctions in Hall Creek area, Montana; Compatible with catastrophic extinction: Science, v. 223, p. 1177–1179.
Sundquist, E. T., 1986, Geologic analogs; Their value and limitations in carbon dioxide research, *in* Trabalka, J. R., and others, eds., The changing carbon cycle, a global analysis: New York, Springer Verlag, p. 371–402.
Supko, P. R., Perch-Nielsen, K., and others, 1977, Initial reports of the Deep Sea Drilling Project: Washington, D.C.: U.S. Government Printing Office, v. 39, 1139 p.
Thierstein, H. R., 1982, Terminational Cretaceous plankton extinction; A critical assessment: Geological Society of America Special Paper 190, p. 385–400.
Tschudy, R. H., Pillmore, C. L., Orth, C. L., Gilmore, J. S., and Knight, J. D., 1984, Disruption of the terrestrial plant ecosystem at the Cretaceous–Tertiary boundary: Science, v. 225, p. 1030–1032.
Varekamp, J. C., and Thomas, E., 1982, Chalcophile elements in Cretaceous–Tertiary boundary sediments; Terrestrial or extraterrestrial?: Geological Society of America Special Paper 190, p. 461–468.
Wolbach, W. S., Lewis, R. S., and Anders, E., 1985, Cretaceous extinctions; Evidence of wildfires and search for meteoric material: Science, v. 230, p. 167–170.
Worsley, T. R., 1974, Cretaceous–Tertiary boundary event in the ocean: Society of Economic Paleontologists and Mineralogists Special Publication 20, p. 94–125.

Manuscript Accepted by the Society April 4, 1988

Typeset by WESType Publishers Service, Inc., Boulder, Colorado
Printed in U.S.A. by Imperial Printing Co., St. Joseph, Michigan

Index

[Italic page numbers indicate major references]

acoustic units, 39
Alaska, western, 11
algae, coralline, 101
Algerian continental margin, 96
Alicia, Hurricane, *13*
alluvium, 29, 35, 112
  erosion of, 28
Alsek Valley, 117, 124
Ape Canyon time, 25
appendages, crustacean, 109
Aquilapollenites, 152
arenite, quartz, 94
arkose, lithic, 94
ash, Mazama, *47*, 51, *52*, *55*
  vitric, 51
Astoria Canyon, 51, 54
Astoria Fan, 51, 54
Atlantic Ocean, western, 11
avalanche, debris, 28, 31, 35, 46
Azores, 80

Baffin Bay, 89
Balearic Abyssal Plain, *96*
ball-bearing bed, 33, 34, 35
bars, *120*, 124
basin plains, *93*
beach recovery, 18
beachrock, calcareous, 101
Beaufort, North Carolina, 95
bed, ball-bearing, 33, 34, 35
bedforms, *65*, *72*, *83*, 90, *104*, 117, 120
Bermuda, 80
Big Pine Creek, 116
Black Shell Turbidite, 93, 95
Black Shell Turbidity Current, 95, 96
Blake Bahama Abyssal Plain, 97
Blanco Saddle, 51, 54
bolide-impact, 151
Brazilian shelf, 20
Brazos-Colorado Delta, 17
British Columbia, 51
Bryan Mound, *14*

cables, breaks, *80*, *87*
caldera
  acoustic units, *39*
  collapsed, 39
  Mount Mazama, *39*
California, southern, 127
Camille, Hurricane, 12, *13*
Camp Baker, 30
Canadian, eastern continental margin, 89
Cape Breton Island, 80
Caravaca, Spain, 145, 152
carbon, 145
Cascade Range, 47
Cascadia Basin, 47, 51, 54, 55
Cascadia Channel, 51
Castle Creek eruptive period, 26
Castle Lake, 23
Cayuga, 122
cement, calcite, 104
Central Valley, 78, 88
Cerithium Limestone, 143
Channeled Scabland, *67*, 74
Chaski Bay, 46
Chaski Slide, 46, 55
chutes, *61*, 70
clasts, coral, 109
clay, 3, 17, 24, 30, 31, 121, *143*, 149, 151, 152
  marsh, 151
Cleetwood backflow, 46, 47, 49
Coal Bank Bridge, 28
Coldwater Lake, 23
Colorado, 149
Columbia Plateau, 67
Columbia River, 26, 51, 54
Columbia River drainage, 51, 54, 55
Congo Submarine Canyon, 72
continental margin, 9, 20, 59, 77, 89, 93
  Algerian, 96
  Spanish, 96
continental slope, 87, 90
Conus, 104
convulsive event,
  categories, 2
  defined, 1
corals, 10, 101
  aragonitic, 104
  fragments, 104, 109
  honeycomb, 101
Coriolis effect, 9, 10
Cowlitz River, 26, 35
Crater Lake, Oregon, 38, 43, 46, 51, 55
craters, 3, 43
  transient, 3
Cretaceous/Tertiary boundary, 145, 147, 148, 149
currents, 17, 20
  bottom, *12*, 13, 14
  turbidity, 74, *80*, 83, 87, *88*, *93*, 99, 135, 141
Cyclamina incisa, 132
cycle, tidal, 11
cyclones
  extratropical, *8*, *10*
  tropical, *8*, *10*
Cypraea, 104

debris flow, 27, 30, 93
debris, marine, 102
Deep Sea Drilling Project. *See* DSDP sites
deforestation, 151
Delphi, Indiana, 112
deposit, fallout, 151
deposition, depths, 12
Deschutes River, 51
Devil's Kitchen syncline, *132*, 135, 136, 139, 140
dinoflagellates, 143
discharge, river, 72
DSDP sites, 145, 147, 148

Eagle Rest Peak, 139, 140
earthquakes, 19, 29, 80, 89, 140, 141
  Alaska (1964), 129
  Grand Banks (1929), *77*, *80*
  Hebgen Lake (1959), 129
Eastern Valley, 78, 80, 83, *88*
  bedforms, *83*
Ekman veering, 9, 14
Elk Rock Lake, 28
energy, kinetic, 9
erosion, 30, 55, *102*, 121
  alluvium, 28
  nearshore, 12
  sand, 19

fan, washover, 8
Fish Clay, 143, 145
fish remains, 143
fiskeler, 143
flood wave, 27
floods, *111*, *123*, 125
flow
  debris, 27, 30, 93
  turbidity, 27, 30, 93
foraminifera, 102, 109, 132, 145
Fort Wayne Moraine, 120, 124

Galveston Island, 15
Giant Turbidite, 96
glacial scour, 33
glaciers, 112, 116
glass, volcanic, 52
glauconite, 146
Globigerina eugubina, 145, 149
  zone, 145
Graham Ridge, *129*, 139, 140
Grand Banks, Newfoundland, 77
  earthquake, *77*, *80*
  slump, 96

Grand Banks Valley, 78, 80, 88
gravel, 32, 80, 83, *102*, 105, 108, 111
  boulder, 101, 117, 121
  cobble, 112, 117, 121, 124
Great Alaska earthquake (1964), 129
Grenada Basin, 97
Grenada Basin Plain, 97
Gubbio, Italy, 145, 152
Guembelitria cretacea zone, 145
Gulf Coast, hurricane, 10, 13
Gulf of Mexico, 9, 11, 12, 20
  hurricane, 10

Halifax, Nova Scotia, 80
Hatteras Abyssal Plain, *93*, 95
Hatteras Canyon, 95
Hatteras Canyon system, 95
Hatteras Fan, 95
Hatteras Outer Ridge, 95
Hatteras Submarine Canyon system, 95
Hawaiian Islands, 101, 106, 108
Hawaiian Ridge, 107
heating, solar, 8
Hebgen Lake earthquake, 129
Hess Rise, 148
Hilo, Hawaii, 106
Hispaniola, 97
Hispaniola Caicos Abyssal Plain, 97
Hulopoe Bay, 102
Hulopoe Gravel, 101, *102*, 106, 108
Hulopoe Waves, 106
Huntington, 112
Huron-Erie Lobe, 124
hurricanes
  Alicia, *13*
  Camille, 12, *13*
  Gulf of Mexico, 10
  Gulf Coast, 10, 13
  western Atlantic, 10

ice, 112, 113, 123, 125
  grounded, 78
  sheet, 116, 125
iceberg, melting, 90
ignimbrite, 49, 55
Illinois, 125
Independence Fan, 116
Indiana, 112
iridium, 3, 146, 149, 150
Ishigaki Island, Japan, 109

Jackson Creek Lake, 23
jökulhlaup, *112*, 117, 120, 122, 124

Kahului, Maui, 106
Kapihaa Bay, 102, 105
Kawaiu Gulch, 104
Klamath Mountain region, 54
Klintar, 121, 122
Knik Glacier, 116
Knik Valley, 124
Krakatoa explosion, 3
K-T boundary. *See* Cretaceous-Tertiary boundary

Labrador, central, 89
Lafayette, 117
lahar bulking factor. *See* LBF
lahars, *26*, *30*
  deposits, 30
  lake-breakout, 35
  Mount St. Helens, *23*
  North Fork, 30
  Pine Creek age, 25, 34
  sequence, *24*
Lake Douglas, 125
Lake George, 116, 124
Lake Maumee, 121
Lake Missoula, 105
Lake Pontiac, 125
Lake Watseka, 125
Lake Wauponsee, 125
lakes, intraglacial, 113
Lanai islands, 101, 104, 106, 107
Lanai slide, 107
Lanai Volcano, 109
landslide, submarine, 107, 109
lapilli, pumice, 51
Laurential Channel, 78, 80, 87
Laurentian Fan, 70, 78, 80, 83, 87, 89, 90
LBF, 28, 30
  defined 28
Line Island, 104
lithic blocks, 47
lithic fragments, 47, 49
Lituya Bay, Alaska, 109
Llao Rock lava flow, 49
Lower Var Canyon, 59, 62, 63, *70*
  bedforms, *65*

Madison Canyon, 129
Madison Canyon rockslide, 129
margins, continental, 9, 20, 59, 77, 89, 93
Marshall Island, 104
mass movement, submarine, 136, 138
Maumee flood, 120, 122, 123
Maumee Torrent, *120*, 122, 124
Mazama ash, *47*, 51, *52*, 55
Median Valley, 59, 62
Mediterranean Basin, 96
megaripples, 122
meltwater, 112, 113, 115, 123, 124, 125
  flood surges, 25, 28
Merriam Cone, 39
microfossils, 109
Mid-Atlantic U.S. Continental Rise, 95
Mississippi Delta, 12, 13
model, nearshore storm processes, 18
Montana, 129
Moon Bay, British Columbia, 109
Morgan Creek, 149
Mount Baker, 24
Mount Mazama, *38*, 54
  caldera acoustic units, *39*
Mount Ontake, Japan, 28
Mount Rainier, 24
Mount St. Helens, 23, 26, 28, 35, 46, 54
mud, shelf, 17
Mulina lateralis, 95

nannofossils, 147
Naridad Basin, 99
Nice Canyon, 60, 86
Nice, France, 59, 109
North America, western, deforestation, 151
North Dakota, 125
North Fork Touttle River, 26, 35
  lahar, 30
North Fork valley, 26
North Pacific abyssal plain, 149
Nucola proxima, 95

Ohio, 124
ooze, calcareous, 145, 149, 153
outwash, 90, 116, 121
  Wabash Valley, 111

Pacific Northwest, 38, 39, 47, 55
Pacific Ocean
  Hess Rise, 148
  northeast, 38
Paillon Canyon, 59, 62, 72
Perrysville, 120
Petriccio, Italy, 145
Pine Creek age, 23, 24, 25, 27, 28, 35
Placentia Bay, 80
plains, basin, *93*
planktons, 153
Pleito Formation, 127, 130, 131, 132, 136
Pleito sequence, 140, 141
Plinian column, 54
Port Royal, Jamaica, 109
Puerto Rico, 97
Puerto Rico Trench, *96*
pumice, 47, 49, 52
  blocks, 47

quartz, 3

radiation, solar, 8
Raton Basin, New Mexico, 149, 151
Reed Canyon Siltstone Member, 128
reefs, coral, 101
Rich valley, 120
ring fracture, Mount Mazama, 43, 51
Rio Grande Rise, 147
ripples, 66, 67, 70, 87, 117, 118
  Scabland, 88

river discharge, 72
rockslides, 129
  Tecuya, 140
Rogue River drainage, 49, 51
rotation, of earth, 10

Sagami Bay, Japan, 109
St. Pierre Bank, 78, 80, 87
St. Pierre Slope, 78, 80, 83, 87
St. Pierre Valley, 80
Salamonie Moraine, 112
Salt Creek, 128, 139, 140
San Emigdio Canyon, 135, 141
San Emigdio Creek, 140
San Emigdio Formation, 127, 141
San Emigdio Range, 127, 129, 140
San Emigdio sequence, 140
San Joaquin basin, 127, 129, 136, 139, 140, 141
Sangamon, 106
Saskatchewan, Canada, 47, 125, 149
Scabland ripples, 88
Scaglia rossa, 145
Scotian Slope, 83, 87, 89
scour, sea-floor, 12
SeaMARC I sonar system, *60*
sedimentation, hemipelagic, 93
Seward, Alaska, 109
shear stresses, 9, 13, 17
shelf
  Brazilian, 20
  Texas, 17, 20
shells, 17
  basal, 17
  fragments, 95, 101
  molluscan, 104
siderophiles, 150
Silver Lake, 26, 35
Silver Lake lahar, 23
Siphogenerina nodifera, 132
slide, submarine, 59, 74, 136, 138
slope, continental, *80*
slumping, 83, 89
smectite, 152
snails, 109
Soft Turbidite, 97
Sohm Abyssal Plain, 78, 83, 87, 89, 90, 99
soil
  fossil, 102
  stripping, 106
solar heating, 8
solar radiation, 8
soot, 145, 181
South Atlantic Ocean, 147
South Nation River landslip, 83
Spanish continental margin, 96
spike, fern-spore, 151
Spirit Lake, 23, 24, 27, 30
Spisula solidissima, 95
Stevns Klint, Denmark, 143, 151
storm beds, *17*, *20*
storms, 7
  geologic past, *9*
  intensity, *9*
  size, *9*
  surge, 11, 15
Strangelove Ocean, 148
stripping, soil, 106
Strombus, 104
  dentatus, 104
  gibberulus gibbosus, 104
sturzstrom, 129, 138, 141
  defined, 129
submarine landslide, 107, 109
submarine mass movement, 136, 138
submarine slide, 59, 74, 136, 138
surge, storm, 11, 15

Tecuya conglomerate, 139
Tecuya Creek, 128
Tecuya Formation, 127, *128*, 130
Tecuya rockslide, 140
Tehachapi range, 129
Tejon Formation, 128, 140
Temblor Formation, 127
temperatures, global, 10
Terre Haute, 112, 155
Texas shelf, 20
  sediments, 17
thunderstorms, 8
tides, 12
Tippecanoe Fan, 112, 113, 115
Tippecanoe River, 115
tornadoes, 8
Toutle River, 24, 26
Toutle River valley, 26
Toutle-Cowlitz River system, 23, 25, 26
Tower, 26
tree wells, 29
tropical storm Delia, *13*
tsunamis, 3, 59, 80, 106, 109
Tufts Abyssal Plain, 51
turbidite, Black Shell, 93
turbidites, 55, 83, 89, 90, 135
turbidity current, 74, *80*, 83, 87, *88*, *93*, 99, 135, 141
turbidity flow, 72, 73
Typhoon Tip, 10
typhoons, northern Pacific, 10

Umpqua River, 51
Union City Moraine, 112
United States, southeast coast, 11
Upper Var Canyon, 59, 62, 65, *70*

Valdez, Alaska, 109
Var Canyon, 59
  bedforms, *65*, *72*
Var Prodelta, 59, 61, 70
Var River, 59
Vermillion River, 120
Virgin Platform, 97
volcanic glass, 52
volcanism, 38, 55, 151
volcanoes
  Cascade Range, 24
  Mount Mazama, 38
vortex
  hurricane, 8
  vertical, 8

Wabash Moraine, 112
Wabash River, 112, 124
Wabash Valley, 112, 113, 116, 120, 123, 124
Wabash Valley outwash, 111
Waimanalo terrace, 106
Walvis Ridge, 147, 148
Washington, 51, 67, 74
water
  intraglacial, 116
  pore, 51
waves, 3, 17, *66*, 71, *102*
  flood, 27
  gravel, 67, *86*, 88, 89, 90, 105
  great, 101
  heights, 10
  Hulapoe, 106
  mud, 66
  sand, 66
  storm, 139
  train, *106*
Western Boundary Undercurrent, 94, 95
Western Valley, 78, 88, 89
Whale Bank, 78
Whittier, Alaska, 109
Willapa Canyon, 51
Williamsport, 120
wind
  shears, 8
  systems, 8
  velocities, 10
Wineglass Welded Tuff, 39, *47*, 54
Wizard Island, 39
wood
  carbonized, 49
  dated, 26